AN INTRODUCTION TO
INVERSE PROBLEMS
IN PHYSICS

AN INTRODUCTION TO
INVERSE PROBLEMS
IN PHYSICS

M Razavy

University of Alberta, Canada

World Scientific

NEW JERSEY · LONDON · SINGAPORE · BEIJING · SHANGHAI · HONG KONG · TAIPEI · CHENNAI · TOKYO

Published by

World Scientific Publishing Co. Pte. Ltd.

5 Toh Tuck Link, Singapore 596224

USA office: 27 Warren Street, Suite 401-402, Hackensack, NJ 07601

UK office: 57 Shelton Street, Covent Garden, London WC2H 9HE

Library of Congress Cataloging-in-Publication Data

Names: Razavy, Mohsen, author.

Title: An introduction to inverse problems in physics / M. Razavy, University of Alberta, Canada.

Description: New Jersey : World Scientific, [2020] | Includes bibliographical references and index.

Identifiers: LCCN 2020021972 (print) | LCCN 2020021973 (ebook) | ISBN 9789811221668 (hardcover) |
    ISBN 9789811221675 (ebook) | ISBN 9789811221682 (ebook other)

Subjects: LCSH: Inverse problems (Differential equations) | Differential equations--Numerical solutions. |
    Mathematical physics.

Classification: LCC QC20.7.D5 R39 2020 (print) | LCC QC20.7.D5 (ebook) | DDC 530.15/5357--dc23

LC record available at https://lccn.loc.gov/2020021972

LC ebook record available at https://lccn.loc.gov/2020021973

British Library Cataloguing-in-Publication Data

A catalogue record for this book is available from the British Library.

For any available supplementary material, please visit
https://www.worldscientific.com/worldscibooks/10.1142/11860#t=suppl

Printed in Singapore

To Haide, Maryam and Rabeé

Preface

In this book we consider a number of inverse problems from classical dynamics of vibrating linear chains of particles connected with springs to the quantum scattering of particles, and from inverse scattering applied to the construction of local nuclear potentials to the determination of forces in atom-atom collisions. While different techniques for inversion are reviewed, the main emphasis is on the well-known methods developed by Gel'fand–Levitan, and by Marchenko for cases where the input information is for fixed partial waves and those advanced by Newton and Sabatier, and by Lipperheide and Fiedeldey where the available information is given at fixed energy. In all these cases when the spectral function is assumed to be a rational function of the variable of the problem, the complexity of the solution is greatly reduced. A large section of the book is devoted to the method based on the continued fraction expansion method, where it is applied to a number of problems of classical wave propagation, nuclear scattering, torsional vibration and inverse problem of geomagnetic induction.

For more than thirty years, my colleagues and I have been working on different inverse problems with mixed results. There were times that we were successful in finding solutions for certain problems and there were times when the results were not promising. In the latter group, the problems were either analytically too complicated, or different numerical techniques were unstable mainly due to the accumulation of the round-off errors.

In writing this book, I am indebted to a number of friends and collaborators who worked with me on and off on inverse problems, among them Dr. Marie Hron, Professor W. van Dijk and in particular, Professor M.A. Hooshyar. In addition, I was fortunate to have the assistance of a number of graduate and undergraduate physics students. The support of my wife, Ghodssi, was essential in writing this book, and I am grateful to her for her encouragement.

Edmonton, Alberta, Canada.
January 2020

Contents

Introduction

In a vast body of problems, called inverse problems of physics, we can find two distinct categories. In the first one, we have a well-defined physical system, and the solution of the direct problem is unambiguous, stable and in general unique. Now supposing that the final solution or the result of a theory is known, can we reverse the process and find the initial question or find out the law(s) governing the physical system from the known results? The second category which contains the majority of interesting problems in physics consists of a group of problems where we have a set of laws or rules and some input data, and from these we can calculate certain results. These results may or may not be empirically observables. Now within the confines of these laws or set of rules, can we interchange the role of the input and the output, and thus replace the direct problem with the inverse one? In other words we want to know whether it is possible to begin with the empirically determined results and find the underlying basic law?

Let us illustrate the first category by the following examples:

(1) Consider a system composed of identical masses connected linearly to each other by springs with different spring constants. In the direct problem we can calculate the characteristic frequencies of such a system. This is achieved by diagonalizing a tri-diagonal matrix. For the inverse problem the question is that if the the total mass and all characteristic frequencies are known, is it possible determine each one of the spring constants?

(2) A very important set of inverse problems in mathematical physics can be stated in the following way: Suppose that we are given a second order linear differential equation such as the Schrödinger equation for a fixed partial wave and subject to specific boundary conditions. We can solve the direct problem when the local potential $V(r)$ is known and we can calculate the bound state energies and the scattering phase shift, the latter as a function of energy. The inverse problem is to find a way to determine a local, energy independent $V(r)$ from the phase shifts and the binding energies.

For most of the important physical applications that we encounter, a precise statement of the mathematical problem cannot be given, and we have to resort to approximate models and methods to describe the system, particularly when we are trying to solve inverse problems. The direct problem is, in general, well-posed, and the results are not very sensitive to the small variations of the input data. On

1

the other hand "inverse problems" are typically ill-posed problems, meaning that a small error in the initial parameters can result in large errors in the final results. Thus these problems are often unstable. For instance consider the simple case of the matrix eigenvalue problem. Here the direct problem refers to the calculation of the eigenvalues of a symmetric matrix, and the inverse eigenvalue problem is that of the construction of a symmetric matrix when the set of eigenvalues are known. Here the direct problem is well-posed, but not the inverse matrix eigenvalue problem.

We say a physical problem is well-posed if it has the following properties [1]:

(1) A solution for it is guaranteed to exist.

(2) The solution must be unique.

(3) The solution has to depend continuously on the data.

In physical sciences we find two distinct sets of inverse problems. First those problems which can be precisely stated but the answer, in general, may not be unique. For instance in classical dynamics, the inverse problem of vibration of a linear system composed of masses and springs, Sec. 1.6, or in quantum theory the inverse problem of bound states in a confining potential, Sec. 2.1 deal with examples of this type. In the second group of inverse problems we can make a mathematically well-defined formulation if and only if we make a number of simplifying assumptions so that the problem can be formulated without ambiguity. These assumptions can be about the nature of the physical system, or an extension of the range of validity of the observables or the input data beyond what is physically accessible or measurable. For instance, in quantum scattering theory, consider the question of the determination of a potential function from the energy dependence of the phase shift. In this problem we have to make a number of assumptions, such as whether the potential should be local or nonlocal and/or whether the potential should be energy (or velocity) dependent or independent.

Inverse methods can be used to investigate different aspects of problems that we encounter in theoretical physics. Among them we will consider the followings:

(a) For exactly solvable problems we can use these methods to study the uniqueness of the results that we have found from the solution of the direct problem. Interesting examples belonging to this group are Wigner's derivation of the general form of the commutation relation for the harmonic oscillator [2], [3] and the Bargmann construction of phase equivalent potentials [4], and the fact that having the standard quantum mechanical angular momentum eigenvalues, $\hbar^2 \ell(\ell+1)$, do not necessarily imply the existence of a central force law [5].

(b) The elegant method of Gel'fand–Levitan [6], [7] and also the interesting method of Marchenko [8] to solve the inverse problem of wave equation, i.e. the construction of a local energy independent short range potential for the Schrödinger equation from the information known about the energy dependence of the phase shift, is the central part of any discussion of the inverse scattering theory, particularly in nuclear and particle physics. A good review entitled "Inverse Problems in Classical and Quantum Physics" by A.A. Almasy discusses a number of wide ranging problems including problems from particle physics and field theory.

(c) Except for a limited set of examples, we cannot find analytical solutions to the inverse problems, therefore for practical applications we have to resort to numerical methods. Unfortunately, unlike the direct problems, most inverse problems are ill-posed and are unstable. While many different ways of obtaining numerical solutions have been proposed, some more accurate than the others, in general, these solutions should be used as guides to the form of the output. In this book we will try to sample a number of different inverse problems numerically. Nearly all of the numerical solutions reported in this text have been done in 1980's, with the numerical techniques and computers available in that decade. Today solutions of these problems may yield much better results.

This book begins with a review of a number of inverse problems of classical dynamics including a brief discussion of construction of the Lagrangian and the Hamiltonian from the equations of motion and also the conditions under which such a construction is possible. Then the application of the inverse method in semiclassical formulation of quantum mechanics for certain physical systems is studied. This is followed by a short account of the Heisenberg formulation. The next chapter is devoted to the inverse problem for the Schrödinger equation using the Gel'fand–Levitan formulation in one dimension as well as for the radial equation for spherically symmetric potential in spherical polar coordinates. In the latter case it is assumed that for a given partial wave one knows the phase shifts for all energies and also the bound states energies. Marchenko's approach is studied in the following chapter, where it is applied to solve a number of physically interesting problems.

The Newton–Sabatier method for determining the potential from scattering data given at fixed energy but for large number of partial waves is discussed next. Then few applications such as finding solvable cases of the Fokker–Planck equation are presented. Inverse problems for classical wave propagation are the subject of the next few chapters. Here continued fractions are introduced as a powerful way of getting approximate solutions for inverse problems of various types. This approach combined with the input data given in the form of a rational fraction provides a reliable technique for solving many problems in atomic, nuclear and geophysical inverse problems numerically. Among a number of applications, the solution of the inverse problem of torsional vibration is discussed. The next two chapters are devoted to the approximate determination of nuclear forces from the empirical knowledge of the phase shifts and binding energies. The inverse problem of electrical conductivity which is of great interest in geophysics is the subject which is presented in the following chapter. Finally, there is a brief review of the inverse problem of reflection from a moving target.

I must also acknowledge the great books of Newton, [9], by Chadan and Sabatier [8] and by Zakhariev and Suzko [11] that paved the way for my studies of this subject over the years, and were very helpful in guiding me in different ways. Only when I had completed the writing of the book that I came across the very interesting dissertation entitled "Inverse Problems in Classical and Quantum Physics" by A.A. Almasy who discusses a number of wide-ranging problems including problems from particle physics and field theory [12].

References

[1] J. Hadamard, Sur les Problémes aux Dérivées Partielles et Leur Signification Physique, Princeton University Bulletin, pp. 49-52.

[2] E.P. Wigner, Phys. Rev. 77, 711, (1950).

[3] M. Razavy, *Heisenberg's Quantum Mechanics*, (World Scientific, Singapore, 2011) Chapter 5.

[4] V. Bagamann, Phys. Rev. 75, 301 (1949).

[5] M. Razavy, Phys. Letts. 88A, 215 (1982).

[6] I.M. Gel'fand and B.M. Levitan, Dokl. Akad. Nauk SSSR, 77, 557 (1951).

[7] I.M. Gel'fand and B.M. Levitan, Isvest. Akad. Nauk SSSR, 15, 309 (1951).

[8] V.A. Marchenko, Dokl. Akad. Nauk SSSR, 72, 457 (1950).

[9] R.G. Newton, *Scattering Theory of Waves and Particles*, (Second Edition, 1980).

[10] K. Chadan and P.C. Sabatier, *Inverse Prblems in Quantum Scattering Theory*, Second Edition, (Springer-Verlag, 1989).

[11] B.N. Zakhariev and A.A. Suzko, *Direct and Inverse Problems: Potentials in Quantum Scattering*, (Springer-Verlag, Berlin 1990).

[12] A.A. Almasy, *Inverse Problems in Classical and Quantum Physics*, Ph.D. Dissertation, Johannes Gutenberg-Universität in Mainz (2007), arXiv.0912.0455 math-ph (2009).

Chapter 1

Inverse Problems in Classical Dynamics

Some of the very interesting problems of classical mechanics belong to the category of inverse problems. While these are elementary in nature, and are exactly solvable, they provide a very good introduction to the physics of the inverse problems.

1.1 Inverse Problem for Trajectory

We will start our review of the inverse problems in classical mechanics by considering a problem that goes back to I. Newton [1]. That is the derivation of the inverse-square law of planetary motion from empirical laws of Kepler.

First we note that Kepler's second law, i.e. the area swept out by the position vector in unit time is constant, in polar coordinates, can be written as

$$mr^2\dot{\theta} = \ell = \text{constant}, \tag{1.1}$$

where dot denotes the time derivative, and where m is the mass of the planet. In addition we know that the orbit is an ellipse with the origin at one of the two foci. Thus the orbit in polar coordinate is expressible as

$$r = \frac{a}{1 + \epsilon \cos \theta}, \quad \epsilon < 1. \tag{1.2}$$

By differentiating r twice with respect to time and then using ℓ to eliminate $\dot{\theta}$, we find

$$\dot{r} = \frac{a\epsilon \sin \theta}{(1 + \epsilon \cos \theta)^2}, \tag{1.3}$$

and

$$\ddot{r} = \frac{\ell\epsilon}{a}(\cos\theta)\dot{\theta} = \frac{\ell^2\epsilon}{mar^2}\cos\theta, \qquad (1.4)$$

Eliminating $\cos\theta$ between (1.2) and (1.4), we can write $\ddot{r}$ as a function of r only

$$\ddot{r} = \frac{\ell^2}{mar^2}\left(\frac{a}{r} - 1\right). \qquad (1.5)$$

But we know that the radial component of the acceleration in polar coordinates a_r is given by

$$a_r = \left[\ddot{r} - r\left(\dot{\theta}\right)^2\right] = \frac{1}{m^2}\left(\frac{\ell^2}{r^3} - \frac{\ell^2}{ar^2} - r\frac{\ell^2}{r^4}\right) = -\frac{\ell^2}{m^2a}\left(\frac{1}{r^2}\right). \qquad (1.6)$$

Therefore the radial component of the force is given by $F_r = -\frac{\ell^2}{ma}r^{-2}$.

1.2 Determination of the Shape of the Potential Energy from the Period of Oscillation

Another classical problem which is of interest is to find the shape of a confining potential (i.e. a potential $V(x)$ that goes to $+\infty$ as $x \to \pm\infty$) from the energy dependence of its period of oscillations [2].

For the one-dimensional motion the total energy, E, of the particle is given by

$$E = \frac{1}{2}m\dot{x}^2 + V(x), \qquad (1.7)$$

where x is the position of the particle and $V(x)$ is a confining potential with the condition $V(0) = 0$. Equation (1.7) is a first order differential equation for x and can easily be solved for t as a function of x

$$t = \sqrt{\frac{1}{2}m}\int\frac{dx}{\sqrt{E - V(x)}} + t_0. \qquad (1.8)$$

The two constants of motion are E, which is the total energy and t_0. The limits of the motion are the two turning points where the potential energy is equal to the total energy, $E = V(x)$. We only consider confining potentials where the motion is bounded by two such points. Now let us consider the coordinate x as function of V. This function $x(V)$ is double-valued, i.e. for a given value of V we find two different values of x. Therefore, the integrand in (1.8) can be divided into two parts, one from $x = x_1$ to $x = 0$ and the other from $x = 0$ to $x = x_2$. We denote the function $x(V)$ in these two regions by $x = x_1(V)$ and $x = x_2(V)$ respectively. We also note that the limits of integration with respect to V are E and zero, therefore the period is given by

$$T(E) = \sqrt{2m} \int_0^E \frac{dx_2(V)}{dV} \frac{dV}{\sqrt{E-V}} + \sqrt{2m} \int_E^0 \frac{dx_1(V)}{dV} \frac{dV}{\sqrt{E-V}}$$

$$= \sqrt{2m} \int_0^E \frac{[dx_2(V) - dx_1(V)]}{dV} \frac{dV}{\sqrt{E-V}}. \tag{1.9}$$

Now we divide both sides of (1.9) by $\sqrt{\mathcal{E} - E}$, where $\mathcal{E}$ is a parameter, and then integrate the result with respect to E from 0 to $\mathcal{E}$. Thus we find Abel's integral equation [3]–[4]

$$\int_0^{\mathcal{E}} \frac{T(E)dE}{\sqrt{\mathcal{E}-E}} = \sqrt{2m} \int_0^{\mathcal{E}} \int_0^E \left(\frac{dx_2}{dV} - \frac{dx_1}{dV} \right) \frac{dV \, dE}{\sqrt{(\mathcal{E}-E)(E-V)}}. \tag{1.10}$$

By changing the order of integration in Eq. (1.10) we have

$$\int_0^{\mathcal{E}} \frac{T(E)dE}{\sqrt{\mathcal{E}-E}} = \sqrt{2m} \int_0^{\mathcal{E}} \left(\frac{dx_2}{dV} - \frac{dx_1}{dV} \right) dV \times \int_V^{\mathcal{E}} \frac{dE}{\sqrt{(\mathcal{E}-E)(E-V)}}$$

$$= \pi\sqrt{2m} \left[x_2(\mathcal{E}) - x_1(\mathcal{E}) \right], \tag{1.11}$$

where we have substituted the value of the last integral in Eq. (1.11) and the value is π. Now by replacing $\mathcal{E}$ by V, we obtain

$$x_2(V) - x_1(V) = \frac{1}{\pi\sqrt{2m}} \int_0^V \frac{T(E)}{\sqrt{V-E}} dE. \tag{1.12}$$

Thus by knowing $T(E)$ we can find the difference $x_2(V) - x_1(V)$, but not $x_2(V)$ and $x_1(V)$ separately. This means that there are infinitely many curves $V = V(x)$ which gives us the same $T(E)$, and that these functions, $V(x)$, differ from each other in such a way that $x_2(V) - x_1(V)$ is the same for every curve. By imposing some other conditions on $V(x)$, such as its symmetry about the V axis we find a unique potential which is given by [2]–[6]

$$x(V) = \frac{1}{2\pi\sqrt{2m}} \int_0^V \frac{T(E)dE}{\sqrt{V-E}}. \tag{1.13}$$

As an example if $T(E) \propto E^{-\frac{1}{4}}$, then from (1.13) we find

$$x(V) \propto \int_0^V \frac{dE}{E^{\frac{1}{4}}\sqrt{V-E}} = V^{\frac{1}{4}} \int_0^1 \frac{dx}{x^{\frac{1}{4}}\sqrt{1-x}}$$

$$= V^{\frac{1}{4}} \beta\left(\frac{3}{4}, \frac{1}{4} \right), \tag{1.14}$$

where β is the β function and is related to Γ function by [7],

$$\beta(z, w) = \beta(w, z) = \int_0^1 t^{z-1}(1-t)^{w-1} dt = \frac{\Gamma(z)\Gamma(w)}{\Gamma(z+w)}. \tag{1.15}$$

Thus taking the symmetry about the V axis we find that the potential is $V(x) \propto |x|^4$, i.e. the motion is that of a classical anharmonic oscillator.

Abel's Integral Equation — The integral equation (1.9) is an Abel integral equation and is a special case of a class of the general integral equations of the form [3], [6]

$$f(x) = \frac{1}{\Gamma(\alpha)} \int_0^x \frac{\eta(\xi)d\xi}{(x-\xi)^{1-\alpha}}, \qquad 0 < \alpha < 1, \tag{1.16}$$

where $\Gamma(\alpha)$ is the Gamma function. In this integral equation $f(x)$ is a known function and we want to find $\eta(\xi)$. The formal solution of (1.16) which is obtained by inversion is given by

$$\eta(\xi) = \frac{1}{\Gamma(1-\alpha)} \frac{d}{d\xi} \int_0^\xi \frac{f(x)}{(\xi-x)^\alpha} dx$$
$$= \frac{1}{\Gamma(1-\alpha)} \left\{ \frac{f(0)}{t^\alpha} + \int_0^\xi \frac{f'(x)}{(\xi-x)^\alpha} dx \right\}. \tag{1.17}$$

Equation (1.16) has a continuous solution in the interval $a \le \xi \le b$ subject to the following conditions:
(a) $f(x)$ must be continuous in this interval,
(b) $f(0) = 0$ and
(c) $\int_0^\xi f(x)(\xi-x)^{\alpha-1}dx$ must have a continuous derivative in the interval [6], [5].
In a number of problems we find Abel's integral equation of the form

$$f(x) = \int_x^b \frac{\eta(\xi)d\xi}{\sqrt{\xi^2 - x^2}}, \tag{1.18}$$

where $0 \le x \le \xi \le b \le \infty$. The solution of this integral equation is

$$\eta(\xi) = -\frac{2}{\pi} \frac{d}{d\xi} \int_\xi^b \frac{x f(x) dx}{\sqrt{x^2 - \xi^2}}, \tag{1.19}$$

or [6]

$$\eta(\xi) = \frac{2\xi}{\pi} \left\{ \frac{f(b)}{b^2 - \xi^2} - \int_\xi^b \frac{f'(x)dx}{\sqrt{x^2 - \xi^2}} \right\}. \tag{1.20}$$

An elegant way of solving Abel's integral equation is by utilizing the Laplace transform method. This method is given, for example in Arfken's book [8].

1.3   Action Equivalent Hamiltonians

Let us solve the inverse one-dimensional motion of particle for the simplest case where the period $T = \frac{2\pi}{\omega}$ is independent of energy. Then from (1.12) we conclude that

$$x_2(V) - x_1(V) = \sqrt{\frac{2V}{m}} \frac{1}{\omega}. \tag{1.21}$$

This relation shows that V is a quadratic function of x. Let us write it as

$$V(x) = \begin{cases} \frac{1}{2} m\omega_+^2 x^2 & \text{for } x > 0 \\[2mm] \frac{1}{2} m\omega_-^2 x^2 & \text{for } x < 0 \end{cases}, \tag{1.22}$$

and this solution satisfies the condition $V(0) = 0$. From (1.21) and (1.22) it follows that the angular frequency of the motion, ω, is related to ω_+ and ω_- by

$$\omega = \frac{2\omega_+\omega_-}{\omega_+ + \omega_-}. \tag{1.23}$$

Thus for all non-zero values of the parameters ω_+ and ω_- of the potential the resulting angular frequency is ω, and therefore in an action-angle formulation of the problem the Hamiltonian is given by [9], [10]

$$H = \omega J_x. \tag{1.24}$$

A Two-Dimensional Motion with Closed Orbit — Let us study the direct problem for the motion of a particle of unit mass in two dimensions x and y, where the Hamiltonian is given by

$$H = \frac{1}{2} \left(p_x^2 + \omega^2 x^2 \right) + \frac{1}{2} \left(p_y^2 + \omega^2 y^2 \right). \tag{1.25}$$

If we transform this Hamiltonian to the action-angle variables, then we have [9]–[11].

$$H = \omega(J_x + J_y). \tag{1.26}$$

According to what we have found earlier for one-dimensional motion the simplest two-dimensional motion for the Hamiltonian (1.26) given in terms of the canonical variables is

$$H = \frac{1}{2} \left[p_x^2 + p_y^2 + \Omega_x(x)x^2 + \Omega_y(y)y^2 \right]. \tag{1.27}$$

where

$$\Omega_x(x) = \omega \left[\frac{\theta(x)}{1 + \lambda_1} + \frac{\theta(x)}{1 - \lambda_1} \right]. \tag{1.28}$$

and

$$\Omega_y(y) = \omega \left[\frac{\theta(y)}{1 + \lambda_2} + \frac{\theta(y)}{1 - \lambda_2} \right]. \tag{1.29}$$

In these relations $\theta(x)$ is the step function

$$\theta(x) = \begin{cases} 1 & \text{for } x > 0 \\[2mm] 0 & \text{for } x < 0 \end{cases}, \tag{1.30}$$

and λ_1 and λ_2 are two arbitrary parameters $|\lambda_{1,2}| < 1$.

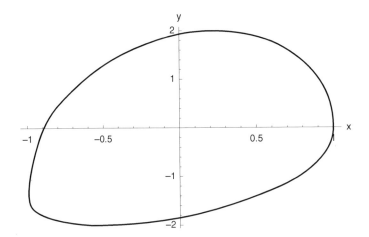

Figure 1.1: The closed orbit found from the Hamiltonian (1.27) for $m = 1$, $\omega = 1$, $\lambda_1 = 0.4$ and $\lambda_2 = 0.2$. Apart from the energy there are no apparent symmetries or conserved quantities for this motion [11].

In conclusion we observe that the symmetry of the Hamiltonian (1.25) and the symmetry of the Hamiltonian (1.27), which we found by solving the inverse problem, are not the same, i.e. (1.25) remains invariant under the transformation $x \rightleftharpoons y$, but (1.27) does not, even though in both cases the orbits are closed. This is particularly significant once we consider the question of the quantization.

We will come back to this important aspect of using the results of solving the inverse problem later in connection with the quantum mechanical inverse problems.

1.4   Abel's Original Inverse Problem

An inverse problem of historical interest has been called "determination of the shape of a hill from the travel time measurement" [12]. If a particle slides up and down a hill with no friction and with the initial energy E, assuming that we can measure the travel time for up and down motion $T(E)$ as a function of the initial velocity of the particle, then under certain conditions we can find the shape of the hill. In the direct problem, we know that the potential energy of the particle at the height y is given by $mgy(s)$, where s measures the arc length along the hill. The total energy of the particle is the sum of kinetic and potential energies:

$$E = \frac{1}{2}m \left(\frac{ds}{dt} \right)^2 + V(s). \tag{1.31}$$

At $t = 0$ we assume that the particle is at $s(0) = 0$, and has a velocity $\left(\frac{ds}{dt} \right)_{t=0} = v_0$. Solving (1.31) for $\frac{ds}{dt}$, choosing the positive square root since $v_0 > 0$, and then integrating (1.31) we obtain

$$t = \sqrt{\frac{m}{2}} \int_0^s \frac{ds'}{\sqrt{E - V(s')}}. \tag{1.32}$$

Thus starting at $s = 0$, the particle moves up and reaches the classical turning point s_1, where $E = V(s_1)$. The corresponding value of t at this point is $\frac{1}{2}T(E)$, i.e.

$$T(E) = \sqrt{2m} \int_0^{s_1(E)} \frac{ds'}{\sqrt{E - V(s')}}. \tag{1.33}$$

At the time $t = \frac{1}{2}T(E)$, the velocity of the particle becomes zero and after that the velocity becomes negative and the particle will start descending. Therefore the solution for $t \geq \frac{1}{2}T(E)$, is obtained if we take the negative square root in (1.31). The result in this case is

$$t = \frac{1}{2}T(E) + \sqrt{\frac{m}{2}} \int_0^{s_1(E)} \frac{ds'}{\sqrt{E - V(s')}}, \quad \frac{1}{2}T(E) \leq t. \tag{1.34}$$

Now if we set $s_1 = 0$ in the above equation we find the time that takes the particle to move back to the point $s = 0$ is just the same $\frac{1}{2}T(E)$ given by (1.33). If $V(s)$ is not a monotonically increasing function of s, we can divide the upward (or downward) motion of the particle into different segments and then find the total time [12]. Here, for the sake of simplicity, we assume that $V(s)$ is a monotonic function of s. With this condition the solution of the Abel integral equation is given by

$$\frac{ds(V)}{dV} = \frac{T(0)}{\pi \sqrt{2mV}} + \frac{1}{\pi \sqrt{2m}} \int_0^V \frac{dT(E)}{dE} \frac{dE}{\sqrt{V - E}}, \tag{1.35}$$

where, because of the initial conditions, $T(0) = 0$. Finally we integrate (1.35) from 0 to V to get $s_1(V)$

$$s_1(V) = \frac{1}{\pi \sqrt{2m}} \int_0^V \frac{T(E)dE}{\sqrt{V - E}}. \tag{1.36}$$

Once we have $s_1(V)$, by inverting, we find $s(V)$ and since $V(s) = mgy(s)$ and $\left(\frac{dx}{ds}\right)^2 + \left(\frac{dy}{ds}\right)^2 = 1$, therefore we have a parametric equation for $x(s)$ and $y(s)$:

$$x(s) = \int_0^s \sqrt{1 - \left[\frac{dy(s')}{ds'}\right]^2} \, ds', \quad \text{and} \quad y(s) = \frac{V(s)}{mg}. \tag{1.37}$$

Even when the profile of the hill cannot be described by a monotonic function $y = y(s)$, still there are cases that the problem can be solved [12].

1.5   Inverse Scattering Problem in Classical Mechanics

As another example of application of inverse problems in classical mechanics we want to consider the inverse scattering problem of a projectile of mass m from a fixed (or a massive) target. The projectile is coming from infinity, and has a total energy $E = \frac{1}{2}mv_\infty^2$ and we assume that the interaction potential is repulsive and goes to zero as $r \to \infty$. The trajectory of the particle is a hyperbola and its shape is determined by the impact parameter b which is equal to the closest distance to the center that the particle will reach if the potential were zero everywhere. However because of the repulsive nature of the potential the nearest point that the projectile will reach the target is $r = r_0$. The hyperbolic path of the particle has two asymptotes corresponding to $t = -\infty$ and $t = +\infty$, and the interaction causes a defection of the asymptotes by an angle θ which is the scattering angle. We use the polar coordinates (r, ϕ) to describe the position of the particle as a function of time. Since the force is central, from the laws of motion we find two conserved quantities, the angular momentum ℓ and the energy E:

$$\ell = -bmv_\infty, \tag{1.38}$$

and

$$E = \frac{1}{2}m\dot{r}^2(t) + \frac{\ell^2}{2mr^2} + V(r). \tag{1.39}$$

In these relations dot means derivative with respect to time, ℓ is the angular momentum and b is the impact parameter. Since the angular momentum is conserved we have

$$\dot{\phi}(t) = \frac{\ell}{mr^2} = -\frac{b}{r^2}\sqrt{\frac{2E}{m}} < 0, \tag{1.40}$$

so that $\phi(t)$ is a monotonically decreasing function of t. From Eq. (1.39) we find $\dot{r}(t)$ to be:

$$\dot{r}(t) = \pm\sqrt{\frac{2}{m}}\left[\left(1 - \frac{b^2}{r^2}\right)E - V(r)\right]^{\frac{1}{2}}. \tag{1.41}$$

Now if at the time $t = t_0$ the projectile reaches the nearest distance from the target, i.e. $r_0 = r(t_0)$, then we choose the minus sign in (1.41) for $-\infty < t < t_0$, and the plus sign for $t_0 < t < +\infty$. Thus the value of r_0 is determined from

$$E\left(1 - \frac{b^2}{r_0^2}\right) - V(r_0) = 0. \tag{1.42}$$

For finding the asymptotes, we have to know the trajectory. This is obtained by eliminating t between (1.40) and (1.41):

$$d\phi = -\sqrt{\frac{2E}{m}}\frac{b}{r^2}dt = -\frac{1}{r^2}\left[\left(1 - \frac{V(r)}{E}\right)\frac{1}{b^2} - \frac{1}{r^2}\right]^{-\frac{1}{2}}dr. \tag{1.43}$$

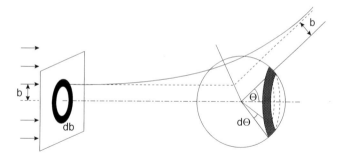

Figure 1.2: Scattering of a particle by a repulsive force.

Thus by integrating this equation and noting that the scattering angle, $\theta = \pi - 2\phi$, we find

$$\theta = \theta(b) = \pi - 2 \int_{r_0}^{\infty} \frac{dr}{r^2} \left[\left(1 - \frac{V(r)}{E} \right) \frac{1}{b^2} - \frac{1}{r^2} \right]^{-\frac{1}{2}}. \qquad (1.44)$$

This is the solution of the direct problem, i.e. if we know the potential we can calculate the scattering angle as a function of energy and of the impact parameter.

In the direct problem, the important concept is the cross section which is a function of θ. Thus we consider a mono-energetic beam of particles incident in the annular ring of area $2\pi b\, db$ bounded by the impact parameters b and $b + db$, and that the number of scattered particles through the solid angle $d\Omega = 2\pi \sin\theta d\theta$ is $\sigma(\theta)$ (see Fig. 1.2). By equating the number of incoming and outgoing particles we find [10]

$$\sigma(\theta) = -\frac{b}{\sin\theta} \frac{db}{d\theta}. \qquad (1.45)$$

The Inverse Scattering Problem in Classical Mechanics — In order to solve the inverse problem, we assume that $\theta(b)$ is known and we want to find $V(r)$. Following the work of Firsov and of Keller [13], [14], we introduce the new variables:

$$x = \frac{1}{b^2}, \quad u = \frac{1}{r}, \quad \theta(b) = \bar{\theta}(x), \quad V(r) = \bar{V}(u), \qquad (1.46)$$

and

$$\beta(x) = \frac{1}{2} \left(\pi - \bar{\theta}(x) \right). \qquad (1.47)$$

With these changes, Eq. (1.44) can be written as

$$\beta(x) = \int_{u=0}^{u_0} \left[x \left(1 - \frac{\bar{V}(u)}{E} \right) - u^2 \right]^{-\frac{1}{2}} du, \qquad (1.48)$$

where $u_0 = \frac{1}{r_0}$. Next we make the following substitution:

$$v(u) = 1 - \frac{\bar{V}(u)}{E}, \quad w = \frac{u^2}{v} \quad \text{and} \quad g(w) = \frac{1}{\sqrt{v}} \frac{du}{dw}. \qquad (1.49)$$

With these changes (1.48) becomes the standard Abel integral equation

$$\beta(x) = \int_{w=0}^{x} \frac{g(w)dw}{\sqrt{x-w}}, \qquad 0 \le x \le x_{max}, \tag{1.50}$$

where $x_{max} = b_{min}^{-2}$. In Eq. (1.50) when the square root is zero, i.e. for $x = w$, then $u = u_0$. Now Eq. (1.48) is the Abel integral equation for which we know what the solution is (see Eq. (1.17))

$$g(w) = \frac{1}{\pi} \frac{d}{dw} \int_0^w \frac{\beta(x)dx}{\sqrt{w-x}}. \tag{1.51}$$

Having obtained $g(w)$ we want to determine $V(r)$. Now from (1.49), we have $u = \sqrt{vw}$, therefore

$$g(w) = \frac{\sqrt{w}}{2v} \frac{dv}{dw} + \frac{1}{2\sqrt{w}}. \tag{1.52}$$

We solve this equation for v as a function of w;

$$v = \bar{v}(w) = \exp \int_0^w \left[2g(\bar{w}) \bar{w}^{-\frac{1}{2}} - \bar{w}^{-1} \right] d\bar{w}. \tag{1.53}$$

Let us note that $u = 0$ implies $w = 0$, and that $\bar{V}(0) = V(\infty) = 0$. In this way we obtain a parametric representation of $V(r)$, viz, from (1.48) and (1.49) we find

$$V = E\left(1 - \bar{v}(w)\right), \qquad r = \frac{1}{\sqrt{w\bar{v}(w)}}, \tag{1.54}$$

for $0 \le w \le x_{max} = b_{min}^{-2}$.

Another way to simplify our formulation is to integrate (1.51) by parts, and the integration yields the following result:

$$g(w) = \frac{d}{dw} \left[\sqrt{w} - \frac{1}{\pi} \int_0^w \sqrt{w-x} \left(\frac{d\theta}{dx} \right) dx \right.$$

$$= \frac{1}{2\sqrt{w}} + \frac{1}{2\pi} \int_0^w \frac{d\theta}{dx} \frac{dx}{\sqrt{w-x}} \bigg]. \tag{1.55}$$

Now by changing the integration variable to θ, and noting that $\theta(0) = 0$, we find

$$g(w) = \frac{1}{2\sqrt{w}} + \frac{1}{2\pi} \int_0^{\theta(w)} \frac{d\theta}{\sqrt{w-x(\theta)}}. \tag{1.56}$$

Using this result Eq. (1.53) becomes

$$v = \exp \left\{ \frac{1}{\pi} \int_0^w \frac{dw'}{\sqrt{w'}} \int_0^{\theta(w')} \frac{d\theta}{\sqrt{w'-x(\theta)}} \right\}. \tag{1.57}$$

Now we will use these results to solve the inverse problem of Rutherford scattering.

Inverse Problem of the Rutherford Scattering — We know that in the well-known Rutherford scattering, the cross section is given by [10]

$$\sigma(\theta) = \frac{A}{4 \sin^4 \frac{\theta}{2}}, \tag{1.58}$$

where $A = \frac{e^2}{4E^2}$. Using Eq. (1.45) we can express b^2 in terms of the scattering angle θ

$$-\frac{b\,db}{\sin\theta d\theta} = \frac{A}{4 \sin^4 \frac{\theta}{2}}, \tag{1.59}$$

or simplifying we find

$$\frac{1}{b^2} = x(\theta) = A^{-1} \tan^2 \frac{\theta}{2}, \quad \theta = \tan^{-1} \sqrt{Ax}. \tag{1.60}$$

Now we substitute $x(\theta)$ in (1.57) to find v:

$$v = \exp \left\{ \frac{1}{\pi} \int_0^w \frac{dw'}{\sqrt{w'}} \int_0^{2\tan^{-1}\sqrt{Aw'}} \frac{d\theta}{\left[w' - A^{-1}\tan^2 \frac{\theta}{2}\right]^{\frac{1}{2}}} \right\}. \tag{1.61}$$

To evaluate the θ integral, we change the variable θ to ϕ, where

$$\tan \frac{\theta}{2} = \sqrt{Aw'} \sin \phi. \tag{1.62}$$

The last integral in (1.61) can be calculated [7]

$$\int_0^{2\tan^{-1}\sqrt{Aw'}} \frac{d\theta}{\left[w' - A^{-1}\tan^2 \frac{\theta}{2}\right]^{\frac{1}{2}}}$$

$$= 2\sqrt{A} \int_0^{\frac{\pi}{2}} \frac{d\phi}{1 + Aw' \sin^2 \phi} = \frac{\pi\sqrt{a}}{\sqrt{1+Aw'}}. \tag{1.63}$$

By substituting this integral in (1.52) we can carry out the integration over w' and obtain v;

$$v = 1 + 2Aw + 2\sqrt{Aw(2+Aw)}. \tag{1.64}$$

Since $u = \frac{1}{r} = \sqrt{vw}$, from Eqs. (1.46) and (1.49), we find

$$V(r) = 2E\sqrt{A}u = eu = \frac{e}{r}, \tag{1.65}$$

which is the Coulomb potential.

1.6 Inverse Problem of a Linear Chain of Masses Coupled to Springs

A simple and much studied example of an inverse problem in classical dynamics is that of the determination of masses and different spring constants for a finite chain from the knowledge of the characteristic frequencies of the system [15], [16]. In order to state the questions relating to the input and output of this problem clearly, we need to study certain aspects of the theory of matrices first. Therefore we begin this section with the review of the mathematical tools needed for solving this particular problem [16].

Rayleigh Quotient— Let $\mathbf{x}$ be a column $n \times 1$ matrix (or a vector with n components) and A_n be a symmetric $n \times n$ matrix, and let us consider the following quotient called Rayleigh quotient [17]

$$R = \frac{\mathbf{x}^T \cdot A_n \mathbf{x}}{\mathbf{x}^T \cdot \mathbf{x}}. \tag{1.66}$$

In this relation $\mathbf{x}^T$ is the transpose of $\mathbf{x}$ and dot denotes the scalar product of two vectors. We note that by replacing $\mathbf{x}$ by $b\mathbf{x}$, with b a constant R will not be affected, since it is the ratio of two quadratic forms. Thus we can write R as the value of $\left(\mathbf{x}^T \cdot A\mathbf{x} \right)$ subject to the normalization condition $\mathbf{x}^T \cdot \mathbf{x} = 1$. This constraint on $\mathbf{x}$ can be enforced by introducing a Lagrange multiplier λ, and then finding the condition for R to be stationary. Now we consider the condition under which the quadratic form

$$R = \mathbf{x}^T \cdot A_n \mathbf{x} - \lambda \mathbf{x}^T \cdot \mathbf{x}, \tag{1.67}$$

becomes stationary. By differentiating R with respect to a matrix element x_i and setting it equal to zero we get

$$\frac{\partial R}{\partial x_i} = 2(a_{1i}x_1 + a_{2i}x_2 + \cdots + a_{ni}x_n) - 2\lambda x_i = 0. \tag{1.68}$$

If we collect all the terms $\frac{\partial R}{\partial x_1}, \frac{\partial R}{\partial x_2}, \cdots, \frac{\partial R}{\partial x_n}$, we find the equation

$$A_n \mathbf{x} - \lambda \mathbf{x} = 0, \tag{1.69}$$

and this shows that λ is an eigenvalue of the matrix A.

From this direct problem we find a set of eigenvalues $\lambda_1, \lambda_2, \cdots, \lambda_n$, but knowing this set is not sufficient to solve the inverse problem since in the latter we need n masses, $m_1, m_2, \cdots, m_n$ and n spring constants $k_1, k_2, \cdots, k_n$. The rest of the needed information can be obtained from the eigenvalues $\mu_1, \mu_2, \cdots, \mu_{n-1}$ of the matrix A_{n-1} which will be defined below and in addition the total mass of the system $\sum_{i=1}^{n} m_i$. If we delete the row n and the column n from the matrix we find $(n-1) \times (n-1)$ matrix A_{n-1}, with $n-1$ eigenvalues. Whereas the matrix $(A_n - \lambda)\mathbf{x} = 0$ represents the motion of the system shown in Fig. 1.3 (the top figure) with all particles moving, the eigenvalue equation $(A_{n-1} - \mu)\mathbf{x} = 0$ describes the system shown in the same figure (below) where the nth mass is attached to the wall and is not moving. Since $x_1 = 0$ is equivalent to $\mathbf{x}^T \cdot \mathbf{e}1 = 0$, where $\mathbf{e}_1 = [0, 0, \cdots, 1]$, we can introduce two Lagrange multiplier λ and 2μ and consider under what conditions

$$R = \mathbf{x}^T \cdot A_n \mathbf{x} - \lambda \mathbf{x}^T \cdot \mathbf{x} - 2\mu \mathbf{x}^T \cdot \mathbf{e}_n, \tag{1.70}$$

will become stationary. The condition on R is

$$\frac{\partial R}{\partial x_i} = 2 \sum_{j=1}^{n} a_{ij} x_j - 2\lambda x_j - 2\mu \delta_{i1} = 0 \qquad (1.71)$$

which can be written as

$$A_n \mathbf{x} - \lambda \mathbf{x} - \mu \mathbf{e}_1 = 0. \qquad (1.72)$$

Since $\mathbf{x}_1, \mathbf{x}_2, \cdots, \mathbf{x}_n$ span R_n, we can write

$$\mathbf{x} = \sum_{i=1}^{n} \beta_i \mathbf{x}_i, \qquad (1.73)$$

and by substituting this expression in (1.72) we find

$$\sum_{i=1}^{n} (\lambda_i - \lambda) \beta_i \mathbf{x}_i - \mu \mathbf{e}_1 = 0. \qquad (1.74)$$

From the orthogonality of the eigenvectors $\mathbf{x}_i$, i.e. $\mathbf{x}_j^T \cdot \mathbf{x}_i = \delta_{ij}$ we obtain

$$(\lambda_j - \lambda) \beta_j = \mu \mathbf{x}^T \cdot \mathbf{e}_1 = \mu x_{1j}, \qquad (1.75)$$

where x_{1j} denotes the first component of $\mathbf{x}_j$. Thus we have

$$\beta_j = \frac{\mu x_{1j}}{(\lambda_j - \lambda)}. \qquad (1.76)$$

Now by imposing the constraint condition $\mathbf{x}_1 = 0$ we get

$$f(\lambda) \equiv \sum_{i=1}^{n} \frac{|x_{1i}|^2}{\lambda_i - \lambda} = 0. \qquad (1.77)$$

This relation shows that the eigenvalues $\mu_1, \mu_2, \cdots, \mu_{n-1}$ are the roots of $f(\lambda)$, and that these eigenvalues interlace $\lambda_1, \lambda_2, \cdots, \lambda_n$, i.e.

$$\lambda_1 < \mu_1 < \lambda_2 < \mu_2, \cdots, \mu_{n-1} < \lambda_n. \qquad (1.78)$$

If x_{ni} is normalized i.e. $\sum_{i=1}^{n} |x_{ni}|^2 = 1$, and if we know all the elements x_{n1}, $x_{n2}, \cdots, x_{nn}$ then from Eq. (1.77) we can find $\mu_1, \mu_2, \cdots, \mu_{n-1}$. Thus in the direct problem knowing the matrix A_n enables us to determine the set of μ s. But this way of looking at the problem is useful since we can also solve the inverse problem. That is if we are given $(n-1)$ μ s, and we know they interlace λ s, then we can find the elements $x_{n1}, x_{n2}, \cdots, x_{nn}$.

Using the relation for the spectral decomposition of a given symmetric matrix A,

$$\text{adj} (\lambda_i I - A) = \prod_{k=1, k \neq i}^{n} (\lambda_i - \lambda_k) x_i x_i^T, \qquad (1.79)$$

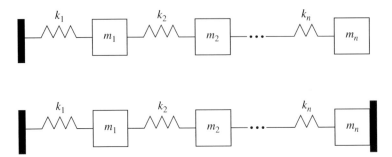

Figure 1.3: Linear chains of masses and springs. The top figure shows the case where all n masses can move, and the bottom shows the case where the last mass m_n is rigidly attached to the wall [16].

where I is the unit matrix, we can evaluate both sides of this relation at (n, n) position to obtain

$$\det(\lambda_i I_{n-1}) = x_{ni}^2 \prod_{k=1, k \neq i}^{n} (\lambda_i - \lambda_k).$$ (1.80)

In this relation x_{ni} is the last entry of $\mathbf{x}_i$, and A_{n-1} denotes the $(n-1)(n-1)$ matrix. Solving for x_{ni}^2, we get

$$|x_{1i}|^2 = \frac{\prod_{k=1, k \neq i}^{n-1} (\mu_k - \lambda_i)}{\prod_{k=1, k \neq i}^{n} (\lambda_k - \lambda_i)},$$ (1.81)

and this expression gives us the matrix elements of A. Thus we conclude that if the eigenvalues $\mu_1, \mu_2, \cdots, \mu_{n-1}$ are known then we know the first entries in the vectors $\mathbf{x}_1, \mathbf{x}_2, \cdots, \mathbf{x}_n$. If we know $\lambda_1, \lambda_2, \cdots, \lambda_n$ and $x_{n1}, x_{n2}, \cdots, x_{nn}$ we have $2n + 1$ quantities. Because $\mathbf{x}_n$ is normalized x_n s are not independent.

The Jacobi Matrix — The Jacobi matrix is a square $n \times n$ matrix with just three diagonal elements being nonzero. A typical Jacobi matrix has the following form [18], [19]

$$K_n = \begin{bmatrix} a_1 & b_1 & 0 & 0 & \cdots & 0 \\ b_1 & a_2 & b_2 & 0 & \cdots & 0 \\ 0 & b_2 & a_3 & b_3 & \cdots & 0 \\ \hdotsfor{6} \\ \cdots & \cdots & \cdots & \cdots & \cdots & b_{n-1} \\ \cdots & \cdots & \cdots & \cdots & b_{n-1} & a_n \end{bmatrix}$$ (1.82)

Since changing the sign of b_j has no effect on the eigenvalues, we can assume that $b_1, b_2, \cdots, b_{n-1}$ are all negative. We can collect the n column vector equations $(A - \lambda_i I)\mathbf{x}_i = 0$ into one matrix equation

$$A\mathbf{X} = \mathbf{X}\Lambda,$$ (1.83)

where

$$\mathbf{X} = [\mathbf{x}_1, \mathbf{x}_2, \cdots, \mathbf{x}_n]. \tag{1.84}$$

Let us now denote the transpose of $\mathbf{X}$ by $\mathbf{U}$, i.e. $\mathbf{X}^T = \mathbf{U}$, then $A\mathbf{U}^T = \mathbf{U}^T\Lambda$ and on transposing we have

$$\mathbf{U}A = \Lambda\mathbf{U}, \tag{1.85}$$

or in particular if $A = J$ is a Jacobi matrix we have

$$[\mathbf{u}_1, \mathbf{u}_2, \cdots, \mathbf{u}_n]J = \Lambda[\mathbf{u}_1, \mathbf{u}_2, \cdots, \mathbf{u}_n], \tag{1.86}$$

where $\mathbf{U} = [\mathbf{u}_1, \mathbf{u}_2, \cdots, \mathbf{u}_n]$. Now let us consider a method for determining the elements a_j, b_j of this J matrix from the set of eigenvalues λ_j and μ_j. There is a wealth of literature on the inversion of symmetric Jacobi matrices and the reader is referred to the following papers where in these works, further references can be found to other works [16]–[25]. Let us consider a simple way of this inversion.

The first column of Eq. (1.86) is

$$a_1\mathbf{u}_1 + b_1\mathbf{u}_2 = \Lambda\mathbf{u}_1. \tag{1.87}$$

Since $\mathbf{u}_1, \mathbf{u}_2, \cdots, \mathbf{u}_n$ form an orthogonal set, from (1.87) we find

$$a_1 = \mathbf{u}_1^T \cdot \Lambda\mathbf{u}_1. \tag{1.88}$$

To find b_1 we write (1.87) as

$$b_1\mathbf{u}_2 = (\Lambda - a_1)\mathbf{u}_1 = \mathbf{z}_2. \tag{1.89}$$

The vector $\mathbf{z}_2$ is known from its definition and $\mathbf{u}_2$ is a unit vector. Thus

$$b_1^2 = \|\mathbf{z}_2\|^2, \quad b_1 = \|\mathbf{z}_2\|, \tag{1.90}$$

and

$$\mathbf{u}_2 = \frac{1}{b_1}\mathbf{z}_2. \tag{1.91}$$

Now let us consider the second column of (1.86);

$$b_1\mathbf{u}_1 + a_2\mathbf{u}_2 + b_2\mathbf{u}_3 = \Lambda\mathbf{u}_2. \tag{1.92}$$

Again we multiply this equation by $\mathbf{u}_2^T$, using the fact that $\mathbf{u}_2^T \cdot \mathbf{u}_1 = \mathbf{u}_2^T \cdot \mathbf{u}_3 = 0$ and get

$$a_2 = \mathbf{u}_2^T\Lambda\mathbf{u}_2, \tag{1.93}$$

and

$$b_2\mathbf{u}_3 = \mathbf{z}_3 = (\Lambda - a_2)\mathbf{u}_2 - b_1\mathbf{u}_1. \tag{1.94}$$

Since every term in the right-hand side of this equation is known therefore $\mathbf{z}_3$ can be found. From the last relation we deduce that

$$b_2 = \|\mathbf{z}_3\|, \quad \mathbf{u}_3 = \frac{\mathbf{z}_3}{b_2}. \tag{1.95}$$

We continue this process to find all a_i s and b_i s.

Lanczos Algorithm — The procedure of finding a_i and b_i, $i = 1, 2 \cdots n$ is called Lanczos algorithm [20]. The original Lanczos algorithm concerns with the solution of the following problem:

Given a symmetric matrix A and a normalized vector $\mathbf{x}_1$ (i.e. $\mathbf{x}_1^T \cdot \mathbf{x}_1 = 1$), compute a Jacobian matrix J and an orthogonal matrix $X = [\mathbf{x}_1, \mathbf{x}_2, \cdots, \mathbf{x}_n]$ such that $A = XJX^T$. Variants of Lanczos's algorithm, and different ways of finding the elements in a Jacobi matrix from the eigenvalues are given in Refs. [21]–[28].

Eigenfrequencies of the Mass-Spring Chain — Now let us consider the problem of n masses $m_1, m_2, \cdots, m_n$ connected by springs with spring constants k_1, $k_2, \cdots, k_n$ to each other. For the system composed of masses and springs, Fig. 1.3 (top), we can write this eigenvalue equation as

$$
\begin{bmatrix}
(k_1 + k_2) & -k_2 & \cdots & 0 \\
-k_2 & (k_2 + k_3) & \cdots & 0 \\
\cdots\cdots\cdots\cdots\cdots\cdots\cdots\cdots\cdots \\
\cdots & \cdots & \cdots & -k_n \\
\cdots & \cdots & -k_n & k_n
\end{bmatrix}
\begin{bmatrix}
x_1 \\ x_2 \\ \cdots \\ x_{n-1} \\ x_n
\end{bmatrix}
$$

$$
= \lambda
\begin{bmatrix}
m_1 & 0 & \cdots & 0 \\
0 & m_2 & \cdots & 0 \\
\cdots\cdots\cdots\cdots\cdots\cdots \\
\cdots & \cdots & m_{n-1} & 0 \\
\cdots & \cdots & \cdots & m_n
\end{bmatrix}
\begin{bmatrix}
x_1 \\ x_2 \\ \cdots \\ x_{n-1} \\ x_n
\end{bmatrix},
\tag{1.96}
$$

where $\lambda = \omega^2$, and $\omega_1, \omega_2 \cdots \omega_n$ are the natural frequencies of the system.

Here we want to find the characteristic frequencies ω_i s of such a system under two different boundary conditions: In the first one the mass m_1 is attached with a spring k_1 to a rigid wall, but m_n is free to move, as shown in Fig. 1.3 at the top. The second is when the last mass m_n is fixed by attaching it to the wall, as shown in the lower figure Fig. 1.3. The equations of motion for the system at the top can be written as

$$
(K - \lambda M)\mathbf{u} = 0, \quad \lambda = \omega^2,
\tag{1.97}
$$

where K is the Jacobi matrix with the elements related to the spring constants k_i as shown in (1.97). If we want to write (1.97) in the standard form of an eigenvalue equation we introduce the matrix D;

$$
M = D^2, \quad D^{-1}KD^{-1} = J, \quad D\mathbf{u} = \mathbf{x},
\tag{1.98}
$$

and then (1.97) can be written as

$$
(J - \lambda I)\mathbf{x} = 0,
\tag{1.99}
$$

where I is a unit $n \times n$ matrix. Using Lanczos algorithm we find the elements of J, and then we try to find matrices for K and M from J [16], [28]. As we can see from (1.98) the matrices K and A are related to each other $K = DJD$. By inspection we see that

$$K[1, 1, \cdots, 1] = k_1 \mathbf{e}_1, \tag{1.100}$$

where $\mathbf{e}_1 = [1, 0, \cdots, 0]$. Thus

$$DAD[1, 1, \cdots, 1] = k_1 \mathbf{e}_1, \tag{1.101}$$

and therefore

$$A[d_1, d_2, \cdots, d_n] = D^{-1} k_1 \mathbf{e}_1 = d_1^{-1} k_1 \mathbf{e}_1. \tag{1.102}$$

Since the off diagonal elements of the K matrix are negative, Eq. (1.96), when we solve

$$K[x_1, x_2, \cdots, x_n] = \mathbf{e}_1, \tag{1.103}$$

then the solution $x_1, x_2, \cdots, x_n$ will have positive components. The solution of the last equation in (1.98) is $\mathbf{d} = c\mathbf{x}$, where c can be found from the total mass of the system

$$m = \sum_{i=1}^{n} m_i = \sum_{i=1}^{n} d_i^2 = \| \mathbf{d} \|^2 = c^2 \| \mathbf{x} \|^2 . \tag{1.104}$$

Therefore, knowing m we can calculate $\| \mathbf{x} \|^2$ then find c and $\mathbf{d}$, and hence the masses $m_i = d_i^2$. Having found d_i s we can compute $K = DAD$ and find the spring constants $k_1, k_2, \cdots, k_n$ [15], [16].

An Analytically Solvable Spring-Mass Problem — Let us consider the special case of Eq. (1.96) when $n = 2$. Then the eigenvalues are the roots of the quadratic equation

$$\begin{vmatrix} k_1 + k_2 - \lambda m_1 & -k_2 \\ -k_2 & k_2 - \lambda m_2 \end{vmatrix} = 0. \tag{1.105}$$

In the direct problem we know k_1, k_2, m_1 and m_2, and the eigenvalue equation (1.105) can be solved for λ_1 and λ_2, the two roots of this equation. In the inverse problem for this case we are given the eigenvalues plus m_1 and m_2 and from (1.105) we obtain k_1 in terms of k_2:

$$k_1 = \frac{1}{k_2} (m_1 m_2 \lambda_1 \lambda_2), \tag{1.106}$$

where k_2 is the root of the quadratic equation

$$\left(\frac{m_1 + m_2}{m_1 m_2} \right) k_2^2 - (\lambda_1 + \lambda_2) k_2 + m_2 \lambda_1 \lambda_2 = 0. \tag{1.107}$$

This equation has a real positive root if and only if

$$(\lambda_1 - \lambda_2)^2 > \frac{4m_2}{m_1} \lambda_1 \lambda_2. \tag{1.108}$$

This result indicates that, in general, for the accuracy of the numerical solution of the inverse problem the eigenvalues should not be close to each other [16].

Some Other Solvable Mass-Spring Systems — There are a number of other interesting problems closely related to the mass-spring system that we have studied. Among them is the question of recovery of masses and the spring constants of a finite system when we have the natural frequencies of N masses connected with springs with different spring constants plus the frequencies of the system when it is perturbed by modifying one mass and adding one spring to the system [29]. Another inverse problem of interest is the determination of a finite element model of longitudinally vibrating rod with one end fixed and the other end supported on a spring. This case has been studied by Wei and Dai [30].

Let us note that as long as the number n of masses (and springs) are finite the motion is reversible (see e.g. Ref. [31]).

1.7 Direct Problem of Non-exponential and of Exponential Decays in a Linear Chain

One of the natural generalizations of the mass-spring system is to study mechanical models where both the number of particles and springs go to infinity. The exact solution of such a problem is if interest, since the motion becomes irreversible i.e. once the system starts from a given configuration it will not come back to its original state.

First let us study a very special type of inverse problem for this case. Consider a linear chain of particles each of mass M_j, attached to each other by means of springs with spring constants K_j. Assume that initially all of the particles are at rest in their equilibrium position except one particle which is displaced and released with zero initial velocity. How we should choose different masses and different spring constants so that the motion of this particle decays exponentially?

The Decay of Motion in the Schrödiger Chain — Before discussing this question, let us investigate the solution of the direct problem for the case where masses are all equal and the connecting springs all have the the same spring constant, assuming that there are infinite number of masses and springs. This problem was first studied and solved by Schrödinger [32]. If we denote the mass of each particle by m and the spring constant by k, then the motion of the jth particle in the chain is given by

$$m\ddot{\xi}_j = k(\xi_{j+1} + \xi_{j-1} - 2\xi_j), \quad j = 0, \pm 1, \pm 2, \cdots, \tag{1.109}$$

where ξ_j is the displacement of the jth particle from the equilibrium and k is the spring constant which we write in terms of another constant ν with the dimension of wave number,

$$k = \frac{1}{4}m\nu^2. \tag{1.110}$$

Let assume that initially the central particle $j = 0$ is displaced by a distance A, and then released with zero initial velocity, while all of the other particles are initially at rest in their equilibrium position. Thus at $t = 0$ we have the initial conditions

$$\xi_0(t = 0) = A, \quad \xi_j(t = 0) = 0, \quad j \neq 0, \tag{1.111}$$

and

$$\dot{\xi}_0(t = 0) = 0, \quad j = 0, \pm 1, \pm 2 \cdots. \tag{1.112}$$

A method for solving the system of equations (1.109) is to construct the generating function $G(z, t)$ which we define by [33], [34]

$$G(z, t) = \sum_{j=-\infty}^{+\infty} \xi_j(t) z^{2j}. \tag{1.113}$$

We multiply (1.109) by z^j, we sum over all j s, and use (1.113) to obtain the following equation for $G(z, t)$:

$$\frac{d^2 G(z, t)}{dt^2} = \frac{1}{4} \nu^2 \left(\frac{1}{z} - z \right)^2 G(z, t). \tag{1.114}$$

We can find the initial conditions for this differential equation from (1.111)–(1.113).

$$G(z, 0) = A, \quad \left(\frac{dG(z, t)}{dt} \right)_{t=0} = 0. \tag{1.115}$$

With these conditions we find the solution of (1.113) to be

$$G(z, t) = A \cosh \left[\frac{1}{2} \nu t \left(z - \frac{1}{z} \right) \right]. \tag{1.116}$$

This $G(z, t)$ is the generating function for the Bessel function of even order [35];

$$\cosh \left[\frac{1}{2} \nu t \left(z - \frac{1}{z} \right) \right] = \sum_{j=-\infty}^{+\infty} J_{2j}(\nu t) z^{2j}, \tag{1.117}$$

and therefore

$$\xi_j(t) = A J_{2j}(\nu t). \tag{1.118}$$

Thus in this model the energy of the central particle is lost to the other particles in the system, and the rate of energy loss is

$$\left(\frac{dE_0}{dt} \right)_{t \to \infty} \to \left(\frac{m\nu^2 A^2}{2\pi t} \right) \cos(2\nu t), \tag{1.119}$$

i.e. the energy transfer between this particle and the rest of the system has an oscillatory behaviour and goes to zero as t^{-1}.

1.8   Inverse Problem of Dynamics for a Non-uniform Chain

Let us first consider the propagation of a disturbance along a non-uniform chain. If M_j denotes the mass of the jth particle in the chain and the spring constant between the particle j and $j+1$ is K_j, then we have the set of equations of motion

$$M_j \ddot{x}_j = K_j(x_{j+1} - x_j) + K_{j-1}(x_{j-1} - x_j), \quad j = 1, \, 2 \cdots N, \tag{1.120}$$

where x_j is the displacement of the jth particle from its position of equilibrium. Following the work of Dyson we introduce the variable ξ_j s by [36]

$$\xi_j = M_j^{\frac{1}{2}} x_j, \tag{1.121}$$

where all M_j s are greater than zero. Also let us introduce the set of constants $\omega_1, \omega_2, \cdots, \omega_{2N-2}$ by the following relations

$$\omega_{2j-1}^2 = \frac{K_j}{M_j}, \quad \omega_{2j}^2 = \frac{K_j}{M_{j+1}}, \tag{1.122}$$

then the equations of motion take the form

$$\ddot{\xi}_j = (\omega_{2j-1}\omega_{2j})\xi_{j+1} + (\omega_{2j-3}\omega_{2j-2})\xi_{j-1} - \left(\omega_{2j-1}^2 + \omega_{2j-2}^2\right)\xi_j. \tag{1.123}$$

Now we define the variables $z_1, z_2, \cdots, z_{N-1}$ by

$$\dot{z} = \omega_{2j}\xi_{j+1} - \omega_{2j-1}\xi_j, \tag{1.124}$$

and then we can rewrite (1.123) as

$$\dot{\xi}_j = \omega_{2j-1}z_j - \omega_{2j-2}z_{j-1}. \tag{1.125}$$

By combining (1.124) and (1.125) we arrive at a single set of first order equations:

$$\dot{X}_j = \omega_j X_{j+1} - \omega_{j-1} X_{j-1}, \tag{1.126}$$

where $X_1, X_2, \cdots, X_{2N-1}$ are defined by

$$X_{2j-1} = \xi_j, \quad X_{2j} = z_j. \tag{1.127}$$

The characteristic frequencies ν_j of the chain are the eigenvalues of a $(2N-1) \times (2N-1)$ matrix Λ whose elements are given by

$$\Lambda_{j+1,j} = -\Lambda_{j,j+1} = i\omega_j. \tag{1.128}$$

Here we want to choose the parameters M_j s and K_j s in such a way that in the limit of $N \to \infty$ we can find an exact solution for the motion of the particles, i.e. a solvable many-body problem. Let us consider a model where

$$\omega_j = j\lambda, \tag{1.129}$$

where λ is a positive constant. For this choice of w_j, the ratio of the masses of two neighbouring particles according to Eq. (1.122) is

$$\frac{M_{j+1}}{M_j} = \frac{(2j-1)^2}{4j^2}, \quad j = 1,\, 2, \cdots. \tag{1.130}$$

By substituting (1.121), (1.129) and (1.130) in (1.120), we find the equation of motion for x_j;

$$\ddot{x}_j = \lambda^2(2j-1)^2(x_{j+1} - x_j) + 4\lambda^2(j-1)^2(x_{j-1} - x_j). \tag{1.131}$$

Assuming that initially only the particle $j = 1$ is displaced by one unit of length and with no initial velocity, while all other particles in the chain are at rest with no initial displacement, i.e.

$$x_1(0) = 1, \quad \dot{x}_j(0) = 0, \quad j = 1,\, 2, \cdots, \tag{1.132}$$

and

$$x_j(0) = 0, \quad j \neq 1, \tag{1.133}$$

then we can write (1.131) with the initial conditions (1.132), (1.133) as a set of equations

$$\dot{X}_j = \lambda[jX_{j+1} - (j-1)X_{j-1}], \tag{1.134}$$

subject to the initial conditions

$$X_1(0) = M_1^{\frac{1}{2}}, \quad X_j(0) = 0, \quad j \neq 1. \tag{1.135}$$

We can solve this problem by the same method of constructing the generating function that we discussed earlier. Let us assume that the generating function is of the form

$$G(z,t) = \sum_{j=1}^{\infty} X_j(t)z^{j-1}, \tag{1.136}$$

then by differentiating $G(z,t)$ we find

$$z\frac{\partial}{\partial z}(zG(z,t)) = \sum_{j=1}^{\infty}(j-1)X_{j-1}z^{j-1}, \tag{1.137}$$

and

$$\frac{\partial}{\partial z}G(z,t) = \sum_{j=1}^{\infty} jX_{j+1}z^{j-1}. \tag{1.138}$$

Now if we multiply (1.134) by z^{j-1} and sum over all j values we find the following first order partial differential equation

$$\frac{\partial G(z,t)}{\partial t} = \lambda\left[\frac{\partial G(z,t)}{\partial z} - z\frac{\partial}{\partial z}(zG(z,t))\right]. \tag{1.139}$$

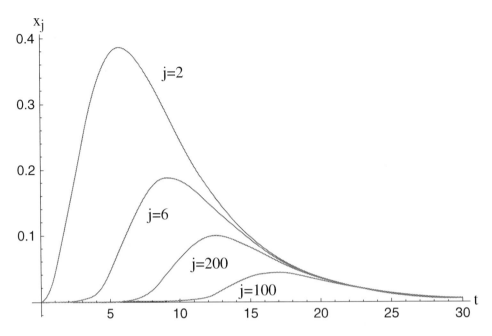

Figure 1.4: The displacement of the jth particle in the non-uniform chain as a function of time, Eq. (1.143), for $j = 2$, 6, 20, 100.

The initial condition on $G(z, t)$ can be found from Eqs. (1.135) and (1.136), i.e.

$$G(z, t = 0) = M_1^{\frac{1}{2}}. \tag{1.140}$$

The solution of (1.134) satisfying the boundary condition (1.140) can be expressed as

$$G(z, t) = \frac{M_1^{\frac{1}{2}}}{\cosh \lambda t}(1 + z \tanh \lambda t)^{-1}. \tag{1.141}$$

Now to find $X_j(t)$, we expand $G(z, t)$ in powers of z and compare the results with (1.136). Thus we find $X_j(t)$ to be

$$X_j(t) = \frac{M_1^{\frac{1}{2}}(-1)^{j-1}}{\cosh \lambda t}(\tanh \lambda t)^{j-1}. \tag{1.142}$$

From $X_j(t)$ we can find the displacement of the jth particle in this model;

$$M_j^{\frac{1}{2}} x_j(t) = \frac{M_1^{\frac{1}{2}}}{\cosh \lambda t}(\tanh \lambda t)^{2j-2}. \tag{1.143}$$

In particular the motion of the first particle is given by

$$x_1(t) = \frac{1}{\cosh \lambda t}, \tag{1.144}$$

which shows that for times $t > \frac{1}{\lambda}$ the displacement decays exponentially in time [37]. Since we have found the exact solution of motion according to the second law of dynamics, the time-reversed motion is also a solution of the problem. The damped motion of the first particle results from its coupling to a many-particle system.

1.9 Direct and Inverse Problems of Analytical Dynamics

In classical mechanics the Newton second law of motion establishes a relation between the motion of the bodies and the forces acting between them. Thus at first it seems that the direct problem should be the derivation (or construction) of the Lagrangian and the Hamiltonian from the second and the third laws of motion. Historically this was the direction of the development of analytical mechanics taken by Lagrange, Hamilton, Poisson and Jacobi. They showed how useful the analytical formulation of the classical mechanics (i.e. Lagrangian and Hamiltonian formulations) can be, both as a concise way of writing different mechanical problems and also as a very powerful method of finding approximate and exact solutions [38]–[41].

It is also possible to start with a Hamiltonian (or Lagrangian) and obtain the equations of motion using the method of calculus of variations. That is, if we choose the classical Hamiltonian function, $H(p_j, q_j, t)$, as the basic mathematical way of expressing a given motion, then the direct method shows that we can obtain a unique set of equations from the canonical equations of motion,

$$\frac{dq_j(t)}{dt} = \frac{\partial H}{\partial p_j}, \quad \text{and} \quad \frac{dp_j(t)}{dt} = -\frac{\partial H}{\partial q_j}, \quad j = 1, \, 2 \cdots N. \qquad (1.145)$$

In these relations $q_j(t)$ is the coordinate of the jth particle at the time t, $p_j(t)$ is its canonical momentum conjugate to $q_j(t)$, and N is the number of degrees of freedom of the system. By eliminating $p_j(t)$ s between the two equations in (1.145) we find the equations of motion for $q_1(t), q_2(t), \cdots, q_n(t)$. Here we expect to get a unique set of differential equations for $q_j(t)$ s. Thus the direct problem is well-defined for those problems for which the Hamiltonians (or Lagrangians) are known.

For the direct problem we start with Eqs. (1.145) and determine the motion in $2N$ dimensional phase space $p_1(t) \cdots p_N(t); q_1(t) \cdots q_N(t)$, or we find the equations of motion in the coordinate space:

$$H(p_j, \, q_j, t) \rightarrow \begin{cases} m_j \ddot{q}_j = F_j(q_1, \cdots, q_N; p_1, \cdots, p_N) \\ \\ \begin{cases} m_j \dot{q}_j = Q_j(q_1, \cdots, q_N; p_1, \cdots, p_N) \\ m_j \dot{p}_j = P_j(q_1, \cdots, q_N; p_1, \cdots, p_N) \end{cases} \end{cases} \qquad (1.146)$$

Thus if we start with $H(p_j, q_j, t)$, find the equations of motion either in the coordinate space or in the phase space, and then use these equations to obtain the first integrals, i.e. to solve the inverse problem. We will not always recover just $H(p_j, q_j, t)$, but among the pq-equivalent Hamiltonian, there will be one Hamiltonian that we started with. That is, we have a set of Hamiltonians giving us the correct equations of motion in coordinate space;

$$m_j\ddot{q}_j = F_j(q_1, \cdots, q_N; p_1, \cdots, p_N; t) \to K(p_j, q_j; t), \quad (q\text{-equivalent Hamiltonians}),$$
(1.147)

and another, smaller set generating the motion in phase space;

$$\begin{cases} m_j\dot{q}_j = Q_j(q_1, \cdots, q_N; p_1, \cdots, p_N; t) \\ m_j\dot{p}_j = P_j(q_1, \cdots, q_N; p_1, \cdots, p_N; t), \end{cases} \to H_1(p_j, q_j),$$
(1.148)

the latter is pq equivalent Hamiltonian. As a very simple example, let us consider the Hamiltonian for the damped motion of a single particle moving along the x axis:

$$H = \frac{p^2}{2m}e^{-\lambda t}.$$
(1.149)

From this H we obtain the equation of motion in the phase space

$$\frac{dp}{dt} = \frac{\partial H}{\partial q} = 0, \quad \frac{dq}{dt} = \frac{p}{m}e^{-\lambda t},$$
(1.150)

or [43]

$$\frac{d^2q}{dt^2} = -\lambda\frac{dq}{dt}.$$
(1.151)

This equation of motion can also be found from the more complicated Hamiltonian H_1, where

$$H_1 = \frac{1}{3mp_0}(2p_0p)^{\frac{3}{2}}e^{-\lambda t},$$
(1.152)

and p_0 is a constant.

Nonuniqueness of the Hamiltonian for Conservative Many-particle System — Starting with the equations of motion in the coordinate space for a conservative system we find nonuniqueness in determination of the Lagrangian and the Hamiltonian functions. For instance let us consider a system of N interacting particles with the equations of motion

$$m_j\frac{d^2\mathbf{r}_j}{dt^2} = -\nabla_j V(\mathbf{r}_1, \cdots, \mathbf{r}_N), \quad j = 1, 2, \cdots, N.$$
(1.153)

If we sum (1.153) over all j s we find that

$$\frac{d}{dt}\sum_{j=1}^{N} m_j\mathbf{v}_j = 0,$$
(1.154)

which implies the conservation of total momentum of the system. We can construct a Galilean invariant Lagrangian by adding a term

$$\frac{1}{2\mu} \sum_{j=1}^{N} (m_j \mathbf{v}_j)^2 , \qquad (1.155)$$

to the standard Lagrangian which is:

$$L(\mathbf{r}, \dot{\mathbf{r}}) = \sum_{j=1}^{N} m_j \mathbf{v}_j^2 - V. \qquad (1.156)$$

Thus if μ is a constant having the dimension of mass, then we write $L(\mu)$ as

$$L_\mu(\mathbf{r}, \dot{\mathbf{r}}) = \sum_{j=1}^{N} m_j \mathbf{v}_j^2 + \frac{1}{2\mu} \sum_{j=1}^{N} (m_j \mathbf{v}_j)^2 - V. \qquad (1.157)$$

The equations of motion derived from (1.157) are given by (1.153), since the time derivative of the total momentum is zero, Eq. (1.154). However the relation between the canonical momentum $\mathbf{p}_k$ and the velocity $\mathbf{v}_k$ is now more complicated, i.e.

$$\mathbf{p}_k = \frac{\partial L_\mu(\mathbf{r}, \dot{\mathbf{r}})}{\partial \mathbf{v}_k} = m_k \mathbf{v}_k + \left(\frac{m_k}{\mu}\right) \sum_{j=1}^{N} m_j \mathbf{v}_j. \qquad (1.158)$$

From this Lagrangian we find a Hamiltonian which also depends on the parameter μ;

$$H_\mu(\mathbf{r}, \mathbf{p}) = \sum_{j=1}^{N} \frac{\mathbf{p}_k^2}{2m} - \frac{(\sum_{j=1}^{N} \mathbf{p}_j)^2}{2\mu + \sum_{j=1}^{N} m_j} + V, \qquad (1.159)$$

where $\mathbf{p}_k$ s are given by (1.158).

1.10 From the Classical Equations of Motion to the Lagrangian and Hamiltonian Formulations

For most conservative systems it is easy to write down the standard Lagrangian (or the Hamiltonian) describing the motion. However if the forces acting on particles in a conservative system are known, we can ask whether it is possible to construct a unique Lagrangian (or Hamiltonian) for the motion or not? But when the system is dissipative, then even the question of existence of any Lagrangian (or Hamiltonian) is questionable.

Helmholtz's Conditions for Existence of a Lagrangian — The non-uniqueness of the Lagrangians function for dynamical systems will be studied in detail later.

But first let us examine the conditions for the existence of a Lagrangian function and how it can be constructed [38]–[41].

The laws governing the motion of a dynamical system can be formulated according to Hamilton's principle and this principle states that there is a function $L(x_1, \cdots x_N; \dot{x}_1 \cdots \dot{x}_N, t)$ and that this L satisfies the following condition [2]:

If the system at the instant t_1 has the coordinates $x_i^{(1)}$ and at t_2 has the coordinates $x_i^{(2)}$ then this condition implies that the functional differential δS defined by

$$\delta S = \delta \int_{t_1}^{t_2} L(x_i, \dot{x}_i, t)\, dt = \int_{t_1}^{t_2} \sum_{j=1}^{N} \frac{\delta L(x_i, \dot{x}_i, t)}{\delta x_j} \delta x_j(t) dt$$

$$= \int_{t_1}^{t_2} \sum_{j,k} \mu_{jk}(x_i, \dot{x}_i, t)\left[\ddot{x}_k - \frac{1}{m} f_k(x_i, \dot{x}_i, t) \right] \delta x_j(t) dt,$$

$$j = 1, 2, \cdots N, \tag{1.160}$$

must vanish identically for any continuous variation of the path consistent with the requirement

$$\delta x_j(t_1) = \delta x_j(t_2) = 0. \tag{1.161}$$

In Eq. (1.160) $\mu_{jk}(x_i, \dot{x}_i, t)$ is a positive definite matrix which may be viewed as an integrating factor [38]–[42]. Thus the requirement of the extremum of action S implies the vanishing of the integrand between any two arbitrary instances t_1 and t_2. Since we have imposed the condition of positive definiteness on the integrating factor μ_{jk} and since the variation $\delta x_j(t)$ is arbitrary (apart from the continuity condition and the variational conditions of (1.161)), therefore we conclude that

$$\ddot{x}_k(t) = \frac{1}{m} f_k(x_i, \dot{x}_i, t), \quad k = 1, 2, \cdots N, \tag{1.162}$$

which is Newton's second law of motion.

The Lagrangian function determined from the requirement of $\delta S \equiv 0$ is not unique. We can add the total time derivative of any function $\Phi(x, \dot{x}, t)$ to the Lagrangian without affecting the resulting equations of motion [43]. For the special case of $\mu_{jk} = \delta_{jk}$ we find the Lagrangian to be the same as that given by (1.156).

The second variation of S with respect to $\delta x_j(t)$ and $\delta x_k(t')$ must satisfy the identity:

$$\frac{\delta^2 S}{\delta x_j(t)\, \delta x_k(t')} = \frac{\delta^2 S}{\delta x_k(t')\, \delta x_j(t)}. \tag{1.163}$$

By imposing this condition on (1.160) we find the following set of equations which μ_{jk} s have to satisfy:

$$\mu_{jk} = \mu_{kj}, \tag{1.164}$$

$$\frac{\partial \mu_{kj}}{\partial \dot{x}_l} = \frac{\partial \mu_{jl}}{\partial \dot{x}_k} = \frac{\partial \mu_{lk}}{\partial \dot{x}_j}, \tag{1.165}$$

$$\hat{D}\mu_{jk} + \frac{1}{2m} \sum_{l} \left(\mu_{jl} \frac{\partial f_l}{\partial \dot{x}_k} + \mu_{kl} \frac{\partial f_l}{\partial \dot{x}_j} \right) = 0, \tag{1.166}$$

and

$$\hat{\mathcal{D}} \sum_k \left(\mu_{ik} \frac{\partial f_k}{\partial \dot{x}_j} - \mu_{jk} \frac{\partial f_k}{\partial \dot{x}_i} \right) - 2 \sum_k \left(\mu_{ik} \frac{\partial f_k}{\partial \dot{x}_j} - \mu_{jk} \frac{\partial f_k}{\partial \dot{x}_i} \right) = 0, \tag{1.167}$$

where in Eqs. (1.166) and (1.167) $\hat{\mathcal{D}}$ denotes the following differential operator

$$\hat{\mathcal{D}} = \frac{\partial}{\partial t} + \sum_k \left[\dot{x}_k \frac{\partial}{\partial x_k} + \frac{1}{m} f_k (x_i, \dot{x}_i, t) \frac{\partial}{\partial \dot{x}_k} \right]. \tag{1.168}$$

These are the Helmholtz conditions [38], [39].

We can construct the Lagrangian from the equations of motion

$$m\ddot{x}_k = f_k (x_i, \dot{x}_i, t), \quad k = 1, 2, \cdots N, \tag{1.169}$$

by solving a linear partial differential equation. To this end we first write the following identity:

$$\frac{\delta L}{\delta x_k} = \frac{\partial L}{\partial x_k} - \frac{d}{dt} \left(\frac{\partial L}{\partial \dot{x}_k} \right) \equiv \frac{\partial L}{\partial x_k} - \frac{\partial^2 L}{\partial t \partial \dot{x}_k}$$

$$- \sum_j \dot{x}_j \frac{\partial^2 L}{\partial x_j \partial \dot{x}_k} - \sum_j \ddot{x}_j \frac{\partial^2 L}{\partial \dot{x}_j \partial \dot{x}_k} = 0, \tag{1.170}$$

and then we substitute for $\ddot{x}$ from the equation of motion to obtain

$$\frac{\partial L}{\partial x_k} - \frac{\partial^2 L}{\partial t \partial \dot{x}_k} - \sum_j \dot{x}_j \frac{\partial^2 L}{\partial x_j \partial \dot{x}_k} - \frac{1}{m} \sum_j \frac{\partial^2 L}{\partial \dot{x}_j \partial \dot{x}_k} f_j (x_i, \dot{x}_i, t) = 0, k = 1 \, 2, \cdots N. \tag{1.171}$$

This equation together with the set of equations (1.164)–(1.167) determine both $L(x, \dot{x}, t)$ and $\mu(x, \dot{x}, t)$.

For another method of constructing Lagrangians and Hamiltonians for non-linear systems the reader is referred to a paper by Musielak *et al.* [44].

In the case of a one-dimensional motion Eq. (1.171) simplifies and later we will use it to find the Lagrangian for simple damped systems. But there are other ways of constructing the Lagrangian for motions confined to one dimension [45], [46]. For instance we can start with the equation of motion

$$m\ddot{x} = f (x, \dot{x}), \tag{1.172}$$

and write the Lagrangian for this motion as

$$L (x, \dot{x}) = \dot{x} \int^{\dot{x}} \frac{1}{v^2} G(v, x) dv, \tag{1.173}$$

where $G(v, x)$ is a function to be determined later. By substituting $L (x, \dot{x})$ in the Euler–Lagrange equation we find

$$\frac{d}{dt} \left(\frac{\partial L}{\partial \dot{x}} \right) - \frac{\partial L}{\partial x} = \frac{\ddot{x}}{\dot{x}} \left(\frac{\partial G (\dot{x}, x)}{\partial \dot{x}} \right) + \frac{\partial G (\dot{x}, x)}{\partial x} \equiv m\ddot{x} - f (\dot{x}, x). \tag{1.174}$$

From Eq. (1.174) it follows that $G(\dot{x},x)$ is the solution of the following differential equation

$$\frac{1}{m\dot{x}}f(x,\dot{x})\frac{\partial}{\partial\dot{x}}G(\dot{x},x) = -\frac{\partial}{\partial x}G(\dot{x},x),\qquad(1.175)$$

provided that

$$\frac{1}{\dot{x}}\frac{\partial}{\partial\dot{x}}G(\dot{x},x) \neq 0.\qquad(1.176)$$

Now let us consider two examples using this method of construction of the Lagrangian. When the equation of motion for a damped system is of the form

$$\ddot{x} + \frac{\dot{x}}{\left(\frac{dg(\dot{x})}{d\dot{x}}\right)}\frac{dV(x)}{dx} = 0,\qquad(1.177)$$

then the Lagrangian which is found by solving Eq. (1.175) is

$$L(x,\dot{x}) = \dot{x}\int^{\dot{x}}\frac{1}{v^2}F\left[z(x,v)\right]dv,\qquad(1.178)$$

where F is an arbitrary function of its argument and

$$z(v,x) = g(v) + V(x).\qquad(1.179)$$

However the condition

$$\frac{1}{v}\frac{dg(v)}{dv}\frac{dF(z)}{dz} \neq 0,\qquad(1.180)$$

must be satisfied. Thus for the very simple case of $\ddot{x} + \lambda\dot{x} = 0$ and for $F(z) = z$ we find

$$L(x,\dot{x}) = \dot{x}\ln\dot{x} - \lambda x.\qquad(1.181)$$

As a second example let us consider the damped harmonic oscillator

$$\ddot{x} + \lambda\dot{x} + \omega_0^2 x = 0.\qquad(1.182)$$

By calculating $G(\dot{x},x)$ we find that $G(\dot{x},x) = F(z)$, where now

$$z = \frac{1}{2}\frac{\left[x + \left(\frac{\lambda}{2} - i\omega\right)\dot{x}\right]^{\nu^*}}{\left[x + \left(\frac{\lambda}{2} + i\omega\right)\dot{x}\right]^{\nu}} = z^*,\qquad(1.183)$$

with z^* being the complex conjugate of z and

$$\nu^2 = \frac{\frac{\lambda}{2} + i\omega}{\frac{\lambda}{2} - i\omega},\qquad(1.184)$$

with $\omega^2 = \omega_0^2 - \frac{\lambda^2}{4}$.

These q-equivalent Hamiltonians are acceptable as generators of motion in analytical dynamics, and depending on the integrating factors used, there are infinitely many Hamiltonians. However as we will see in Sec. 3.3 they are not admissible for quantizing a particular system. Since their classical phase space trajectories are very different from those of the standard Hamiltonians they may not be also admissible even in classical statistical mechanics.

The Inverse Problem for Liouville Equation — The inverse problem of the classical Liouville equation can be described as the problem of finding a time-dependent Hamiltonian H which generates a given phase space density ρ, and both the Hamiltonian H and the density ρ satisfy the same Liouville equation [6]. Frisch and collabortors have shown that, subject to some physically well-defined subsidiary conditions, the solution exists and is unique, when ρ is chosen to be the solution of the standard Fokker–Planck equation when the particles are free, are in a constant external field, or are harmonically bound.

Let us consider a system of particles and denote the coordinate and the canonical momentum of the jth particle by q_j and p_j respectively. As we have seen earlier we can derive the equation of motion of these particles from a time-dependent Hamiltonian $H(q, p, t) \equiv H(q_i, p_i, t)$, Eq. (1.145). We assume the following initial conditions for the first order equations of motion:

$$q_j(0) = q_j, \quad p_j(0) = p_j, \quad H(0) = H_0. \tag{1.185}$$

In classical statistical mechanics one wants to find the time-dependent density

$$\rho(q, p, t) \equiv \rho \tag{1.186}$$

corresponding to a given $H = H(q, p, t)$ which satisfies Liouville's equation

$$\frac{\partial \rho}{\partial t} + \{\rho,\, H\} = \frac{\partial \rho}{\partial t} + \sum_j \left(\frac{\partial \rho}{\partial q_j} \frac{\partial H}{\partial p_j} - \frac{\partial \rho}{\partial p_j} \frac{\partial H}{\partial q_j} \right) = 0, \tag{1.187}$$

subject to the initial condition

$$\rho(0) = \rho_0(q, p). \tag{1.188}$$

Now we define the dynamical inverse problem for the Liouville equation by the following procedure [6]:

(1) We assume a marginally consistent, time-dependent point process given by $\rho(q, p, t)$ and its marginal distributions.

(2) We try to find a "Hamiltonian" function $H(q, p, t)$ corresponding to the ρ given in (1.186) which is a solution of Eq. (1.187).

By considering (1.187) as a linear partial differential equation for $H = H(q, p, t)$, we have the characteristic equations for H

$$\frac{dp_j}{-\dfrac{\partial \rho}{\partial q_j}} = \frac{dq_j}{\dfrac{\partial \rho}{\partial p_j}} = \frac{dH}{\dfrac{\partial \rho}{\partial t}}. \tag{1.189}$$

These equations are not identical with Hamilton's canonical equations (1.145) for H except when ρ is judiciously chosen [6]. Solution of (1.189) only determines the general solution of (1.187). If one specifies the initial condition on H and in addition provides information about H on a characteristic strip or singular region of phase space, then one can show the existence of the Cauchy problem for H with very weak conditions on the continuously differentiable density ρ [48].

1.11   Langevin and Fokker–Planck Equations

If a particle of mass m is immersed in a fluid with the friction coefficient γ, its equation of motion (in one dimension) is given by

$$\frac{dv}{dt} + \gamma v = -K(x) + A(t). \tag{1.190}$$

In this equation v is the speed of the particle, x is its position, $K(x)$ is the acceleration due to mechanical forces and $A(t)$ is the acceleration due to random forces.

The Fokker–Planck equation for the distribution in phase space of the single particle, $\mathcal{P}(x, v, t)$ is given by

$$\frac{\partial \mathcal{P}}{\partial t} + v\frac{\partial \mathcal{P}}{\partial x} - K(x)\frac{\partial \mathcal{P}}{\partial v} = \gamma\frac{\partial(v\mathcal{P})}{\partial v} + D\gamma^2\frac{\partial^2 \mathcal{P}}{\partial v^2}. \tag{1.191}$$

The solution of (1.191) subject to the boundary condition

$$\mathcal{P}(x, v, 0) = \delta(x - x_0)\delta(v - v_0), \tag{1.192}$$

can be obtained for the following forms of $K(x)$:

$$K(x) = \begin{cases} 0 & \text{free particle} \\ F_0 & \text{a constant force} \\ \omega^2 x & \text{harmononically bound particle} \end{cases}. \tag{1.193}$$

Solutions of the Fokker–Planck equation for these cases are given in [6], and the general Hamiltonian for a the motion of a particle in a viscous medium with frictional force βv is discussed in detail in [49]–[51].

References

[1] I. Newton, *The Principia: Mathematical Principles of Natural Philosophy*, (University of California Press, 1999).

[2] L.D. Landau and E.M. Lifshitz, *Mechanics*, (Pergamon Press. Oxford, 1960), Chap. III.

[3] N.H. Abel, J. Reine Angew. Math. 1, 13 (1826).

[4] A.H. Carter, Am. J. Phys. 68, 698 (2000).

[5] M. Bocher, *An Introduction to the Study of Integral Equations*, Cambridge Tracts in Mathematics and Mathematical Physics, No. 10 (Hefner, New York, 1960).

[6] R. Gorenflo and S. Vessella, *Abel Integral Equations, Analysis and Applications*, (Springer-Verlag, Berlin, 1991).

[7] I.S. Gradshteyn and I.M. Ryzhik, *Tables of Integrals, Series, and the Products*, (Academic Press, New York, 1965) p. 875.

[8] G. Arfken, *Mathematical Methods for Physicists*, Third Edition, (Academic Press, New York, 1985) p. 875.

[9] E.T. Whittaker, *A Treatise on the Analytical Dynamics of Particles and Rigid Bodies*, Fourth Edition (Cambridge University Press, London 1960).

[10] H. Goldstein, *Classical Mechanics*, (Addison-Wesley, Reading, 1956).

[11] J.F. Marko and M. Razavy, Lettere Al Nuovo Cimento, 40, 533, 1984.

[12] J.B. Keller, Am. Math. Monthly, 83, 107 (1976).

[13] O.B. Firsov, Zh. Eksp. Teor. Fiz. 24, 279 (1953). Firsov's solution is described in L.D. Landau and E.M. Lifshitz, *Mechanics*, (Pergamon Press, Oxford, 1960) p. 52.

[14] J.B. Keller, I. Kay and J. Shmoys, Phys. Rev. 557, 102 (1956).

[15] M.T. Chu and G.H. Golub, *Inverse Eigenvalue Problems: Theory, Algorithm and Applications*, (Oxford University Press, London, 2005).

[16] G.M.L. Gladwell, *Inverse Problems in Vibration*, Second Edition, Volume 119 of *Solid Mechanics and Its Applications*, (Dordrecht, Kluwer, 2004).

[17] See for instance *The Symmetric Eigenvalue Problem*, (SIAM, Classics in Applied Mathematics, 1988).

[18] D. Belkić, in *Advances in Quantum Chemistry* V. 145, Chapter 4 (2011).

[19] I.M. Gel'fand and S.V. Fornin, *Calculus of Variation*, translated by R.A. Silverman, (Dover, New York, 1991).

[20] C. Lanczos, J. Res. Natl. Bur. Stand. 45, 255 (1950), ibid. 49, 33 (1958).

[21] For other methods of inversion of a Jacobi matrix see D.N. Ghosh Roy. *Methods of Inverse Problems in Physics*, (CRC Press Boca Raton, 1991) Chapter 11.

[22] G. Meurant, SIAM J. Matrix Anal. Appl. 13, 707 (1992).

[23] X. Wu and E. Jiang, J. Shanghai University, 11, 27 (2007).

[24] G.M.I. Gladwell and N.B. Williams, Inverse Problems, 4, 1013 (1988).

[25] R.A. Usmani, Computer Mat. Appl. 27, 59 (1994).

[26] J.H. Wilkinson, *The Algebraic Eigenvalue Problem*, (Oxford University Press, Oxford, 1965) p. 388.

[27] A.S. Housholder, *The Theory of Matrices in Mathematical Analysis*, (Blaisdell, New York, 1964) p. 254.

[28] G.H. Golub and G. Meurant, *Matrices, Moments and Quadrature with Applications*, (Princeton University Press, Princeton, 2010), Chapter 4.

[29] R. del Rio and M. Kudryavtsev, Inverse Problems, 28, 055007 (2012).

[30] Y. Wei and H. Dai, J. Comp. Appl. Math, 300, 172 (2016).

[31] H.J. Kreuzer, *Nonequilibrium Thermodynamics and Its Statistical Foundations*, (Oxford University Press, Oxford, 1981).

[32] E. Schrödinger, Ann. Phys. (Leipzig), 44, 916 (1914).

[33] R. Bellman and K.L. Cook, *Differential-Difference Equtions*, (Academic Press, New York, 1963).

[34] M. Razavy, Can. J. Phys. 57, 1731 (1979).

[35] I.S. Gradshtetyn and I.M. Ryzhik, *Table of Integrals, Series and Products*, (Academic Press, New York, 1965) p. 973.

[36] F.J. Dyson, Phys. Rev. 92, 1331 (1953).

[37] M. Razavy, Can. J. Phys. 58, 1019 (1980).

[38] P. Havas, Nuovo Cimento Supp. 5, 363 (1957).

[39] H. Helmholtz, J. Reine Angew. Math. 100, 137 (1887).

[40] R.M. Santilli, *Foundations of Theoretical Mechanics: The Inverse Problem in Newtonian Mechanics*, (Springer-Verlag, New York, 1978).

[41] V. Dodonov, V.I. Man'ko and V.D. Skarzhinsky, Had. J. 4, 1734 (1981).

[42] C. Lanczos, *The Variational Principles of Mechanics*, Fourth Edition, (Dover, New York, 1986).

[43] For a number of similar examples see M. Razavy *Classical and Quantum Dissipative Systems*, Second Edition (World Scientific, Singapore, 2017).

[44] Z.E. Musielak, D. Roy and L.D. Swift, Chaos, Solitons & Fractals, 38, 894 (2008).

[45] J.A. Kobussen, Acta Phys. Austriaca, 59, 293 (1979).

[46] S. Okubo, Phys. Rev. A 23, 2776 (1981).

[47] H.L. Frisch, G. Forgacs and S.T. Chui, Phys. Rev. A 20, 561 (1979).

[48] H. Bateman, *Differential Equations*, (Chelsea, New York, 1966).

[49] M. Razavy, Can. J. Phys. 56, 311 (1978).

[50] M. Razavy, Can. J. Phys. 56. 1372 (1978).

[51] M. Razavy, *Classical and Quantum Disiipative Motion*, Second Edition, (World Scientific, Singapore, 2017).

Chapter 2

Inverse Problems in Semiclassical Formulation of Quantum Mechanics

To find the solution to a large number of quantum mechanical problems, the semi-classical approximation either simplifies the calculation of bound and scattering problems, and in some others like tunneling of particles it is a powerful and most of the times the only way to formulate and obtain approximate solutions [1]. In this chapter we consider the inverse problem for the bound states of one-dimensional confining potentials, and in Chapter 12 we study the question of tunneling.

2.1 Quantum Mechanical Bound States for Confining Potentials

As we will see later, for a one-dimensional motion or for a three-dimensional spherically symmetric potentials, we can use Gel'fand–Levitan [2], [3] equation or Abraham–Moses [4] approach to find the potential exactly, but here we show how it is possible to use Abel's integral equation and determine the potential approximately [5]. The semiclassical or WKB approximation is well-known and the conditions for its validity can be found in most of the textbooks on quantum mechanics. Here we will only consider the inverse problem for confining potentials, i.e. those for which $\lim_{x \to \pm\infty} V(x) \to \infty$ or $\lim_{r \to \pm\infty} V(r) \to \infty$. For the one-dimensional potentials that we want to consider first, the WKB quantization condition is

$$\int_{x_1}^{x_2} \sqrt{E - V(x)}\, dx = (n + \epsilon)\, \frac{\pi \hbar}{\sqrt{2m}}. \tag{2.1}$$

In this relation ϵ is a parameter which depends on the shape of the potential, e.g. for the harmonic oscillator it is $\frac{1}{2}$ whereas for an infinite square well it is zero. The two limits of the integral x_1 and x_2 are the classical turning points, $V(x_1) = V(x_2) = E$. We want to consider one-dimensional potentials and we write the quantization rule as

$$\Phi(E) = \int_0^a \sqrt{E - V(x)}\, dx = \left(n + \frac{1}{2}\right) \frac{\pi \hbar}{2\sqrt{2m}}, \tag{2.2}$$

where a is the classical turning point $V(a) = E$. Now if we differentiate (2.2) with respect to E, change the variable from x to V, and choose the energy scale so that $V(0) = 0$, then we have an approximate form of the potential having the given eigenvalues (see Table 2.1).

$$\frac{d\Phi(E)}{dE} = \frac{1}{2} \int_0^E \frac{\left(\frac{dx}{dV}\right)}{\sqrt{E - V}}\, dV. \tag{2.3}$$

This is exactly an Abel integral equation the we have seen before Eq. (1.16). We invert (2.3) to find $\left(\frac{dx}{dV}\right)$

$$\frac{dx}{dV} = \frac{2}{\pi} \frac{d}{dV} \int_0^V \frac{\left(\frac{d\Phi(E)}{dE}\right)}{\sqrt{V - E}}\, dE. \tag{2.4}$$

Table 2.1: Eigenvalues for some solvable potentials have been used to find the shapes of the corresponding potentials [5].

$E(n)$	$V(x)$	Shape of the Potential
$n^{\frac{2}{3}}$	$\propto |x|$	Ramp
n	$\propto x^2$	Harmonic oscillator
$n^{\frac{4}{3}}$	$\propto x^4$	Anharmonic oscillator
n^2	....	Infinite square well

Finally we integrate (2.4) to obtain x as a function of V:

$$x(V) = \frac{2}{\pi} \int_0^V \frac{\left(\frac{d\Phi(E)}{dE}\right) dE}{\sqrt{V-E}}. \tag{2.5}$$

By solving this equation for $V(x)$ we find the confining potential [5].

2.2 Semiclassical Formulation of the Inverse Scattering Problem

In scattering theory when a large number of partial waves contribute to the scattering amplitude, $f(\theta)$, and the phase shifts are large, rather than summing over many partial waves, it is simpler to use the semiclassical approximation (WKB method) and express the scattering amplitude as an integral over the impact parameter, as we will discuss in this section.

We can start with the WKB approximation for both the angular part, viz, Legendre equation, and for the radial part, i.e. for the radial Schrödinger equation, and then use the resulting expressions and write the scattering amplitude in terms of an integral over the impact parameter. This interesting approach is discussed in an elegant way in the book "Quantum Mechanics" by Landau and Lifshitz [6].

First, let us consider the semiclassical solution of the Legendre differential equation

$$\frac{d^2 P_\ell}{d\theta^2} + \cot\theta \frac{dP_\ell}{d\theta} + \ell(\ell+1)P_\ell = 0, \tag{2.6}$$

where by substituting

$$P_\ell(\cos\theta) = \frac{\chi(\theta)}{\sqrt{\sin\theta}}, \tag{2.7}$$

in (2.6) we can eliminate the first derivative and the resulting equation for $\chi(\theta)$ becomes

$$\chi'' + \left[\left(\ell+\frac{1}{2}\right)^2 + \frac{1}{4}\csc^2(\theta)\right]\chi(\theta) = 0. \tag{2.8}$$

In this equation prime denotes differentiation with respect to θ. The de Broglie wavelength λ associated with this motion is

$$\lambda = \left[\left(\ell+\frac{1}{2}\right)^2 + \frac{1}{4}\csc^2\theta\right]^{-\frac{1}{2}}. \tag{2.9}$$

The requirement that the rate of change of λ with respect to θ should be small for the validity of semiclassical approximation, for the angular part of the wave function gives the inequalities

$$\theta\ell \gg 1, \quad (\pi - \theta)\ell \gg 1. \tag{2.10}$$

With these conditions the second term in the bracket in (2.8) can be ignored, except for the range of angles very close to zero or π. Thus we get

$$\chi''(\theta) + \left(\ell + \frac{1}{2}\right)^2 \chi(\theta) = 0. \tag{2.11}$$

The solution of this equation is

$$\chi(\theta) = A \sin\left[\left(\ell + \frac{1}{2}\right)\theta + \alpha\right], \tag{2.12}$$

where A and α are the constants of integration. Substituting (2.12) in (2.7) we get

$$P_\ell(\cos\theta) = A\frac{\sin\left[\left(\ell + \frac{1}{2}\right) + \alpha\right]}{\sqrt{\sin\theta}}. \tag{2.13}$$

To determine A and α we observe that for angles $\theta \ll 1$ in Eq. (2.6) we can put $\cot\theta \approx \frac{1}{\theta}$ and also replace $\ell(\ell+1)$ by $\left(\ell + \frac{1}{2}\right)^2$, and thus obtain

$$\frac{d^2 P_\ell}{d\theta^2} + \frac{1}{\theta}\frac{dP_\ell}{d\theta} + \left(\ell + \frac{1}{2}\right)^2 P_\ell = 0. \tag{2.14}$$

This equation has the form of the Bessel function of zero order, therefore the approximate solution of (2.6) becomes

$$P_\ell(\cos\theta) = J_0\left[\left(\ell + \frac{1}{2}\right)\theta\right], \quad \theta \ll 1. \tag{2.15}$$

Since we have $P_\ell(\theta = 0) = 1$, therefore we have set the constant multiplicative factor equal to one. For $\theta\ell \gg 1$, the Bessel function can be replaced by its asymptotic expression for large values of the argument. Thus

$$P_\ell(\cos\theta) \approx \sqrt{\frac{2}{\pi\ell}}\frac{\sin\left[\left(\ell + \frac{1}{2}\right)\theta + \frac{1}{4}\pi\right]}{\sqrt{\theta}}. \tag{2.16}$$

Comparing (2.16) with (2.15) we find

$$A = \sqrt{\frac{2}{\pi\ell}}, \quad \text{and} \quad \alpha = \frac{\pi}{4}. \tag{2.17}$$

We conclude that in the semiclassical limit the Legendre polynomial can be approximated by

$$P_\ell(\cos\theta) \approx \sqrt{\frac{2}{\pi\ell}}\sin\left[\left(\ell + \frac{1}{2}\right)\theta + \frac{\pi}{4}\right]. \tag{2.18}$$

Semiclassical Expression for the Scattering Amplitude — We start with the exact expression for the scattering amplitude $f(\theta)$ which is

$$f(\theta) = \frac{1}{2ik} \sum_{\ell=0}^{\infty} (2\ell + 1) P_\ell(\cos\theta) e^{2i\delta_\ell}. \tag{2.19}$$

In the semiclassical regime the phases δ_ℓ are large, and then the value of the sum in (2.19) is determined by the terms with large integers ℓ. Therefore we can use the asymptotic expression for $P_\ell(\cos\theta)$ which we write as

$$P_\ell(\cos\theta) \approx \frac{i}{\sqrt{2\pi\ell \sin\theta}} \left\{ \exp\left[i\left(\ell + \frac{1}{2}\right)\theta + \frac{i\pi}{4} \right] - \exp\left[-i\left(\ell + \frac{1}{2}\right)\theta - \frac{i\pi}{4} \right] \right\}, \tag{2.20}$$

to find $f(\theta)$. By substituting (2.20) in (2.19) we have

$$f(\theta) = \frac{1}{k} \sum_{\ell=0}^{\infty} \sqrt{\frac{\ell}{2\pi \sin\theta}} \left\{ \exp\left[i\left(2\delta_\ell - \left(\ell + \frac{1}{2}\right)\theta - \frac{\pi}{4} \right) \right] \right.$$
$$\left. - \exp\left[i\left(2\delta_\ell + \left(\ell + \frac{1}{2}\right)\theta + \frac{\pi}{4} \right) \right] \right\}. \tag{2.21}$$

Now we observe that each term in (2.21), because of their large phases, are rapidly oscillating functions of ℓ. The major contribution to the sum over ℓ comes from the values of ℓ from the neighborhood of the roots of the equation

$$2\frac{d\delta_\ell}{d\ell} \pm \theta = 0. \tag{2.22}$$

Around this extremum, there are a large number of terms in the series for which the exponential factors have almost the same value and they do not cancel each other. Next we have to find the phases δ_ℓ in the same classical limit that we have obtained $P_\ell(\cos\theta)$. Applying the same technique to solve the radial Schrödinger equation, we find that the semiclassical phase shifts (WKB) are related to the potential $V(r)$ and to the energy $E = \frac{\hbar^2 k^2}{2m}$ by [6]

$$\delta_\ell = \int_{r_0}^{\infty} \left\{ \frac{1}{\hbar} \left[2m(E - V(r)) - \frac{\hbar^2 \left(\ell + \frac{1}{2}\right)^2}{r^2} \right]^{\frac{1}{2}} - k \right\} dr$$
$$+ \frac{\pi}{2} \left(\ell + \frac{1}{2}\right) - k r_0. \tag{2.23}$$

In this relation r_0 is the classical turning point which is dependent on ℓ. Now we introduce the angular momentum of the incident particle $mvb = \hbar\left(\ell + \frac{1}{2}\right)$, where b is the impact parameter and v is the velocity of the particle at infinity. By calculating $\frac{d\delta_\ell}{d\ell}$ we find that the scattering angle θ from (2.22) is given by

$$\frac{1}{2}(\pi \mp \theta) = \int_{r_0}^{\infty} \frac{mvb\,dr}{r^2 \left[2m(E - V(r)) - \left(\frac{mvb}{r}\right)^2 \right]^{\frac{1}{2}}}, \tag{2.24}$$

and this is the same expression that we found for classical scattering Eq. (1.44). In order to have a root for b, for a repulsive field we choose the minus sign and for an attractive field we use the plus sign in front of θ in (2.24) [6].

Equation (2.24) is identical with the classical equation giving us the scattering angle for the potential $V(r)$ (compare with Eq. (1.44)). Therefore for the semiclassical scattering we can use the same method of inversion that we discussed in Sec. 1.5 for the inverse scattering problem in classical mechanics.

The Semiclassical (WKB) Method of Inversion — The deflection function, whose magnitude is equal to the center of mass scattering angle, is of great value in the interpreting the features of the differential cross section and also for solving inverse problem of semiclassical scattering. We found that the derivative of the "classical" deflection angle with the respect to $\ell + \frac{1}{2} = \lambda$ is given by Eq. (2.22). For this case the Able integral equation can be inverted to yield the quasi-potential $Q(\sigma)$:

$$
\begin{aligned}
Q(\sigma) &= \frac{2E}{\pi} \int_{\sigma}^{\infty} \frac{\theta(\lambda)}{\sqrt{\lambda^2 - \sigma^2}} d\lambda \\
&= \frac{4E}{\pi} \frac{1}{\sigma} \frac{d}{d\sigma} \left[\int_{0}^{\infty} \frac{\delta(\lambda)}{\sqrt{\lambda^2 - \sigma^2}} \lambda d\lambda \right].
\end{aligned}
\tag{2.25}
$$

The quasi-potential which is related to the potential which is a function of the radial distance r by Sabtier transformation [7], [8]

$$
V_{WKB}(r) = E \left[1 - \exp\left(-\frac{Q(\sigma)}{E} \right) \right],
\tag{2.26}
$$

provided that there is one-to-one correspondence between r and σ:

$$
r = \frac{1}{k} \sigma \exp\left(\frac{Q(\sigma)}{2E} \right).
\tag{2.27}
$$

This condition is satisfied if

$$
E > V(r) + \frac{1}{2} r \frac{dV(r)}{dr}.
\tag{2.28}
$$

Now the definition of $Q(\sigma)$ shows that in the limit of $\sigma \to 0$, $Q(\sigma) \to \infty$ and in this limit the transform indicates that V tends to E, and r approaches r_0, where r_0 is the classical turning point.

For a detailed discussion of the methods of inversion based on WKB approximation see Chapter 12.

References

[1] M. Razavy, *Quantum Theory of Tunneling*, Second Edition, (World Scientific, Singapore, 2014) Chapter 24.

[2] I.M. Gel'fand and B.M. Levitan, Dokl. Akad. Nauk SSSR, 77, 557 (1951).

[3] I.M. Gel'fand and B.M. Levitan, Isvest. Akad. Nauk SSSR, 15, 309 (1951).

[4] P.B. Abraham and H.E. Moses, Phys. Rev. A22, 1333 (1980).

[5] A.H. Carter, Am. J. Phys. 68, 698 (2000).

[6] L.D. Landau and E.M. Lifshitz, *Quantum Mechanics: Non-relativistic Theory*, (Pergamon Press, London, 1958) Chapter XIV.

[7] J.B. Keller, I. Kay and J. Shamoys, Phys. Rev. 102, 557 (1956).

[8] P.C. Sabatier, Nuovo Cimento, 37, 1180 (1956).

Chapter 3

Inverse Problems and the Heisenberg Equations of Motion

Before discussing the inverse problem in wave mechanics and the Schrödinger picture, we will briefly consider the inverse problem for the Heisenbeg equations of motion, and whether they lead to new results or not. Since in this case we are dealing with equations involving non-commuting operators, we confine our attention only to one-dimensional motions [1], [2].

In Heisenberg's quantum mechanics the direct problem is to derive the equations of motion from the Hamiltonian operator $H(x, p)$, where x and p are operators satisfying the commutation relation $[p,\ x] = -i\hbar$, and the time derivative of any operator $F(x, p)$, (not explicitly dependent on time), is given by [1]

$$\frac{dF(x, p)}{dt} = \frac{i}{\hbar}[F(x, p),\ H]. \tag{3.1}$$

As in classical dynamics, here we have the equations of motion in the phase space, the Hamiltonian, and in addition we have the canonical commutation relations [1]. Now in the direct problem we assume that the Hamiltonian operator is given in terms of the non-commuting operators p and x, and we want to obtain the equations of motion in the phase space, or starting with the Hamiltonian and eliminating p between $\frac{dp}{dt}$ and $\frac{dx}{dt}$ to obtain the equation of motion for the operator x, i.e. $m\frac{d^2x}{dt^2} = f(x)$, where $f(x)$ is the force acting on the particle.

3.1   Equations of Motion Derived from the Hamiltonian Operator

Let us study the equation of motion for the operators $p(t)$ and $q(t)$ obtained from the Hamiltonian operator via Heisenberg's equations of motion. For simplifying the problem we consider a one-dimensional motion, we set $\hbar = 1$ and write the equations of motion as

$$i\frac{dq(t)}{dt} = [q(t),\ H], \quad i\frac{dp(t)}{dt} = [p(t),\ H]. \tag{3.2}$$

We want to solve these equations as an initial value problem, i.e. to express the operator $q(t)$ as

$$q(t) = \sum_{m,n} c_{m,n}(t) p^m(0) q^n(0), \tag{3.3}$$

where $q(0)$ and $p(0)$ are the initial coordinate and momentum operators respectively and $c_{m,n}(t)$ is a time-dependent functions. Since $p^m(0)q^n(0)$ is not a Hermitian operator we replace it with a Weyl-ordered operator $T_{m,n}(0)$ which is Hermitian and is defined by [3]–[6]

$$T_{m,n}(t) = \left(\frac{1}{2}\right)^n \sum_{k=0}^{n} \frac{n!}{(n-k)!k!} q^k(t) p^m(t) q^{n-k}(t). \tag{3.4}$$

By replacing $p(t) = T_{1,0}(t)$ and $q(t) = T_{0,1}(t)$ in (3.2) we get

$$\begin{cases} i\frac{dT_{0,1}(t)}{dt} = [T_{0,1}(t),\ H] \\ i\frac{dT_{1,0}(t)}{dt} = [T_{1,0}(t),\ H] \end{cases}. \tag{3.5}$$

As Bender and Dunne [3] have shown, the operators $T_{m.n}(t)$ s form an algebra, and that the product $T_{m.n}(t)T_{r,s}(t)$ can be written in terms of $T_{j,k}(t)$ by rearranging the orders of $p(t)$ and $q(t)$;

$$T_{m,n}(t)T_{r,s}(t) = \sum_{j=0}^{\infty} \frac{\left(\frac{i}{2}\right)^j}{j!} \sum_{k=0}^{j} (-1)^{j-k} \binom{j}{k} \frac{n!}{(n-k)!} \frac{m!}{(m+k-j)!}$$

$$\times \frac{r!}{(r-k)!} \frac{s!}{(s+k-j)!} T_{m+r-j,\ n+s-j}(t). \tag{3.6}$$

Using this relation for the products $T_{m,n}(t)T_{r,s}(t)$, we can find the commutation relation and express the result as a linear expression in $T_{j,k}(t)$;

$$[T_{m,n}(t),\ T_{r,s}(t)] = 2 \sum_{j=0}^{\infty} \frac{\left(\frac{i}{2}\right)^{2j+1}}{(2j+1)!} \sum_{k=0}^{2j+1} (-1)^k \binom{2j+1}{k}$$

$$\times \frac{\Gamma(n+1)\Gamma(m+1)\Gamma(r+1)\Gamma(s+1)}{\Gamma(m-k+1)\Gamma(n+k-2j)\Gamma(r+k-2j)\Gamma(s-k+1)}$$

$$\times T_{m+r-2j-1,\ n+s-2j-1}(t). \tag{3.7}$$

Using this commutation relation we can write the time development of any element of the set $\{T_{m,n}(t)\}$. If we assume that the potential $V(q)$ in the Hamiltonian is a polynomial in q, then the Hamiltonian has a simple expression in terms of $T_{m,n}(t)$ s. As an example let us take $V(q)$ to be

$$V(q) = \frac{1}{2}q^2 + \frac{\lambda}{4}q^4, \tag{3.8}$$

then the Hamiltonian can be expressed as

$$H = \frac{1}{2}p^2(t) + \frac{1}{2}q^2(t) + \frac{1}{4}\lambda q^4(t) \tag{3.9}$$

$$= \frac{1}{2}T_{2,0}(t) + \frac{1}{2}T_{0,2}(t) + \frac{\lambda}{4}T_{0,4}(t). \tag{3.10}$$

For this Hamiltonian the time development of $T_{m,n}(t)$ is obtained from the Heisenberg equations

$$i\frac{dT_{m,n}(t)}{dt} = [T_{m,n}(t),\ H] = \left[T_{m,n}(t),\ \frac{1}{2}T_{2,0}(t) + \frac{1}{2}T_{0,2}(t) + \frac{\lambda}{4}T_{0,4}(t)\right]$$

$$= i\left\{nT_{m+1,n-1}(t) - m\left(T_{m-1,n+1}(t) + \lambda T_{m-1,n+3}(t)\right)\right\} \tag{3.11}$$

$$+ \left(\frac{i\lambda m}{4}m(m-1)(m-2)T_{m-3,n=1}\right). \tag{3.12}$$

This relation shows that $T_{0,1}(t) = q(t)$ is coupled to all of the other $T_{m,n}(t)$ s. Thus only for problems of motion in a constant field, a linear potential and a harmonic oscillator we find uncoupled linear equations for $\dot{q}(t) = T_{0,1}(t)$ and $\dot{p}(t) = T_{1,0}(t)$.

For the general case the phase space trajectory for the particle is given by

$$\begin{cases} i\frac{d\langle 0|T_{0,1}(t)|0\rangle}{dt} = \langle 0\,[T_{0,1}(t),\ H]\,0|\rangle \\[2mm] i\frac{d\langle 0|T_{1,0}(t)|0\rangle}{dt} = \langle |0\,[T_{1,0}(t),\ H]\,|0\rangle \end{cases}, \tag{3.13}$$

where $|0\rangle$ is the initial state of the system.

3.2 Determination of the Commutation Relations From the Equations of Motion

The first inverse problem in this category was studied and solved by Wigner for the quantum mechanical harmonic oscillator was the following [2]: Let us consider the following set of rules for the operators replacing the dynamical variables. Setting the mass of the particle m, the Planck's constant $\hbar$ and also the spring constant equal to one, we have:

(1) The generator of the infinitesimal evolution operator is given by the Hamiltonian

$$H = \frac{1}{2}\left(p^2 + q^2\right).$$
(3.14)

(2) The Heisenberg equations of motion defines the time derivatives of q and p.

$$i\dot{q} = [q,\ H],$$
(3.15)

$$i\dot{p} = [p,\ H].$$
(3.16)

By differentiating (3.15) and finding $\ddot{q}$ and then eliminating $\dot{q}$ we have the analogue of the second law

$$\ddot{q} = -[H,\ [H,\ q]] = -q$$
(3.17)

(3) The operators q and p satisfy the canonical commutation relations

$$[p,\ q] = -i, \quad [p,\ p] = [q,\ q] = 0.$$
(3.18)

These fundamental relations are compatible with each other [1]. The direct problem is simple and the solution is given in the well-known book of Landau and Lifshitz [7].

The inverse problem solved by Wigner is as follows: Given the Hamiltonian (3.14), and the Heisenberg equation of motion (3.15) and (3.16), find the most general form of the commutation relation $[p,\ q]$. To this end we consider a representation in which H is diagonal. Now because of the positive-definiteness of the Hamiltonian (3.14), the diagonal elements are all positive. Let us denote these elements by $E_1, E_2, \cdots$, and find the matrix elements of q and p of Eqs. (3.15) and (3.16);

$$p_{n,m} = i(E_m - E_n)q_{n,m},$$
(3.19)

$$q_{n,m} = -i(E_m - E_n)p_{n,m}.$$
(3.20)

From these relations it follows that

$$q_{n,m} = (E_n - E_m)^2 q_{n,m}.$$
(3.21)

Thus the matrix elements $q_{n,m}$ will be nonzero only if

$$E_n - E_m = \pm 1.$$
(3.22)

Also from (3.19) we have $p_{n,m} = 0$ when $q_{n,m} = 0$. Now from Eq. (3.22) it follows that E_n s form an arithmetical series

$$E_n = E_0 + n.$$
(3.23)

If E_n appears in the diagonal elements of H several times, then by means of a unitary transformation we can decompose any system of matrices in which E_n occurs repeatedly without changing H. Thus we can assume that all of the eigenvalues of H are simple. From (3.21) and (3.22) it is clear that only $q_{n,n+1}$ and $q_{n+1,n}$ are nonzero. In addition, since the position operator q is Hermitian, therefore we can assume that its matrix elements must be real. Then from (3.18) we find that all $p_{n,m}$ s are pure imaginary;

$$p_{n,n+1} = -iq_{n,n+1} = -iq_{n+1,n}, \tag{3.24}$$

and

$$p_{n+1,n} = iq_{n+1,n} = -p_{n,n+1}. \tag{3.25}$$

To find the elements of $q_{n,n+1}$ and $q_{n+1,n}$, let us first determine the elements of the diagonal matrix H. From (3.14) we have

$$
\begin{cases}
E_n = E_0 + n = q_{n-1,n}^2 + q_{n,n+1}^2, & n \neq 0 \\[2mm]
E_0 = q_{0,1}^2 & n = 0.
\end{cases} \tag{3.26}
$$

This relation gives us a way of calculating successive values of $q_{n,n+1}$ with the following result

$$
\begin{cases}
q_{n,n+1} = \left(E_0 + \frac{1}{2}n\right)^{\frac{1}{2}}, & \text{for even} \quad n \\[2mm]
q_{n,n+1} = \left(\frac{1}{2} + \frac{1}{2}n\right)^{\frac{1}{2}}, & \text{for odd} \quad n.
\end{cases} \tag{3.27}
$$

Next we find $p_{n,n+1}$ and $p_{n+1,n}$ from (3.24) and (3.25) and substitute these in the expression

$$[p, q]_{n,m} = (pq)_{n,m} - (qp)_{n,m} \tag{3.28}$$

and find that this commutator is diagonal and its diagonal elements are

$$-2iq_{1,0}^2, \quad -2i\left(q_{1,2}^2 - q_{0,1}^2\right), \quad -2i\left(q_{2,3}^2 - q_{1,2}^2\right) \cdots \tag{3.29}$$

Thus we have the result that

$$\left([p, q]_{n,m} + i\right)^2 = -(2E_0 - 1)^2. \tag{3.30}$$

For the special value of $E_0 = \frac{1}{2}$ i.e. the ground state energy of the harmonic oscillator Eq. (3.30) reduces to the well-known result $[p, q] = -i$.

Let us emphasize that this generalized canonical commutation relation is derived for the harmonic oscillator. As Wigner has pointed out we cannot obtain such a relation for potentials of the form $V(x) \propto x^{2n+1}$ (n is zero or an integer).

As we have noticed in Wigner's approach, the canonical solution of the quantum harmonic oscillator is not assumed, instead the compatibility of Hamilton's equations and the Heisenberg's equations are carefully examined. This generalized form of the oscillator called the Wigner oscillator or parabose oscillator plays an important role in the theory of paraboson statistics with applications to quantum field theory and to nuclear physics [8], [9].

3.3   Construction of the Hamiltonian Operator as an Inverse Problem

As we mentioned in the previous section, from the Hamiltonian operator and the equation of motion we can find the general form of the canonical commutation relation for the harmonic oscillator [2]. Now let us suppose that the equation of motion for the position operator is known, $m\ddot{q} = F(q)$, and in addition we are also given the canonical commutation relation $[q, \ p] = i\hbar$. Given these two operator equations is it possible to construct other Hamiltonians in addition to the standard one? In order to differentiate between the standard Hamiltonian of the system $H = p^2/2m + V(q)$ and other generators of motion we denote the latter group by $K(p,q)$. We want to determine $K(p,q)$ with $q = i\partial/\partial p$ (we set $\hbar = 1$) such that

$$\dot{q} = i[K, \ q], \quad \text{and} \quad \ddot{q} = i[K, \ \dot{q}], \tag{3.31}$$

or

$$\ddot{q} = \frac{1}{m}F(q) = -\frac{1}{m}\frac{\partial V(q)}{\partial q} = -[K, \ [K, \ q]]. \tag{3.32}$$

We can use this last equation $\ddot{q} = -[K, \ [K, \ q]]$ to find K. A rather trivial Hamiltonian can be found from H by gauge transformation. Thus if $K(p,q) \equiv H(p,q)$ is a solution of (3.32), then its gauge transformed form H_G is also a solution;

$$H_G = \frac{1}{2m}\left(p + \frac{\partial\chi(q,t)}{\partial q}\right)^2 + V(q) + \frac{\partial\chi(q,t)}{\partial t}. \tag{3.33}$$

In this relation $\chi(q,t)$ is a differentiable but otherwise an arbitrary function of q and t.

Now let us study a more complicated form of K. Here we consider the special form where $K(p,q)$ can be written as a product of two functions $R(p)$ and $S(q)$, i.e.

$$K(p,q) = R(p)S(q), \tag{3.34}$$

with $q = i\frac{\partial}{\partial p}$. With this $K(p,q)$, Eq. (3.32) can be written as

$$\ddot{q} = i\left\{R(p)S(q)\frac{dR(p)}{dp} - \frac{dR(p)}{dp}S(q)R(p)\right\}S(q). \tag{3.35}$$

Let us also assume that $S(q)$ is expressible as a finite or an infinite sum of powers of q, i.e.

$$S(q) = \sum_{k=0}^{\infty} C_k q^k, \tag{3.36}$$

then the operator $S\left(i\frac{d}{dp}\right)R(p)$ can be written as [10], [11]

$$S\left(q = i\frac{d}{dp}\right)R(p) = \sum_{n=0}^{\infty} \frac{(i)^n}{n!}\left(\frac{d^n R(p)}{dp^n}\right)\left(\frac{d^n S(q)}{dq^n}\right). \tag{3.37}$$

Similarly we have

$$S(q)\frac{dR(p)}{dp} = \sum_{n=0}^{\infty} \frac{(i)^n}{n!}\left(\frac{d^{n+1}R(p)}{dp^{n+1}}\right)\left(\frac{d^n S(q)}{dq^n}\right). \tag{3.38}$$

By substituting (3.37) and (3.38) in (3.35) we find

$$\ddot{q} = \frac{1}{m}F(q) = \sum_{n=0}^{\infty} \frac{i^{n+1}}{n!}\left[R(p)\frac{d^{n+1}R(p)}{dp^{n+1}} - \frac{dR(p)}{dp}\frac{d^n R(p)}{dp^n}\right]\frac{d^n S(q)}{dq^n}S(q). \tag{3.39}$$

Now the right-hand side of (3.39) should be independent of p and since $F(q)$ is a real function we have different possibilities, but we will consider only two examples [12].
(1) We can choose the following set of constants for the bracket in (3.39):

$$\left[R(p)\frac{d^{n+1}R(p)}{dp^{n+1}} - \frac{dR(p)}{dp}\frac{d^n R(p)}{dp^n}\right] = \begin{cases} C^{-n-1} & \text{if } n \text{ is an odd integer} \\ 0 & \text{if } n \text{ is an even integer} \end{cases}. \tag{3.40}$$

The solution of this equation is

$$R(p) = \cosh\left(\frac{p}{C} + D\right), \tag{3.41}$$

where C and D are constants. For this choice of $R(p)$ from Eq. (3.39) we find that the force law is

$$F(q) = m\sum_{n=0}^{\infty} \frac{(-1)^{n+1}}{(2n+1)!}\frac{d^{2n+1}S(q)}{dq^{2n+1}}\frac{S(q)}{C^{2n+2}}. \tag{3.42}$$

If $F(q)$ is known, it is difficult to solve (3.42) for $S(q)$ except for very special cases. For instance if we choose $F = -\omega^2 q$, then from (3.42) it follows that $S = C\omega q$, or if we choose $S(q) = \sqrt{\frac{CF_0}{m}}\exp\left(-\frac{1}{2}\mu q\right)$, then $F = F_0 e^{-\mu q}\sin\left(\frac{\mu}{2C}\right)$ [12].
(2) If we set $F(q) = 0$, then

$$R(p)\frac{d^{n+1}R(p)}{dP^{n+1}} - \frac{dR(p)}{dp}\frac{d^n R(p)}{dp^n} = 0. \tag{3.43}$$

From this relation we find

$$K(p, q) = e^{\beta p}S(q), \tag{3.44}$$

where β is a constant.

Note that these Hamiltonians, while in classical dynamics are acceptable, in quantum theory they are not, since they do not generate the motion in phase space correctly.

References

[1] M. Razavy, *Heisenberg's Quantum Mechanics*, (World Scientific, Singapore, 2011).

[2] E.P. Wigner, Phys. Rev. 77, 711, (1950).

[3] C.M. Bender and G.V. Dunne, Phys. Rev. D 40, 2739 (1989).

[4] C.M. Bender and G.V. Dunne, Phys. Rev. D 40, 3504 (1989).

[5] M. Hron and M. Razavy, Phys. Rev. A 51, 4365 (1995).

[6] M. Hron and M. Razavy, Phys. Rev. A 54, 3801 (1996).

[7] L.D. Landau and E.M. Lifshitz, *Quantum Mechanics: Nonrelativistic Theory*, (Pergamon Press, London, 1958) p. 64.

[8] H.S. Green, Phys. Rev. 90, 270 (1953).

[9] E. Jafarov, S. Lievens and J. Van der Jeugt, J. Phys. A: Mathematical and Theoretical, 41, 235301 (2008).

[10] M. Born and P. Jordan, Z. Phys. 34, 858 (1925). The English translation of this article can be found in B.L. van der Waerden, *Sources of Quantum Mechanics*, (North-Holland, Amsredam, 1967).

[11] M. Born and N. Wiener, Z. Phys. 36, 174 (1926).

[12] M. Razavy, Can. J. Phys. 50, 2037 (1972).

Chapter 4

Inverse Scattering Problem for the Schrödinger Equation and the Gel'fand–Levitan Formulation

In Chapters 1 and 2 we studied the classical and semiclassical inverse scattering problems and in both cases we showed that once we know the scattering angle as a function of the impact parameter, the potential can be determined by solving an Abel integral equation. Knowing that the concept of the impact parameter is a classical one, we have to formulate the problem differently, i.e. replace the classical equations of motion by the Schrödinger equation, and the impact parameter by angular momentum or partial waves. In most of the texts on wave mechanics these concepts are discussed in detail. For quantal treatment of scattering problem we need to define some of the basic ideas of scattering theory. The essential ideas and the necessary mathematical techniques can be found in a number of books [1]–[8].

Scattering Amplitude and the Scattering Matrix — Let us assume that the differential cross section $\sigma(k, \theta)$ is given, from this $\sigma(k, \theta)$ we can find the partial wave scattering amplitude $A_\ell(k)$ for the ℓth partial wave where

$$\sigma(k, \theta) = \left| \sum_\ell (2\ell + 1) A_\ell(k) P_\ell(\cos \theta) \right|^2 . \tag{4.1}$$

Once $A_\ell(k)$ is found the phase shift can be calculated from

$$A_\ell(k) = \frac{1}{k} e^{i\delta_\ell} \sin \delta_\ell. \tag{4.2}$$

Now we want to find a local energy-independent potential $v(r) = \frac{2m}{\hbar^2} V(r)$ which is responsible for scattering. There are two ways to formulate the solution of this inverse problem:
(a) We can determine the Jost function and use it as the input, this is the Gel'fand–Levitan approach which we will discuss in this chapter [9]–[12].
(b) We can use the partial wave scattering matrix, S_ℓ, defined by

$$2i A_\ell(k) = S_\ell(k) - 1, \tag{4.3}$$

and solve Marchenko's equation to find $v(r)$ [13]. Marchenko's formulation will be discussed in the next chapter.

4.1   The Jost Solution

We start with the Schrödinger equation for a central short range potential $V(r) = \frac{2m}{\hbar^2} v(r)$ which for the S-state (or $\ell = 0$) partial wave takes the simple form

$$\frac{d^2}{dr^2} \phi(k,r) + k^2 \phi(k,r) = v(r)\phi(k,r). \tag{4.4}$$

In the following discussion we assume that the potential $v(r)$ satisfies the integrability conditions

$$\int_0^\infty |v(r)| dr < \infty. \tag{4.5}$$

$$\int_0^\infty r|v(r)| dr < \infty. \tag{4.6}$$

In the absence of the potential $v(r)$, the solution of (4.4), which is zero at the origin and its derivative is $\phi'(k,0) = 1$, (prime denotes derivative with respect to r) is given by

$$\phi(k,r) = \frac{\sin kr}{k}. \tag{4.7}$$

But first we consider a particularly important solution of (4.4) which we denote by $f(k,r)$ and it is obtained when we impose the boundary conditions:

$$\lim_{r \to \infty} f(k,r) = e^{ikr}, \tag{4.8}$$

$$\lim_{r \to \infty} f'(k,r) = ike^{ikr}. \tag{4.9}$$

We call this solution as the Jost solution [9]–[11]. In Eq. (4.9) we follow the notation used in Newton's book [1]. However in majority of the text books on quantum scattering theory, [2]–[5] the Jost function is defined by

$$\lim_{r \to \infty} \exp(\pm ikr) f(\pm k, r) = 1. \tag{4.10}$$

This function, $f(k, r)$, is called the "Jost solution" of the Schrödinger equation [9]–[11]. From these relations it follows that the functions $f(k, r)$ and $f(-k, r)$ are independent of each other, since according to (4.8) and (4.9) the Wronskian of the two is not zero

$$W[f(k, r),\ f(-k, r)] = -2ik. \tag{4.11}$$

We can find the regular solution of (4.4), i.e. the solution which satisfies the boundary conditions

$$\phi(k, 0) = 0, \quad \text{and} \quad \phi'(k, 0) = 1, \tag{4.12}$$

in the following way:

Using the method of variation of constants we can write the formal solution for $\phi(k, r)$ as a Volterra integral equation [14]

$$\phi(k, r) = \frac{\sin kr}{k} + \int_0^r \frac{\sin\left[k\left(r - r'\right)\right]}{k} V\left(r'\right) \phi\left(k, r'\right) dr'. \tag{4.13}$$

Similarly for the Jost solution, $f(k, r)$, is also the solution of the wave equation (4.4), but now with the boundary conditions (4.8) and (4.9) we obtain

$$f(k, r) = e^{ikr} - \int_r^\infty \frac{\sin\left[k\left(r - r'\right)\right]}{k} V\left(r'\right) f\left(k, r'\right) dr'. \tag{4.14}$$

The exact solution $\phi(k, r)$ satisfying the boundary conditions (4.12) can be written as a linear combination of $f(k, r)$ and $f(-k, r)$, viz,

$$\begin{aligned}
\phi(k, r) &= \left(\frac{1}{2ik}\right) [f(-k, 0) f(k, r) - f(k, 0) f(-k, r)] \\
&= \left(\frac{f(k)}{2ik}\right) [S(k) f(k, r) - f(-k, r)], \\
&\to \frac{f(k)}{2ik} \left[S(k) e^{ikr} - e^{-ikr}\right] \\
&= \frac{f(k) \psi(k, r)}{k} \quad \text{as} \quad r \to \infty. \tag{4.15}
\end{aligned}$$

The function $\psi(kr)$ which is related to $\phi(k, r)$ by

$$\psi(k, r) = \frac{k\phi(k, r)}{f(k)}, \tag{4.16}$$

is the physical wave function. That is the inverse square of the modulus of $f(k)$ measures the probability of finding the particle near $r = 0$ relative to the probability if there were no forces present. Clearly $\phi(k,r)$ satisfies the boundary condition (4.12). In this equation $S(k)$ is the $\ell = 0$ partial wave scattering matrix which is defined by Eq. (4.3) and can be expressed in terms of the Jost function $f(k,0)$ by

$$S(k) = \frac{f(-k,0)}{f(k,0)}. \tag{4.17}$$

Let us also note that $f(k,r)$ has the important property that for real values of k

$$[f(k,r)]^* = f(-k,r). \tag{4.18}$$

As an example consider the S-wave scattering by an exponential potential

$$V(r) = V_0 \exp\left(-\frac{r}{a}\right). \tag{4.19}$$

By changing the variable from r to x where $x = \exp\left(-\frac{r}{a}\right)$, we can find the solution of the Schrödinger equation in terms of the Bessel function of imaginary order. The Jost solution for this potential is found to be

$$f(k,r) = \exp\left[ika\ln\left(a^2 V_0\right)\right] \Gamma(1 - 2iak) J_{-2iak}\left(2a\sqrt{V_0}e^{-\frac{r}{2a}}\right), \tag{4.20}$$

and the Jost function is

$$f(k) = f(k,r=0) = \exp\left[ika\ln\left(a^2 V_0\right)\right] \Gamma(1 - 2iak) J_{-2iak}\left(2a\sqrt{V_0}\right). \tag{4.21}$$

Having obtained $f(k,r)$ and $f(k)$, the physical wave function can be found from (4.16).

4.2 The Jost Function

The function $f(k) \equiv f(k,0)$ is called the Jost function and it plays an important role in quantum scattering theory [7]–[10].

From Eq. (4.18) we find that the Jost function satisfies the relation

$$[f(k)]^* = f(-k). \tag{4.22}$$

For bound states the energy in the Schrödinger equation is negative, and k is pure imaginary, $k = i\gamma$, $\gamma > 0$. The asymptotic form of $\phi(k,r)$ as $r \to \infty$ can be found from (4.15);

$$\phi(i\gamma,r)_{r\to\infty} \to \frac{-1}{2\gamma}\left[f(i\gamma)e^{\gamma r} - f(-i\gamma)e^{-\gamma r}\right]. \tag{4.23}$$

Thus in this limit $\phi(i\gamma, r)$ becomes unbounded unless we set the coefficient of $e^{\gamma r}$ equal to zero, i.e. to require that Jost function be zero for $k = i\gamma$ or in other words $k = i\gamma$ be a pole of the S matrix, i.e. as Eq. (4.17) shows $f(i\gamma) = 0$.

In the case of the exponential potential the bound state is given by the roots of

$$J_{2a\gamma_n}\left(2a\sqrt{V_0}\right). \tag{4.24}$$

Normalization of the Bound State Wave Function — We can relate the wave function normalization to the Jost function and its derivatives using the following method:

We calculate the normalization constant for the bound state by considering the Schrödinger equation for $f(k, r)$,

$$f''(k, r) + k^2 f(k, r) = v(r) f(k, r). \tag{4.25}$$

By differentiating this equation with respect to k and denoting $\left(\frac{\partial f(k,r)}{\partial k}\right)$ by $\dot{f}(k, r)$ we have;

$$\dot{f}''(k, r) + k^2 \dot{f}(k, r) + 2k f(k, r) = v(r) \dot{f}(k, r). \tag{4.26}$$

Now we multiply (4.25) by $\dot{f}(k, r)$ and (4.26) by $f(k, r)$, subtract the two equations and integrate the result with respect to r from zero to infinity to find the following expression

$$\dot{f}'(k, 0) f(k, 0) - \dot{f}(k, 0) f'(k, 0) = 2k \int_0^\infty f^2(k, r) dr. \tag{4.27}$$

For the bound state $f(-i\gamma, r)$ is real and $f(i\gamma, 0) \equiv f(i\gamma) = 0$, and from (4.15) we get

$$\phi(i\gamma, r) = -\left(\frac{1}{2i\gamma}\right) f(-i\gamma) f(i\gamma, r), \tag{4.28}$$

therefore we find the normalization to be

$$\int_0^\infty \phi^2(i\gamma, r) dr = -\frac{\dot{f}(i\gamma)}{2i\gamma f'(i\gamma, 0)}. \tag{4.29}$$

S-Wave Phase Shift — Now if we write $f(k)$ in terms of its norm and its phase we have

$$f(-k) = |f(k)| e^{i\delta(k)}, \tag{4.30}$$

and we define the scattering matrix $S(k)$ by

$$\frac{f(-k)}{f(k)} = e^{2i\delta(k)}, \tag{4.31}$$

where $\delta(k)$, which is the phase shift caused by the potential, is an odd function of k, as can be seen from (4.30)

$$\delta(-k) = -\delta(k). \tag{4.32}$$

Using the properties of $f(\pm k)$ and $f(\pm k, r)$ we can write the asymptotic form of solution of (4.15) as

$$\phi(k,r)_{r\to\infty} = \frac{|f(k)|}{|k|}\sin(kr + \delta(k)), \tag{4.33}$$

and

$$\phi'(k,r)_{r\to\infty} = |f(k)|\cos(kr + \delta(k)). \tag{4.34}$$

In addition the phase shift $\delta(k)$ goes to zero as $k \to \infty$. We will show this asymptotic behaviour of the phase shift for regular potentials, i.e. those satisfying the condition

$$\lim_{r\to0} r^2 v(r) = 0, \quad v(r) = \frac{2m}{\hbar^2}V(r). \tag{4.35}$$

Let us assume that near the origin the potential can be characterized by

$$v(r) \sim v_0(r) = \frac{(\ln r)^n}{r^m}, \quad (m \le 2 \ \text{if} \ \ m = 2, \ n < 0). \tag{4.36}$$

Now for high energies, i.e. large real k values, the phase shift $\delta(k)$ approaches $\delta^B(k)$ which is given by

$$\delta(k) \to \delta^B(k) = -k^{-1}\int_0^\infty v(r)\sin^2(kr)dr. \tag{4.37}$$

We separate the integral into two parts:

$$\delta(k) \sim -k^{-1}\int_0^R v(r)\sin^2(kr)dr - k^{-1}\int_R^\infty v(r)\sin^2(kr)dr, \tag{4.38}$$

where for small R and large k we have [15]

$$k^{-1}\int_R^\infty v(r)\sin^2(kr)dr \sim k^{-1}. \tag{4.39}$$

To evaluate the first term in (4.38) we choose R small enough so that we can use the asymptotic form of the potential for small r;

$$k^{-1}\int_0^R v(r)\sin^2(kr)dr \sim k^{-1}\int_0^\infty v_0\left(\frac{x}{k}\right)\sin^2 x \, dx$$

$$= k^{m-2}\int_0^{kR} x^{-m}\ln^n\left(\frac{x}{k}\right)\sin^2 x \, dx. \tag{4.40}$$

We can see that $\sin^2 x$ in the integrand ensures the convergence of the integral at the lower limit, and thus we may write

$$\delta(k) = -k^{m-2}\int_0^{kR} x^{-m}\ln^n\left(\frac{x}{k}\right)\sin^2 x \, dx \sim -k^{m-2}\int_0^{kR} x^{-m}\ln^n\left(\frac{x}{n}\right)dx. \tag{4.41}$$

Depending on the value of m, there are three different cases that we need to consider:
(1) When $m < 1$, we note that the last integral in (4.41) diverges if the upper limit diverges, and this can be ascertained by partial integration. Thus we find that

$$\delta(k) \to -k^{m-2} \int_0^{kR} x^{-m} \ln^n \left(\frac{x}{k}\right) dx \sim -k^{-1}. \tag{4.42}$$

(2) If $m = 1$, we can evaluate the last integral in (4.41) exactly. The result is

$$k^{-1} \int_0^{kR} x^{-1} \ln^n \left(\frac{x}{k}\right) dx \sim \begin{cases} k^{-1} \ln^{n+1} k & \text{for } n \neq -1 \\ \\ k^{-1} \ln \ln k & \text{for } n = -1. \end{cases} \tag{4.43}$$

From Eqs. (4.39) and (4.43) for $m = 1$ for the asymptotic form of the phase shift we get

$$\delta(k) \sim - \begin{cases} k^{-1} \ln^{n+1} k & \text{if } n > -1 \\ k^{-1} \ln \ln k & \text{if } n = -1 \\ k^{-1} & \text{if } n < -1. \end{cases} \tag{4.44}$$

(3) For the case where $1 < m \leq 2$, the right-hand side of (4.40) converges even in the limit of $kR \to \infty$. Therefore

$$\delta(k) = -k^{m-2} \int_1^{kR} x^{-m} \ln^n \left(\frac{x}{k}\right) dx \to k^{m-2} \ln^n k. \tag{4.45}$$

For nearly all of the problems that we will consider, we assume the asymptotic behaviour given by (4.42).

4.3 The Levinson Theorem

This theorem establishes a relation between the number of bound states in the system and the scattering phase shift at zero and at infinite energy. For simplicity again we consider the S-wave scattering, but the result is valid for any partial wave.

Starting from the definition of the S matrix, $S(k) = e^{2i\delta(k)}$ and differentiating it with respect to k, we find

$$\frac{1}{S(k)} \frac{dS(k)}{dk} = 2i \frac{d\delta(k)}{dk}. \tag{4.46}$$

Noting that $\delta(k)$ is an odd function of k, $\delta(k) = -\delta(-k)$ we write the integral of the left-hand side of (4.46) as

$$I = \int_{-\infty}^{\infty} \frac{1}{S(k)} \frac{dS(k)}{dk} dk = 4i \int_0^{\infty} \frac{d\delta(k)}{dk} dk = 4i[\delta(\infty) - \delta(0)], \tag{4.47}$$

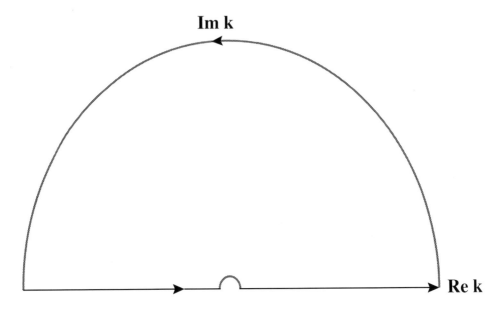

Figure 4.1: The contour for the integration for Eq. (4.49).

where we have regarded the phase shift as a continuous function of k in the range $0 \le k < \infty$.

We also write I in terms of the Jost function

$$I = \int_{-\infty}^{\infty} \frac{d}{dk}[\ln f(-k) - \ln f(k)]dk = 2\int_{-\infty}^{\infty} \frac{d}{dk}\ln f(k)dk. \qquad (4.48)$$

We can evaluate the last integral by contour integration. Let us choose a contour consisting of the following parts: (a) a semi-circle C_1 of radius k_c in the upper half of the k plane, (b) a line joining the points $k = -k_c$ to $k = -\varepsilon$ and from $k = \varepsilon$ to k_c, and (c) a semi-circle of radius ε centered at the origin. Thus we have

$$\frac{1}{2\pi i}\oint_C \frac{d}{dk}[\ln f(k)]dk = \frac{1}{2\pi i}\int_{-k_c}^{k_c} \frac{d}{dk}[\ln f(k)]dk$$

$$+ \frac{1}{2\pi i}\int_C \frac{d}{dk}[\ln f(k)]dk + \frac{1}{2\pi i}\int_\varepsilon \frac{d}{dk}[\ln f(k)]dk. \qquad (4.49)$$

In the limit of $k_c \to \infty$, the contour C will include all of the simple zeros at the points $k_n = i\gamma_n$, $\gamma_n > 0$, $n = 1, 2, \cdots, N$, and these are the poles of the integrand $\frac{d}{dk}\ln[f(k)]$. The first integral on the right-hand side of (4.49) contributes $\frac{1}{2\pi i}\left(\frac{I}{2}\right)$. The second integral (along the semi-circle) goes to zero as the radius of the semi-circle tends to infinity. The last integral along the semi-circle of radius ε contributes an amount πi provided that $f(0) = 0$ (which can happen only for $\ell = 0$

partial wave). Thus from Eqs. (4.47)–(4.49) we conclude that

$$\delta(0) - \delta(\infty) = \begin{cases} \pi\left(N + \frac{1}{2}\right), & \text{if } f(0) = 0 \\ \pi N & \text{otherwise} \end{cases}. \tag{4.50}$$

Here we assume that $r^2 v(r)$ is integrable at infinity. If the potential has a finite range then $\delta(\infty) = 0$, and $\delta(0)$ is determined by the number of bound states.

4.4 The Gel'fand–Levitan Equation

In this section we study a method for constructing a central potential from the known information about the scattering phase shift and also about the bound state energies. For simplicity we confine our discussion to the Schrödinger equation for the S-wave, and we choose the units so that $\hbar = 2m = 1$. In addition we assume that the potential is of short range and satisfies the integrability condition Eqs. (4.5) and (4.6).

We write the Schrödinger equation as

$$\bar{D}(r)\bar{\phi}(E, r) = -k^2\bar{\phi}(E, r), \quad E = k^2 \tag{4.51}$$

where $\bar{\phi}(E, r)$ satisfies the boundary condition (4.12) and $\bar{D}(r)$ denotes the differential operator

$$\bar{D}(r) = \frac{d^2}{dr^2} - \bar{v}(r). \tag{4.52}$$

Now let us define a function $g(r, r')$ which is symmetric in r and r' by the integral

$$g(r, r') = \int_{-\infty}^{+\infty} \bar{\phi}(E, r)\bar{\phi}(E, r') \, dh(k), \tag{4.53}$$

where $h(k)$ is a weight function to be determined later. Because of the symmetry of $g(r, r')$, we have

$$\bar{D}(r)g(r, r') = \bar{D}(r')g(r, r'), \tag{4.54}$$

with the boundary conditions

$$g(r, 0) = g(0, r') = 0. \tag{4.55}$$

Now let us denote the unique solution of the linear integral equation of Gel'fand–Levitan by $K(r, r')$, where

$$K(r, r') = g(r, r') - \int_0^r K(r, r'')g(r'', r') \, dr'', \tag{4.56}$$

and where the kernel $K(r, r')$ does not depend on the energy of the particle. The uniqueness of the solution of this equation is discussed in detail in references [1], [6].

Next we introduce a function $\xi(r, r')$ by

$$\xi(r, r') = D(r)K(r, r') - \bar{D}(r') K(r, r'), \tag{4.57}$$

where $D(r)$ is related to $\bar{D}(r)$ by

$$D(r) = \bar{D}(r) - \Delta v(r), \tag{4.58}$$

and $\Delta v(r)$ is given by

$$\Delta v(r) = -2\frac{d}{dr}K(r, r). \tag{4.59}$$

From the definition of $\xi(r, r')$, Eq. (4.57), and the boundary conditions (4.12) it follows that the function $\xi(r, r')$ satisfies the homogeneous form of the integral equation (4.56). Since the solution of (4.56) is unique therefore the homogeneous solution of this equation is the trivial one $\xi(r, r') \equiv 0$. This result implies that

$$D(r)K(r, r') = \bar{D}(r') K(r, r'). \tag{4.60}$$

In addition from (4.55) we find the boundary conditions for $K(r, r')$:

$$K(r, 0) = K(0, r') = 0. \tag{4.61}$$

Once the solution $K(r, r')$ of Eq. (4.56) is found, we can obtain a wave function $\phi(E, r)$ through the relation

$$\phi(E, r) = \bar{\phi}(E, r) - \int_0^r K(r, r') \bar{\phi}(E, r') \, dr'. \tag{4.62}$$

Next by applying the operator $D(r)$, Eq. (4.58), to the two sides of (4.62), twice integrating by parts and using Eqs. (4.53), (4.60), we find the differential equation for $\phi(E, r)$;

$$D(r)\phi(E, r) = -k^2\phi(E, r). \tag{4.63}$$

To find the boundary condition for $\phi(E, r)$, we note that from (4.61) it follows that $\phi(E, r)$ is the regular solution of the Schrödinger equation, i.e. the solution which satisfies the boundary condition,

$$\phi(E, 0) = 0 \tag{4.64}$$

but now the potential in the wave equation is

$$v(r) = \bar{v}(r) + \Delta v(r). \tag{4.65}$$

Let us try to relate $h(E)$ to the function $\phi(E, r)$. In order to obtain this relationship we observe that the above-mentioned procedure can be reversed. Thus we can start with the equation for $\phi(E, r)$, Eq. (4.63), with the potential $v(r)$, Eq. (4.65), and try to determine the wave equation for $\bar{\phi}(E,r)$ with the potential $\bar{v}(r)$. Since Eq. (4.62) is true in general, there must be a kernel $\bar{K}(r,r')$ such that

$$\bar{\phi}(E,r) = \phi(E,r) + \int_0^r \bar{K}(r,r')\phi(E,r')\,dr', \qquad (4.66)$$

where $\bar{K}(r,r')$ just like $K(r,r')$ is independent of the energy E. Now by substituting $g(r,r')$ given by (4.53) in (4.56), we find that $K(r,r')$ can be written as

$$K(r,r') = \int \mathcal{K}(E,r)\bar{\phi}(E,r')\,dh(E), \qquad (4.67)$$

where $\mathcal{K}(E,r)$ is defined by

$$\mathcal{K}(E,r') = \bar{\phi}(E,r) - \int_0^r K(r,r')\,\bar{\phi}(E,r')\,dr'. \qquad (4.68)$$

By comparing (4.68) with (4.62) we conclude that

$$\mathcal{K}(E,r) = \phi(E,r'), \qquad (4.69)$$

so that

$$K(r,r') = \int \phi(E,r)\bar{\phi}(E,r')\,dh(E). \qquad (4.70)$$

Let us start with the completeness relation for the physical wave function $\psi(E,r)$ which we write as

$$\int_0^\infty \psi(E,r)\psi^*(E,r')\frac{dE}{\sqrt{E}} + \sum_{n=0}^\infty C_n^2\psi(E_n,r)\psi(E_n,r') = \delta(r-r'), \qquad (4.71)$$

where C_n is the normalization constant. This constant can be written either in terms of $\psi(E_n,r)$ or in terms of $\phi(E_n,r)$;

$$C_n = \frac{1}{\sqrt{\int_0^\infty \phi_n^2(E_n,r)dr}}. \qquad (4.72)$$

We also observe that the functions $\bar{\phi}(E,r)$ s, the solutions of the differential equation (4.51) form a complete set of states, i.e.

$$\int_{-\infty}^\infty \bar{\phi}(E,r)\bar{\phi}(E,r')\,d\bar{\rho}(E) = \delta(r-r'). \qquad (4.73)$$

We want to write a similar equation for $\phi(E,r)$ which is also the solution of the Schrödinger equation (4.63) remembering that $\phi(E,r)$ is related to $\psi(E,r)$ by Eq. (4.16). Thus we have

$$\int \phi(E,r)\phi(E,r')\,d\rho(E) = \delta(r-r'). \qquad (4.74)$$

To obtain $\frac{d\rho}{dE}$ we substitute $\psi(E,r)$ in Eq. (4.71) by $\phi(E,r)$ using (4.16)

$$\frac{d\rho(E)}{dE} = \begin{cases} \frac{k}{\pi}|f(k)|^{-2} & \text{for } E > 0 \\ \\ \sum_n C_n^2 \delta(E - E_n) & \text{for } E < 0 \end{cases}, \tag{4.75}$$

where $f(k)$ is the Jost function and (4.75) satisfies the boundary condition $\rho(-\infty) = 0$. The constant of normalization in this case is

$$C_n = \frac{1}{\sqrt{\int_0^\infty \phi_n^2(r)dr}}. \tag{4.76}$$

We will have an expression similar to (4.75) for $\frac{d\rho(\bar E)}{dE}$ with $f(k)$ and C_n being replaced by $\bar f(k)$ and $\bar C_n$ respectively.

Now if we multiply (4.62) by $\bar\phi(E,r')\ d\bar\rho(E)$, and multiply (4.66) by $\phi(E,r')\ d\rho(E)$ and subtract the resulting expressions from each other and then use Eqs. (4.73) and (4.74) we obtain

$$\int d\left[\rho(E) - \bar\rho(E)\right] \phi(E,r)\bar\phi(E,r') = \begin{cases} K(r,r') & r' < r \\ \\ \bar K(r',r), & r' \geq r. \end{cases} \tag{4.77}$$

By comparing (4.77) and (4.70) we find that

$$h(E) = \rho(E) - \bar\rho(E). \tag{4.78}$$

This completes the Gel'fand–Levitan method for the construction of the potential $v(r)$.

To summarise, the potential $v(r)$ which is associated with the spectral function $\rho(E)$ can be obtained by taking the following steps: Suppose that for a potential $\bar v(r)$, the spectral density $\bar\rho(E)$ and the complete set of wave functions $\bar\phi(E,r)$ are known as in (4.75) and (4.51) and we want to construct a potential $v(r)$ which is associated with the spectral function $\rho(E)$ [8]. Since we assume that both $\rho(E)$ and $\bar\rho(E)$ are known functions of E, then from (4.78) we obtain $h(E)$. Knowing $h(E)$ and $\{\bar\phi(E,r)\}$ enables us to calculate $g(r,r')$ from (4.53). Substituting for $g(r,r')$ in (4.56) and then solving the integral equation for $K(r,r')$ we can calculate $\Delta v(r)$ from (4.59), and thus determine $v(r) = \bar v(r) + \Delta v(r)$.

In this book we will not consider the mathematical proofs for the existence and uniqueness of the solution of the inverse problem, but refer the reader to the details of the proofs in the two excellent books [1], [6]. Here we limited our discussion of the Gel'fand–Levitan for the S wave only. For other partial waves we can use a similar approach. The generalization to other partial waves can also be found in the above-mentioned books.

A Solvable Example Obtained from the Gel'fand–Levitan Method — Consider the case where the spectral function $\rho(E)$ differs from the original spectrum $\bar{\rho}(E)$ only by changing one of the normalization constants, viz,

$$\frac{dh(E)}{dE} = \frac{d\rho(E)}{dE} - \frac{d\bar{\rho}(E)}{dE} = (C_0^2 - \bar{C}_0^2)\delta(E - E_0), \qquad (4.79)$$

where $C_0^2 - \bar{C}_0^2$ is the change in the reciprocals of the normalization constants;

$$C_0^2 - \bar{C}_0^2 = \frac{1}{N^2} - \frac{1}{\bar{N}^2} \qquad (4.80)$$

$$= \frac{1}{\int_0^\infty \phi^2(r)\,dr} - \frac{1}{\int_0^\infty \bar{\phi}^2(r)\,dr}. \qquad (4.81)$$

The expression (4.79) for $\frac{dh(E)}{dE}$ yields a simple separable form for $g(r,r')$, Eq. (4.53);

$$g(r,r') = (C_0^2 - \bar{C}_0^2)\bar{\phi}_0(r)\bar{\phi}_0(r'). \qquad (4.82)$$

Since $g(r,r')$ is separable from (4.56) it follows that $K(r,r')$ is also separable

$$K(r,r') = (C_0^2 - \bar{C}_0^2)\phi_0(r)\bar{\phi}_0(r'). \qquad (4.83)$$

In the last two equations $\bar{\phi}_0(r)$ is the wave function of the bound state due to the potential $\bar{v}$, and whose normalization has to be changed and $\phi_0(r)$ is the new wave function. From the integral equation (4.62) we obtain $\phi_0(r)$ to be

$$\phi_0(r) = \frac{\bar{\phi}_0(r)}{1 + (C_0^2 - \bar{C}_0^2)\int_0^r \bar{\phi}_0^{\,2}(r')\,dr'}. \qquad (4.84)$$

By substituting $\phi_0(r)$ in (4.83) we find $K(r,r')$:

$$K(r,r') = \frac{(C_0^2 - \bar{C}_0^2)\bar{\phi}_0(r)\bar{\phi}_0(r')}{1 + (C_0^2 - \bar{C}_0^2)\int_0^r \bar{\phi}_0^2(r')\,dr'}, \qquad (4.85)$$

from which we can calculate $\Delta v(r)$

$$\Delta v(r) = -2\frac{dK(r,r)}{dr}$$
$$= -2\frac{d^2}{dr^2}\ln\left[1 + (C_0^2 - \bar{C}_0^2)\int_0^r \bar{\phi}_0^2(r')\,dr'\right]. \qquad (4.86)$$

The new scattering wave functions are

$$\phi_0(E,r) = \bar{\phi}_0(E,r) - \frac{(C_0^2 - \bar{C}_0^2)\bar{\phi}_0(r)\int_0^r \bar{\phi}_0(r')\bar{\phi}_0(E,r')\,dr'}{1 + (C_0^2 - \bar{C}_0^2)\int_0^r \bar{\phi}_0^2(r')\,dr'}. \qquad (4.87)$$

It is important to observe that the potential $\Delta v(r)$, (4.86), is not symmetric about the vertical axis, i.e. $v(x) \neq v(-x)$. Later, for convenience, we denote the difference between the reciprocals of the normalization constants by c^2

$$(C_0^2 - \bar{C}_0^2) = c^2. \qquad (4.88)$$

4.5   Inverse Problem for One-dimensional Schrödinger Equation

In this section we will consider only those one-dimensional motions where the particle is trapped in a potential well. The inverse problems of transmission and reflection of a particle and the inverse tunneling problem will be postponed to later chapters.

Following the pioneering work of Abraham and Moses [18]–[20] we start with the Hamiltonian

$$H_0 = -\frac{d^2}{dx^2} + v_0(x), \quad -\infty < x < \infty, \tag{4.89}$$

where we have set the mass of the particle equal to $\frac{1}{2}$ and $\hbar = 1$. We assume that this Hamiltonian possesses discrete states and may have a continuous spectra. For discrete states we have a set of real eigenfunctions $\bar{\phi}_n(x)$;

$$H_0\bar{\phi}_n(x) = E_n\bar{\phi}_n(x), \tag{4.90}$$

where $\bar{\phi}_n(x) \to 0$ as $x \to -\infty$, E_n s are discrete eigenvalues and $\bar{\phi}_n(x)$ s are the corresponding eigenfunctions. These eigenfunctions satisfy the normalization condition

$$\bar{C}_n^2 = \frac{1}{\int_{-\infty}^{\infty} \bar{\phi}^2(E_n, x)dx}. \tag{4.91}$$

If in addition to bound states H_0 has a continuous spectrum with eigenvalues E and eigenfunctions $\phi(x, E)$, then

$$H_0\bar{\phi}(E, x) = E\bar{\phi}(E, x). \tag{4.92}$$

The completeness relation for the eigenfunctions can be expressed as

$$\sum_n \bar{C}_n^2 \bar{\phi}_n(x)\bar{\phi}_n(x') + \int \bar{\phi}(x, E)\, \bar{\phi}^*(x', E)\, \rho(E)\, dE = \delta(x - x'). \tag{4.93}$$

Now we show that it is possible to add a potential $\Delta v(x)$ to the Hamiltonian H_0, so that the new Hamiltonian

$$H = H_0 + \Delta v(x), \tag{4.94}$$

has the same continuous spectrum as H_0, except for a finite number of discrete eigenvalues and/or a finite number of normalization constants.

Let us denote the eigenfunctions corresponding to the discrete eigenvalue $\mathcal{E}_n$ of the Hamiltonian H by $\phi_n(x)$, i.e.

$$H\phi_n(x) = \mathcal{E}_n\phi_n(x), \quad \phi_n(x) \to 0 \quad \text{as} \quad x \to -\infty, \tag{4.95}$$

and the wave functions for continuous spectra of H by $\phi(x, \mathcal{E})$, then we can write the completeness relation as

$$\sum_n C_n^2 \phi_n(x) \phi_n(x') + \int \phi(x, \mathcal{E}) \phi^*(x', \mathcal{E}) \rho(\mathcal{E}) d\mathcal{E} = \delta(x - x'), \qquad (4.96)$$

where the weight function $\rho(\mathcal{E})$ has not been changed, since the continuous spectra of H_0 and H are assumed to be the same.

Regarding the completeness relation (4.93) and (4.96) there are two points that we need to remember:

(a) For confining potentials, i.e. those with the property that $v(x) \to +\infty$ as $x \to \pm\infty$, there are no scattering states, therefore we just sum over the discrete eigenvalues, n.

(b) Whether there are scattering states or not, we assume that only a finite number of $\mathcal{E}_n$ s and $\bar{C}_n$ s are not identical with E_n s and C_n s.

With these conditions we apply the Gel'fand–Levitan formulation to determine $\Delta v(x)$. For this type of problems, the input, $g(x, x')$, is defined by

$$g(x, y) = \sum_{n=1}^{N} \left[C_n^2 \phi_n(x) \phi_n(y) - \bar{C}_n^2 \bar{\phi}_n(x) \bar{\phi}_n(y) \right], \qquad (4.97)$$

where N is a finite integer. The function $g(x, y)$ is the analogue of (4.53) that we used in finding the Gel'fand–Levitan kernel $K(r, r')$. In the present case the kernel is the unique solution of the integral equation

$$K(x, y) = g(x, y) - \int_{-\infty}^{x} K(x, z) g(z, y) \, dz. \qquad (4.98)$$

As in the case of S wave problem discussed earlier the potential $\Delta v(x)$ is obtained from

$$\Delta v(x) = -2 \frac{dK(x, x)}{dx}, \qquad (4.99)$$

and the new wave functions for bound and scattering states are given by

$$\phi_n(x) = \bar{\phi}_n(x) - \int_{-\infty}^{x} K(x, y) \bar{\phi}_n(y) dy, \qquad (4.100)$$

and

$$\phi(x, E) = \bar{\phi}(x, E) - \int_{-\infty}^{x} K(x, y) \bar{\phi}(y, E) dy. \qquad (4.101)$$

For the separable form of $g(x, y)$, Eq. (4.97), $K(x, y)$ ia also separable, and the Gel'fand–Levitan equation is solvable.

Let us define an operator $\mathcal{U}$ by

$$\mathcal{U}f(x) = f(x) - \int_{-\infty}^{x} K(x, y) f(y) dy, \qquad (4.102)$$

then we can write (4.100) and (4.101) as

$$\phi_n(x) = \mathcal{U}\bar{\phi}_n(x), \tag{4.103}$$

and

$$\phi(x, E) = \mathcal{U}\bar{\phi}(x, E). \tag{4.104}$$

In the previous section we noticed that we can interchange H_0 and H and then the wave functions $\bar{\phi}_1(E, r)$ and $\phi(\mathcal{E}, r)$ have to be interchanged. Here also if we start with H_0 with wave functions $\bar{\phi}_n(x)$ and $\bar{\phi}(x, E)$, we obtain H and the wave functions $\phi_n(x)$ and $\phi(x, \mathcal{E})$. But if we start with H, $\phi_n(x)$ and $\phi(x, \mathcal{E})$, then we find H_0, $\bar{\phi}_n(x)$ and $\bar{\phi}(x, \mathcal{E})$. In the latter case the kernel $K(x, y)$ is replaced by $\bar{K}(x, y)$ and the operator $\mathcal{U}$ by $\bar{\mathcal{U}}$. If we start with the two wave functions $\bar{\phi}_n(x)$ and $\bar{\phi}(x, E)$, and operate on these with the operator $\mathcal{U}$ we get $\phi_n(x)$ and $\phi(x, \mathcal{E})$. Now if we apply the operator $\bar{\mathcal{U}}$ on these we will recover the original wave functions. Therefore

$$\bar{\mathcal{U}}\mathcal{U} = I, \tag{4.105}$$

where I is the unit operator.

Now let us study the following examples:

Anharmonic Oscillator — Consider the simple harmonic oscillator with unit mass and unit spring constant with the Hamiltonian [19], [22]

$$H = \frac{1}{2}\left(-\frac{d^2}{dx^2} + x^2\right). \tag{4.106}$$

The normalized eigenfunctions and the eigenvalues of H are given by

$$\bar{\phi}_n(x) = \frac{1}{(\sqrt{\pi}2^n n!)} \exp\left(-\frac{1}{2}x^2\right) H_n(x), \quad C_n = 1, \tag{4.107}$$

where H_n is the Hermite polynomial of degree n and where

$$E_n = n + \frac{1}{2}, \quad n = 0, 1\, 2\cdots \tag{4.108}$$

We want to find the Hamiltonian operator and the wave functions for a problem where the ground state with the energy $E_0 = \frac{1}{2}$ is missing. In this case Eq. (4.97) reduces to

$$g(x, y) = -\bar{\phi}_0(x)\bar{\phi}_0(y). \tag{4.109}$$

With this $g(x, y)$ as the input function the Gel'fand–Levitan kernel can be found to be

$$K(x, y) = -\frac{2}{\sqrt{\pi}} \frac{\exp\left[-\frac{1}{2}\left(x^2 + y^2\right)\right]}{\text{erfc}(x)} \tag{4.110}$$

where $\text{erfc}(x)$ is defined by [25]

$$\text{erfc}(x) = \frac{2}{\sqrt{\pi}} \int_x^\infty \exp\left(-t^2\right) dt. \tag{4.111}$$

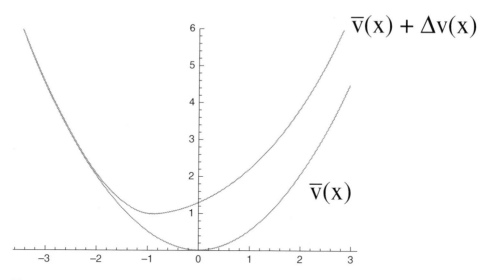

Figure 4.2: Harmonic and anharmonic potentials $\bar{v}(x)$ and $\bar{v}(x)+\Delta v(x)$ are plotted as a function of x. The second potential is found by deleting the ground state energy of $\bar{v}(x)$ using Abraham–Moses method.

From Eq. (4.99) we obtain $\Delta v(x)$ [22], [19]

$$\Delta v(r) = -2\frac{d^2}{dx^2}\ln[\mathrm{erfc}(x)] = \frac{4}{\sqrt{\pi}}\left[\frac{\exp\left(-x^2\right)}{\sqrt{\pi}\mathrm{erfn}(x)} - x\right]$$

$$\times\left(\frac{\exp\left(-x^2\right)}{\mathrm{erfc}(x)}\right). \tag{4.112}$$

We can also calculate $\phi_n(x)$ from (4.101);

$$\phi_n(x) = \bar{\phi}_n(x) - \left(\frac{2}{n\pi}\right)^{\frac{1}{2}}\frac{e^{-x^2}}{\mathrm{erfc}(x)}\bar{\phi}_{n-1}(x), \quad n = 1,\ 2,\ 3\cdots. \tag{4.113}$$

Let us note that while the original potential x^2 is symmetric, $\bar{v}(x) = \bar{v}(-x)$, the potential $\Delta v(x)$ is not (see Fig. 4.2). If we write the potential $v(x) = \bar{v}(x) + \Delta v(x)$ with the factors $\hbar$ and the mass of the particle m and spring constant $m\omega^2$, the potential takes the form

$$V(x) = \frac{1}{2}m\omega^2 x^2 + \hbar\omega^4 w(z)[w(z) - z], \tag{4.114}$$

where z and $w(z)$ are given by

$$w(z) = \frac{\exp\left(-z^2\right)}{\sqrt{\pi}\mathrm{erfc}(z)}, \tag{4.115}$$

and

$$z = \sqrt{\frac{m\omega}{\hbar}}x. \tag{4.116}$$

A Particle in a Box — For the second example let us consider the motion of a particle in a box with the boundaries at $x = -\frac{\pi}{2}$ and $\frac{\pi}{2}$. The Hamiltonian is just the kinetic energy ($\hbar = 2m = 1$),

$$H_0 = -\frac{d^2}{dx^2}, \qquad -\frac{\pi}{2} \le x \le \frac{\pi}{2}, \tag{4.117}$$

and with the normalized eigenfunctions

$$\bar{\phi}_n(x) = \begin{cases} \sqrt{\frac{2}{\pi}}\sin(nx) & n \text{ even} \\ \\ \sqrt{\frac{2}{\pi}}\cos(nx) & n \text{ odd}. \end{cases} \tag{4.118}$$

The eigenvalues of H_0, Eq. (4.117), are

$$E_n = n^2, \quad n = 1, 2, 3 \cdots . \tag{4.119}$$

Here as in the previous example let us consider the effect of deleting the lowest eigenvalue, i.e. $E_1 = 1$, without changing the normalization $\bar{C}_n = 1$, $n = 2, 3$. Again we choose $g(x, y)$ to be

$$g(x, y) = -\frac{2}{\pi}\cos x \cos y, \tag{4.120}$$

and with this $g(x, y)$ as the input we calculate the kernel $K(x, y)$ from the Gel'fand–Levitan equation:

$$K(x, y) = -\frac{4\cos x \cos y}{2x - \pi - \sin(2x)}. \tag{4.121}$$

From $K(x, x)$ we obtain $\Delta v(x)$ which in this case is

$$\Delta v(x) = \frac{8(2x - \pi)}{(2x - \pi - \sin(2x))^2}, \qquad -\frac{\pi}{2} \le x \le \frac{\pi}{2}. \tag{4.122}$$

The new eigenfunctions for the Hamiltonian

$$H = -\frac{d^2}{dx^2} + \Delta v(x), \tag{4.123}$$

are:

$$\phi_n(x) = \bar{\phi}_n(x) + (-1)^n \left(\frac{2\cos x}{2x - \pi - \sin(2x)} \right) \left(\frac{\bar{\phi}_{n+1}(x)}{n+1} + \frac{\bar{\phi}_{n-1}(x)}{n-1} \right), n = 2, 3, \cdots .$$

Other examples can be found in Ref. [20].

4.6 Bargmann Potentials

As an example of exactly solvable potential, we will consider the special case when the Jost function $f(k)$ is a rational function, and for simplicity we choose the S-wave scattering. Then the simplest case is when the Jost function has one zero and one pole. Since $f(k)$ must satisfy the symmetry property (4.22), viz

$$f(-k^*) = (f(k))^*, \qquad (4.124)$$

therefore a simple rational form of $f(k)$ would be

$$f(k) = \frac{k + ia}{k + ib}, \qquad (4.125)$$

where a and b are both real, and $b > 0$.

No Bound State Is Present — When $a > 0$ there is a bound state at $k = -ia$ $(a > 0)$ of energy $-\frac{\hbar^2 a^2}{2m}$. If $a < 0$, there is no bound state, but for $a = 0$ there is a resonance with zero energy. For the Jost function to be analytic in the upper half k-plane b must be positive. This second condition guarantees that $\int_0^\infty r|v(r)|dr < \infty$.

From (4.125) we find

$$|f(k)|^{-2} - 1 = \frac{b^2 - a^2}{k^2 + a^2}, \qquad k \text{ real.} \qquad (4.126)$$

Now according to Eqs. (4.53) and (4.75), $g(r,r')$ of the Gel'fand–Levitan equation when $\bar{v}(r) = 0$ is given by

$$g(r,r') = \int_{-\infty}^{\infty} \bar{\phi}(E,r)\bar{\phi}(E,r')\,dh(E)$$

$$= \int_{-\infty}^{\infty} \bar{\phi}(E,r)\bar{\phi}(E,r')\,[d\rho(E) - d\bar{\rho}(E)]$$

$$= \int_{-\infty}^{\infty} \frac{\sqrt{E}}{\pi}\left[\frac{1}{|f(k)|^2} - 1\right]\bar{\phi}(E,r)\bar{\phi}(E,r')\,dE, \quad E > 0. \qquad (4.127)$$

In the absence of the potential $\bar{v}(r)$, the solution of (4.51) for a free wave function is

$$\bar{\phi}(k,r) = \frac{\sin(kr)}{k}, \qquad (4.128)$$

and the last term in (4.127) is

$$\frac{2}{\pi}\int_0^\infty \frac{\sin(kr)}{k}\frac{\sin(kr')}{k}k^2 dk = \delta(r - r'). \qquad (4.129)$$

By calculating $g(r,r')$ using Eq. (4.127) we find that

$$g(r,r') = \begin{cases} -\left(a^2 - b^2\right)\dfrac{e^{-br}\sinh(br')}{a}, & r \geq r' \\[4mm] -\left(a^2 - b^2\right)\dfrac{\exp(-br')\sinh(br)}{a}, & r \leq r' \end{cases}. \qquad (4.130)$$

By solving the Gel'fand–Levitan integral equation with this $g(r, r')$ we find the kernel $K(r, r')$ to be

$$
K(r, r') = \begin{cases} -\frac{2(b-a)}{1+B\exp(-2br)} e^{-br} \sinh(-br') & r \geq r' \\ \\ -\frac{2(b-a)}{1+B\exp(-2br')} \exp(-br') \sinh(-br) & r' \leq r \end{cases} \tag{4.131}
$$

and the corresponding potential is [21], [22]

$$
v(r) = -8b^2 B \frac{e^{-2br}}{(1 + B\exp(-2br))^2}, \tag{4.132}
$$

where we have set $\frac{\hbar^2}{2m} = 1$ and $B = \frac{b-a}{b+a}$. There are two special cases of this potential which are of interest:
(1) If $b = 0$, then $v(r)$ reduces to

$$
v(r) = \frac{2a^2}{(1 + ar)^2}. \tag{4.133}
$$

This is a potential with long range which does not satisfy the integrability condition $\int_0^\infty |v(r)| r dr < \infty$, and if the phase shift is set to zero at infinity, then $\delta(k) = \left(-\frac{\pi}{2}\right)$, for all k values and therefore the Levinson theorem is violated.
(2) If $a = 0$, we have a resonance at zero energy, and the potential in this case becomes

$$
v(r) = -\frac{2b^2}{\cosh^2(br)}. \tag{4.134}
$$

For this case the phase shift is $\frac{\pi}{2}$, and the cross section is infinite.
From the relation

$$
\frac{f(-k)}{f(k)} = e^{2i\delta_0(k)}, \tag{4.135}
$$

and (4.125) we can determine the phase shift which we can also find by solving the Schrödinger equation with the potential (4.132). This phase shift is given by

$$
k \cot \delta_0(k) = \frac{ab}{b-a} + \frac{1}{(b-a)} k^2. \tag{4.136}
$$

Thus this Bargmann potential yields the phase shift which exactly matches the effective range formula of nuclear theory [23]

$$
k \cot \delta_0(k) = -\frac{1}{a_0} + \frac{1}{2} r_0 k^2. \tag{4.137}
$$

Here a_0 is the scattering length and r_0 is the effective range. The Jost solution found from Eqs. (4.62), (4.128) and (4.131) is given by

$$
f(k, r) = e^{ikr} \left[1 - \left(\frac{2ib}{k + ib} \right) \frac{B\exp(-2br)}{1 + B\exp(-2br)} \right], \tag{4.138}
$$

and the normalized wave function may be found from (4.15);

$$\phi(k,r) = \frac{\sin(kr)}{k} + \left[\frac{b^2 - a^2}{k(k^2 + b^2)}\right] \frac{k \sinh(br)\cos(kr) - \cosh(br)\sin(kr)}{b\cosh(br) + a\sinh(br)}. \quad (4.139)$$

Bargmann Potential with One Bound State — For the presence of the bound state $v(r)$ has to be attractive and a in (4.136) must be negative. Let us replace a by $-\kappa$, $(\kappa > 0)$, then the effective range formula is still exact, but now we have a bound state with the binding energy $-\frac{\hbar^2\kappa^2}{2m}$. For this case we have a one-parameter family of potentials that will give us the same phase shift and the same bound state energy, and at the same time depends on the parameter c. Thus for scattering states we have the wave function $\phi(k,r)$ given by (4.139), and for the bound state we have the normalized wave function [26]

$$\phi_c(r) = 2\left(\frac{c\kappa}{b^2 - \kappa^2}\right)^{\frac{1}{2}} \frac{\sinh(br)}{g_c(\kappa + b, r) - g_c(\kappa - b, r)}, \quad (4.140)$$

where g_c is defined by

$$g_c(k,r) = \frac{1}{k}\left(e^{-kr} + c\sinh(kr)\right). \quad (4.141)$$

The potential now is given by

$$v_c(r) = -4\kappa \frac{d}{dr}\left\{\sinh(br)\frac{g_c(\kappa, r)}{g_c(\kappa + b, r) - g_c(\kappa - b, r)}\right\}. \quad (4.142)$$

Non-uniqueness of the Potential Determined from the Phase Shift — Now we want to show the non-uniqueness of the potential when the the phase shift $\delta_0(k)$ is given for all energies (or wave numbers). For this we consider an Eckart potential [24] depending on three parameters which we denote by $v_{\sigma,\beta,\lambda}$. Earlier we observed that the admissible solution of the Schrödinger equation for the S-wave scattering can be written in terms of the Jost solution as (see Eq. (4.15))

$$\phi(k,r) = \frac{1}{2i|f(k)|}\left[f(-k)f(k,r) - f(k)f(-k,r)\right], \quad (4.143)$$

where $f(k)$ is related to the phase shift by $e^{2i\delta_0(k)} = \frac{f(-k)}{f(k)}$. Now let us consider again the Eckart potential $v(r)$ which we write as

$$v(r) = v_{\sigma,\beta,\lambda} = \frac{-\sigma\lambda^2\beta\exp(-\lambda r)}{(1 + \beta\exp(-\lambda r))^2}, \quad (4.144)$$

where $\lambda > 0$ and $\beta > -1$. We will consider two special cases of this potential:

Let us first choose $\sigma = 2$, i.e. the potential $v_{2,\beta,\lambda}$. For this potential the Jost solution is given by

$$f(-k,r) = \frac{e^{-ikr}(2k + i\mu(r))}{2k - i\lambda}, \quad (4.145)$$

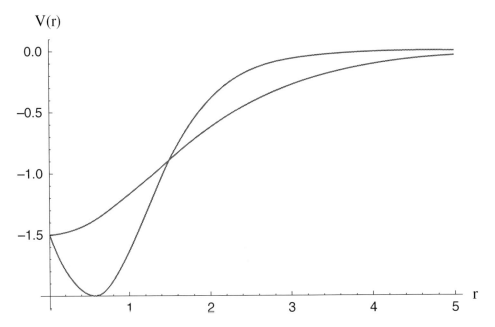

Figure 4.3: Two of the Bargmann phase equivalent potentials are shown in this figure. The one with the minimum is $v_{6,3,2\lambda}(r)$ of the text with the minimum at $a = \frac{\ln 3}{2\lambda}$, and is given by $-2\lambda^2 \cosh^{-2}[\lambda(r - a)]$. The other is $v_{2,1,\lambda}(r) = -\frac{3}{2}\lambda^2 \cosh^{-2}\left(\frac{\lambda r}{2}\right)$ and this potential is a monotonically increasing function of r. Both of these potentials have a very simple Jost function and have a bound state with the energy $E = -\frac{\hbar^2 \lambda^2}{8m}$.

where

$$\mu(r) = \frac{\lambda(\beta e^{-\lambda r} - 1)}{\beta e^{-\lambda r} + 1}. \tag{4.146}$$

By substituting $f(-k, r)$ in the Schrödinger equation we can verify that it is a solution, and has the correct asymptotic behaviour since $\mu(\infty) = -\lambda$. The Jost function $f(-k)$ found from (4.145) is

$$f(-k) = f(-k, 0) = \frac{2k + i\nu}{2k - i\lambda}, \tag{4.147}$$

where

$$\nu = \mu(0) = \frac{\lambda(\beta - 1)}{\beta + 1}. \tag{4.148}$$

Now $f(-i\kappa)$ vanishes for $\kappa = \frac{1}{2}\nu$, so that there is a bound state with the energy $E = -\frac{\hbar^2 \nu^2}{8m}$, provided that $\nu > 0$, i.e. if $\beta > 1$.

A second case that we will consider is the one where $\sigma = 6$ [22]. In this case

$$f(-k, r) = e^{-ikr}\left(\frac{4k^2 + 6ik\mu(r) + \lambda^2 - 3\mu^2(r)}{(2k - i\lambda)(2k - 2i\lambda)}\right), \tag{4.149}$$

and

$$f(-k) = \frac{4k^2 + 6ik\mu(r) + \lambda^2 - 3\nu^2}{(2k - i\lambda)(2k - 2i\lambda)}, \tag{4.150}$$

with $\mu(r)$ and ν as given before. Now consider the potential $v_{6,1,\lambda}(r) = v_1(r)$, with λ arbitrary. Then we find $\nu = 0$ and a very simple Jost function

$$f_1(-k) = \frac{2k + i\lambda}{2k - 2i\lambda}. \tag{4.151}$$

Let us now consider the potential $v_{2,\beta',\lambda'}$ with β' and λ' to be determined later. Then from (4.146) we have

$$f_2(-k) = \frac{2k + i\nu'}{2k - i\lambda'}. \tag{4.152}$$

By setting $f_1(-k) = f_2(-k)$ we obtain

$$2\lambda = \lambda', \quad \text{and} \quad \lambda = \nu' = \lambda' \left(\frac{\beta' - 1}{\beta' + 1} \right), \tag{4.153}$$

which shows that $\beta' = 3$.

Therefore the two potentials

$$v_{2,1,\lambda} = -\frac{6\lambda^2 \exp(-\lambda r)}{(1 + \exp(-\lambda r))^2} = \frac{-3\lambda^2}{2 \cosh^2 \left(\frac{1}{2}\lambda r \right)}, \tag{4.154}$$

and

$$v_{6,3,2\lambda} = -\frac{24\lambda^2 \exp(-2\lambda r)}{(1 + 3\exp(-\lambda r))^2} = \frac{-2\lambda^2}{\cosh^2 \left[\lambda(r - a) \right]}, \tag{4.155}$$

with $a = \frac{\ln 3}{2\lambda}$ lead to the same $f(-k)$, i.e. they are examples of phase equivalent potentials. We also note that these two potentials have the same value of $-\frac{3\lambda^2}{2}$ at $r = 0$, and they give rise to the same bound state energy of $-\frac{\hbar^2 \lambda^2}{8m}$.

4.7   The Jost and Kohn Method of Inversion

A method which can be used if there is no bound state but the Born approximation is not valid, because the phase shifts are large, is due to Jost and Kohn [10]. Our starting point here is the integral equation for the wave function

$$\psi(k, r) = \sin(kr) + \int_0^\infty G(r, r') v(r') \psi(k.r') dr', \tag{4.156}$$

where $G(r, r', k)$ is given by (see also Eq. (13.215))

$$G(r, r', k) = \frac{2}{\pi} \int_0^\infty \frac{\sin(qr)\sin(qr')}{k^2 - q^2} dq \qquad (4.157)$$

$$= -\frac{1}{k} \begin{cases} \sin(kr)\cos(kr') & \text{if } r \leq r', \\ \cos(kr)\sin(kr') & \text{if } r' \leq r. \end{cases} \qquad (4.158)$$

The asymptotic solution of (4.156) is given by

$$\psi(k, r) = \sin(kr) + \tan\delta(k)\cos(kr), \qquad (4.159)$$

where $\delta(k)$ is the S wave phase shift (we omit the subscript 0 in the phase shift). From Eqs. (4.156) and (4.158) we have

$$\tan\delta(k) = -\frac{1}{k} \int_0^\infty \sin(kr)v(r)\psi(k, r)dr. \qquad (4.160)$$

We can solve (4.156) by iteration and substitute the resulting $\sqrt{n}\psi(k, r)$ in (4.160)

$$-k\tan\delta(k) = \int_0^\infty v(r)\sin^2(kr)dr + \sum_{j=1}^\infty \int_0^\infty \cdots \int_0^\infty$$

$$\times \int_0^\infty \sin(kr)v(r)G(r, r_1)v(r_1)dr_1 \cdots$$

$$\times G(r_{j-1}, r_j)v(r_j)\sin(kr_j)dr \cdots dr_j. \qquad (4.161)$$

This equation in turn may be regarded as an equation for $v(r)$ and may be solved by iteration. For this we introduce a parameter μ and write

$$-k\tan\delta(k) = \mu P(k), \qquad (4.162)$$

where P_μ is an infinite series in μ. Also we assume that $v(r)$ has a power expansion in terms of μ, i.e.

$$v(r) = \sum_{\mu=1}^\infty \mu^j v_j(r). \qquad (4.163)$$

By substituting (4.162) and (4.163) in (4.161) and equating to zero the coefficients of μ, $\mu^2 \cdots$ and then putting $\mu = 1$ and by introducing $Q_j(k)$ s as the coefficients of μ^j we have

$$Q_j(k) = \int_0^\infty \sin^2(kr)v_j(r)dr, \qquad (4.164)$$

then we have the following relations:

$$Q_1 = \int_0^\infty \sin^2(kr)v_1(r)dr, \qquad (4.165)$$

and

$$Q_j(k) = -\sum_{j=2}^{\infty} \sum_{\sum \nu_k=m} \int_0^{\infty} \cdots \int_0^{\infty} \sin(k_1 r) v_{\nu_1}(r_1) G(r_1, r_2) v_{\nu_2}(r_2) \cdots$$

$$\times G(r_{j-1}, r_j) v_{\nu_j}(r_j) \sin(kr_j) dr_1 \cdots dr_j, \quad j \geq 2. \tag{4.166}$$

We can find successive terms in the expansion (4.162) and (4.163) by Fourier inversion. We also note that v_m can be expressed in terms of $v_s(s < m)$. Thus we find $v_1(r)$ from the relation

$$rv_1(r) = -\frac{4}{\pi} \int_0^{\infty} \frac{d}{dk} [k \tan \delta(k)] \sin(2kr) dk. \tag{4.167}$$

For $v_j(j > 1)$ we write

$$\int_0^{\infty} \sin^2(kr) v_j(r) dr = \frac{1}{2} \int_0^{\infty} (1 - \cos(2kr)) v_j(r) dr, \tag{4.168}$$

and assume

$$\int_0^{\infty} v_m(r) dr = 0, \quad j > 1. \tag{4.169}$$

Thus we find

$$v_m(r) = \sum_{j=2}^{\infty} \sum_{\sum \nu_j=m} \int_0^{\infty} \cdots \int_0^{\infty} \bar{G}(r, r_1 \cdots r_j) v_{\nu_1}(r_1) \tag{4.170}$$

$$\times v_{\nu_2}(r_2) \cdots v_{\nu_j}(r_j) dr_1 \cdots dr_j, \tag{4.171}$$

where

$$\bar{G}(r, r_1 \cdots r_j) = \frac{4}{\pi} \int_{-\infty}^{\infty} \cos(2kr) \sin(kr_1) G_0(r_1, r_2) \cdots$$

$$\times G_0(r_{j-1}, r_j) \sin(kr_j) dk. \tag{4.172}$$

That this solution is compatible with (4.169) follows from the fact that

$$\int_0^{\infty} \bar{G}_0(r) dr = 0. \tag{4.173}$$

For sufficiently small $\delta(k)$ the series for $v(r)$ converges, but it is difficult to find a general method for determining the radius of convergence.

An Example — Let us find the potential which gives us the effective range formula exactly, i.e. when

$$k \cot \delta(k) = -\frac{1}{a_0} + \frac{1}{2} r_0 k^2, \tag{4.174}$$

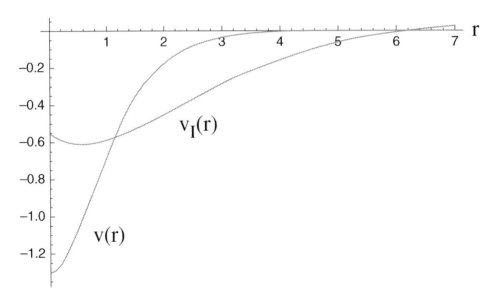

Figure 4.4: The Bargmann potential $v(r)$ giving the exact effective range formula for the neutron-proton scattering is compared with the approximate potential function $v_I(r) = v_1(r) + v_2(r)$ found from inversion using the Jost and Kohn method [10]. The potentials are given in the units of fm^{-2}.

where a_0 is the scattering length and r_0 is the effective range. From (4.173) we obtain

$$- \tan \delta(k) = \frac{-\gamma k^2}{\beta^2 + k^2}, \quad \gamma = \frac{2}{r_0}, \quad \beta^2 = -\frac{2}{r_0 a}, \tag{4.175}$$

where β is a real positive number. Using the Jost and Kohn method we find [10], [11]

$$v_1(r) = -4\beta\gamma e^{-2\beta r}, \tag{4.176}$$

$$v_2(r) = -4\gamma^2 \left[e^{-2\beta r} + \beta r e^{-2\beta r} + e^{-4\beta r} \right]. \tag{4.177}$$

We have seen earlier that the potential which gives the effective range formula is the Bargmann potential, Eq. (4.136). This potential which is of the form given by Eq. (4.132) and also can be written as

$$v(r) = \frac{2\lambda(\lambda + 1)b^2 e^{-br}}{[1 + \lambda(1 - e^{-br})]^2}, \tag{4.178}$$

can be related to the two parameters λ and b to the scattering length a_0 and the effective range r_0, by [27]

$$a = \gamma + \sqrt{\gamma^2 + 4\beta^2} \tag{4.179}$$

$$b = \frac{2}{r_0} \left[-1 + \left(1 - \frac{2r_0}{a_0} \right)^{\frac{1}{2}} \right], \tag{4.180}$$

and

$$\lambda = \left(\frac{\gamma}{2\beta}\right)^2 - \left[\left(\frac{\gamma}{2\beta}\right)^2 + 1\right]^{\frac{1}{2}}. \qquad (4.181)$$

For the singlet scattering of a neutron by a proton at low energies for the parameters in (4.174) we have $a_0 = -23.78$ fm^{-1} and $r_0 = 2.670$ fm^{-1}. Here we have used the exact Bargmann potential $v(r)$ to solve the direct problem. For inversion we have chosen $\mu = 1$ in (4.163). As we can see in Fig. 4.5, the sum of the two leading terms shown by $v_I(r)$ has similar features as the input potential which is $v(r)$. There is another version of this inverse method in which the nth term of the expansion is proportional to λ^n and converges faster to $v(r)$, but is more complicated [10].

The set of Bargmann's potentials that we have found can be enlarged by using the extended Gel'fand–Levitan formulation (see for instance [28]).

References

[1] R.G. Newton, *Scattering Theory of Waves and Particles*, (Second Edition, 1980) Chapter 20.

[2] A.G. Sitenko, *Scattering Theory*, (Springer-Verlag, New York, 1991) p. 104.

[3] C.J. Joachain, *Quantum Collision Theory*, (North-Holland, Amsterdam, 1975) p. 244.

[4] M.L. Goldberger and K.M. Watson, *Collision Theory*, (John Wiley & Sons, New York, 1964) p. 271.

[5] N.F. Mott and H.S.W. Massey, *The Theory of Atomic Collisions*, Third Edition, (Oxford University Press, 1965) p. 138.

[6] K. Chadan and P.C. Sabatier, *Inverse Prblems in Quantum Scattering Theory*, Second Edition, (Springer-Verlag, New York, 1989) Chapter III.

[7] B.N. Zakhariev and A.A. Suzko, *Direct and Inverse Problems: Potentials in Quantum Scattering*, (Springer-Verlag, Berlin 1990).

[8] A thorough discussion of the spectral function as related to the inverse problems can be found in D.N. Ghosh Roy. *Methods of Inverse Problems in Physics*, (CRC Press Boca Raton, 1991) Sec. XVIII.

[9] R. Jost, Helv. Phys. Acta 29, 410 (1956).

[10] R. Jost and W. Kohn, Phys. Rev. 87, 977 (1952).

[11] R. Jost and W. Kohn, Phys. Rev. 88, 382 (1952).

[12] I.M. Gel'fand and B.M. Levitan, Am. Math. Soc. Transl. 1 (1955).

[13] V.A. Marchenko, *Sturm-Liouville Operators and Applications*, Revised Edition (American Mathematical Society, Providence R.I. 2003) Chapters 3 and 4.

[14] Most of the textbooks on integral equations disuss the Volterra integral equations. See for instance A.C. Pipkin, *Integral Equations*, (Springer-Verlag, New York, 2008).

[15] F. Calogero, *Variable Phase Approach to Potential Scattering*, (Academic Press, New York, 1967).

[16] R.G. Newton, J. Math, Phys. 21, 493 (1980).

[17] P.B. Abraham and H.E. Moses, Phys. Rev. A22, 1333 (1980).

[18] H.E. Moses, J. Math. Phys. 20, 2047 (1979).

[19] M.M. Nieto, Phys. Rev. D 24, 1030 (1981).

[20] M. Hron and M. Razavy, Intl. J. Quantum Chem. XXIV, 97 (1983).

[21] V. Bargmann, Phys. Rev. 75, 301 (1949).

[22] V. Bargmann, Rev. Mod. Phys. 21, 488 (1949)

[23] H.A. Bethe, Phys. Rev. 76, 38 (1949).

[24] C. Eckart, Phys. Rev. 35, 1303 (1930).

[25] M. Abramowitz and I.A. Stegun, *Handbook of Mathematical Functions*, (Dover Publications, New York, 1965) p. 297.

[26] R.G. Newton, *Scattering Theory of Waves and Particles*, Second Edition (Springer New York, 1980) p. 433.

[27] N.F. Mott and H.S.W. Massey, *The Theory of Atomic Collisions*, Third Edition, (Oxford University Press, 1965), Chapter VII.

[28] T.A. Weber and D.L. Pursey, Phys. Rev. A 52, 3923 (1995).

Chapter 5

Marchenko's Formulation of the Inverse Scattering Problem

A different method of solving the inverse scattering problem due to Marchenko has the advantage that we need to use directly the partial wave $S_\ell(k)$ matrix rather than the Jost function as the input, and thus avoid the extra step of finding the Jost function from $S_\ell(k)$ [1]–[3].

5.1   Mathematical Preliminaries

We start this section by a theorem due to Titchmarsh [33]. This theorem states that a necessary and sufficient condition that a square integrable function, $F(x) \in L^2(-\infty, \infty)$, should be the limit of an analytic function $F(z)$, $z = x + iy$ as $y \to 0$, in $y > 0$ such that

$$\int_{-\infty}^{\infty} |F(x + iy)|^2 = \mathcal{O}\left(e^{2ay}\right),\tag{5.1}$$

is that

$$\int_{-\infty}^{\infty} F(u)\exp(-ixu)du = 0,\tag{5.2}$$

for all $x < -a$. Thus if $\tilde{F}(q)$ is the inverse Fourier transform of $F(u)$ i.e.

$$\tilde{F}(q) = \frac{1}{\sqrt{2\pi}} \int_{-\infty}^{\infty} F(u)e^{-iqu}du,\tag{5.3}$$

83

then from Titchmarch's theorem it follows that

$$\tilde{F}(q) = 0, \quad \text{for} \quad u < -a. \tag{5.4}$$

Since the Jost solution $f(k, r)$ is a solution of the Schrödinger equation, we can formally write it as a solution of Volterra integral equation [5]

$$f(k, r) = e^{ikr} + \int_r^\infty \frac{\sin [k\,(r - r')]}{k} v\,(r')\,f\,(k, r')\,dr', \quad |k| \neq 0. \tag{5.5}$$

Taking k to be a complex number we write

$$k = k_r + ik_r, \tag{5.6}$$

from the iterative solution of (5.5) we obtain two inequalities:

$$\left| f(k, r) - e^{ikr} \right| \leq \frac{C}{|k|} e^{-k_i r} \int_r^\infty r'\,|v\,(r')|\,dr', \tag{5.7}$$

and

$$\left| \frac{df\,(k, r)}{dr} - ike^{ikr} \right| \leq Ce^{-k_i r} \int_r^\infty |v\,(r')|\,dr' \quad k_i \geq 0, \quad k_i \geq 0 \tag{5.8}$$

where C is a constant. The first inequality gives us the behaviour of $f(k, r)$ when $|k|$ is large, $k_i \geq 0$ and $r \geq 0$ is fixed, and the second gives us the behaviour when r is large and k lies anywhere in the upper half-plane [6]. Having obtained bounds for $f(k, r) - e^{ikr}$, let us consider the function $P(k, r)$ defined by

$$P(k, r) = f(k, r) - e^{ikr}, \tag{5.9}$$

which is analytic in the upper half of the k plane, $k_i > 0$. From Eq. (5.7) it follows that

$$|P(k, r)| \leq \frac{C'}{|k|} \exp(-k_i r), \tag{5.10}$$

where C' is also a constant. Therefore $P(k, r) \in L^2(-\infty, \infty)$, i.e. it is a square integrable function and satisfies the following condition:

$$\int_{-\infty}^\infty |P(k_r + ik_i, r)|^2 dk_r = \mathcal{O}\left(e^{-2k_i r}\right). \tag{5.11}$$

Thus Eq. (5.3) of the Titchmarch's theorem holds, (i.e. Eq. (5.1)) with $a = -r$, and that the Fourier transform $\tilde{P}\,(k_r, r)$ takes the form

$$\tilde{P}\,(k_r, r) = \frac{1}{\sqrt{2\pi}} \int_{-\infty}^\infty P(q, r) \exp\left(-ik_r q\right) dq, \tag{5.12}$$

and it satisfies $\tilde{P}\,(k_r, r) = 0$ for $k_r < r$. Now we introduce the function $A\,(r, k_r)$ by

$$A\,(r, k_r) = \frac{1}{\sqrt{2\pi}} \tilde{P}\,(k_r, r), \tag{5.13}$$

so that

$$A(r, k_r) = \frac{1}{2\pi} \int_{-\infty}^{\infty} P(q, r) e^{-iqk_r} dq, \tag{5.14}$$

and $A(r, k_i) = 0$ for $k_r \leq r$. We invert this Fourier transform to get

$$P(k, r) = \int_{-\infty}^{\infty} A(r, r') \exp(ikr') dr'$$

$$= \int_{r}^{\infty} A(r, r') \exp(ikr') dr', \quad k_r \geq 0. \tag{5.15}$$

Let us note that both sides of the last equation are boundary values of the functions which are analytic in the upper half-plane, $k_r = \text{Im } k \geq 0$. Next from Eqs. (5.9) and (5.15) we find $f(k, r)$

$$f(k, r) = e^{ikr} + \int_{r}^{\infty} A(r, r') e^{ikr'} dr'. \tag{5.16}$$

The Marchenko Equation — We observed that in the Gel'fand–Levitan approach to the inverse scattering the Jost function $f(k)$ entered the basic equation (4.75) for determining the potential. In Marchenko's method the potential is found directly from the S matrix without first finding the Jost function. To avoid complications, at the start we assume that there are no bound states. When this is the case Eq. (4.75) reduces to

$$\frac{2}{\pi} \int_{0}^{\infty} \frac{k^2}{|f(k)|^2} \bar{\phi}(k, r) \bar{\phi}(k, r') \, dk = \delta(r - r'). \tag{5.17}$$

From the symmetry properties of $f(k)$ and the fact that on r-axis $\phi(k, r)$, $\bar{\phi}(k, r')$ and $|f(k)|^2$ are all real we can write (5.17) as

$$\frac{1}{\pi} \int_{-\infty}^{\infty} \frac{k^2}{|f(k)|^2} \bar{\phi}(k, r) \bar{\phi}(k, r') \, dk = \delta(r - r'). \tag{5.18}$$

By introducing $\phi(k, r)$ which is defined by

$$\phi(k, r) = -\frac{1}{2if(k)} [f(-k, r) - S(k)f(k, r)], \tag{5.19}$$

we can express the integrand in (5.18) as

$$\frac{k^2}{|f(k)|^2} \bar{\phi}(k, r) \bar{\phi}(k, r') = \phi(k, r)\phi^*(k, r'). \tag{5.20}$$

Therefore by substituting for $\bar{\phi}(k, r)$ and $\bar{\phi}(k, r')$ in terms of $\phi(k, r)$ in (5.18) and then using (5.19) we find

$$\frac{1}{4\pi} \int_{-\infty}^{\infty} [f(-k, r) - S(k)f(k, r)] [f^*(-k, r') - S^*(k)f^*(k, r')] \, dk = \delta(r - r'). \tag{5.21}$$

Again making use of the analytical properties of $f(k,r)$ and $S(k)$ we can write the above integral as

$$\frac{1}{4\pi} \int_{-\infty}^{\infty} [f(-k,r) - S(k)f(k,r)]f(k,r')\, dk$$

$$+ \frac{1}{4\pi} \int_{-\infty}^{\infty} [f(k,r) - S(-k)f(-k,r)]f(-k,r')\, dk = \delta(r - r'). \tag{5.22}$$

But replacing k by $-k$ changes the second integral into the first, therefore we have

$$\frac{1}{2\pi} \int_{-\infty}^{\infty} [f(-k,r) - S(k)f(k,r)]f(k,r')\, dk = \delta(r - r'). \tag{5.23}$$

Now let us go back to Eq. (5.16) and invert it to get

$$e^{ikr} = f(k,r) + \int_{r}^{\infty} \tilde{A}(r,r')f(k,r')\, dr'. \tag{5.24}$$

This can be done if we take (5.16) as a Volterra integral equation for e^{ikr} with the kernel $\tilde{A}$ and the inhomogeneous term $f(k,r)$. By combining (5.22) and (5.23) we find

$$\int_{-\infty}^{\infty} f(k,r') \left[e^{-ikr} - S(k)e^{ikr} \right] dk = 0, \quad r > r'. \tag{5.25}$$

Now we replace $f(k,r')$ by the equation (5.16) to obtain

$$\int_{-\infty}^{\infty} \left\{ \left[\exp(ikr') + \int_{r'}^{\infty} A(r',r'') \exp(ikr'')\, dr'' \right] \left[e^{-ikr} - S(k)e^{ikr} \right] \right\} dk = 0. \tag{5.26}$$

Expanding this equation we have

$$\int_{-\infty}^{\infty} \exp[ik(r' - r)]\, dk + \int_{r'}^{\infty} A(r',r'') \int_{-\infty}^{\infty} \exp[ik(r'' - r)]\, dk dr''$$

$$- \int_{-\infty}^{\infty} S(k) \exp[ik(r' + r)]\, dk - \int_{r'}^{\infty} A(r',r'') \times$$

$$\left[\int_{-\infty}^{\infty} S(k) \exp[ik(r'' + r)]\, dk \right] dr'' = 0. \tag{5.27}$$

This equation can be simplified if we note that

$$\int_{-\infty}^{\infty} \exp[ik(r' - r)]\, dk = 2\pi\delta(r' - r). \tag{5.28}$$

Let us also define $A_0(r)$ by

$$A_0(r) = \frac{1}{2\pi} \int_{-\infty}^{\infty} [S(k) - 1]e^{ikr}\, dk. \tag{5.29}$$

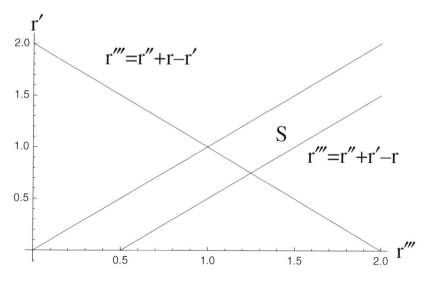

Figure 5.1: The surface integral in Eq. (5.37) is to be evaluated on the area r shown here by S [5].

Then we find that (5.27) can be reduced to the following equation

$$A\left(r',r\right) = A_0\left(r+r'\right) + \int_{r'}^{\infty} A\left(r',r''\right) A_0\left(r''+r\right) dr'', \quad r \geq r'. \tag{5.30}$$

This is Marchenko's integral equation. The known function here is $A_0\left(r+r'\right)$ which is related to the S matrix by (5.29).

When bound states are present, then (5.29) will have additional terms, the contribution of the bound states

$$A_0(r) = \frac{1}{2\pi} \int_{-\infty}^{\infty} [S(k)-1]e^{ikr}dk + \sum_{n=1}^{N} C_n^2 e^{-\gamma_n r}, \tag{5.31}$$

where $-\frac{\gamma_n^2 \hbar^2}{2m}$ is the nth bound state energy, and where C_n^{-2} is the normalization constant for the nth bound state

$$C_n^{-2} = \int_0^{\infty} f^2(-i\gamma_n, r)dr. \tag{5.32}$$

Finding the Potential from the Solution of the Marchenko Equation — We start with the Eq. (5.29) or (5.31) and calculate $A_0\left(r+r'\right)$. By solving the integral equation (5.30) we find the matrix $A\left(r',r\right)$. To obtain the potential we first find the inverse Fourier transform of $P(k,r)$, Eq. (5.9);

$$A(r,r') = \frac{1}{2\pi} \int_{-\infty}^{\infty} P(k,r) e^{-ikr'} dk = \frac{1}{2\pi} \int_{-\infty}^{\infty} e^{-ikr'} dk$$

$$\times \left[\int_0^{\infty} \frac{\sin k\,(r'-r)}{k} v\,(r')\,f\,(k,r')\,dr' \right]$$

$$= \int_{-\infty}^{\infty} e^{-ikr'} J\,(k,r)\,dk, \quad r' \geq r, \tag{5.33}$$

where

$$J\,(k,r) = \int_r^{\infty} \frac{\sin k\,(r'-r)}{k} v\,(r') \left[e^{ikr'} + \int_{r'}^{\infty} A\,(r',r'')\,e^{ikr''}\,dr'' \right] dr'. \tag{5.34}$$

Thus we have integrals of the form

$$\int_0^{\infty} \frac{\sin(ka)\cos(kb)}{k} dk = \begin{cases} \frac{\pi}{2} & a > b \geq 0 \\ \frac{\pi}{4} & a = b > 0 \\ 0 & b > a \geq 0 \end{cases} \tag{5.35}$$

and

$$\int_{-\infty}^{\infty} \frac{\sin k\,(r'-r)}{k} \exp\left[k\,(r'-r'')\right] dk = 2 \int_0^{\infty} \frac{\sin k\,(r'-r)}{k} \cos\left[k\,(r'-r'')\right] dk, \tag{5.36}$$

that we will need in evaluating $A\,(r,r')$. Let us first consider the case where $r'-r'' \leq r'-r$, so that in (5.35) $b \leq a$, and then the integral is π. Now when $r'-r'' < 0$, then $a = r'-r$ and $b = r''-r'$, therefore $a > b$, and for this case $r' > \frac{1}{2}\,(r''+r)$. This means that $r'' \geq \frac{1}{2}\,(r''+r)$, and the integral is π when $r' > \frac{1}{2}\,(r''+r)$. Collecting the various terms we can write (5.33) as the sum of two integrals

$$A\,(r,r'') = \frac{1}{2} \int_{\frac{1}{2}(r''+r)}^{\infty} v\,(r')\,dr' + \frac{1}{2} \int \int_S A\,(r',r''')\,v\,(r')\,dr'dr'''. \tag{5.37}$$

The area S in r',r''' is bounded between $r''' = r'$ and $r''' = r' + r'' - r$ when both $r''' = r'$ and $r''' = r'' + r' - r$ are going to infinity and they also intersect the line $r''' = r - r' + r''$ and the space between these lines is S (see Fig. 5.1). But if we set $r'' = r$ in (5.37), then $r' = r$ and the two parallel lines coincide with the result that $S = 0$ (see Fig. 5.1). Then (5.37) simplifies to

$$A(r,r) = \frac{1}{2} \int_r^{\infty} v\,(r')\,dr'. \tag{5.38}$$

Assuming the continuity of $v(r)$, we find the potential to be

$$v(r) = -2\frac{d}{dr} A(r,r), \tag{5.39}$$

which is the analogue of the expression (4.59) that we found from the Gel'fand–Levitan theory.

One of the other properties of the Marchenko's kernel is that it satisfies a partial differential equation which contains $v(r)$. Noting that the Jost solution $f(k,r)$, Eq. (5.16) satisfies the Schrödinger equation, if we substitute $f(k,r)$ in the latter equation we find

$$\left(\frac{\partial^2}{\partial r^2} - \frac{\partial^2}{\partial r'^2}\right) A\left(r,r'\right) = v(r)A\left(r,r'\right).$$

(5.40)

This equation is subject to the boundary condition [6]

$$-2\frac{d}{dr}A(r,r) = v(r)A(r,r).$$

(5.41)

Solvable Examples Obtained Using Marchenko's Method — We start from a very simple form for the Jost function which is given by [7]

$$f(k) = \frac{k+ia}{k+ib}, \quad \text{Im } a = \text{Im } b = 0,$$

(5.42)

where we assume that $a > 0$, as there is no bound state and $b > 0$ for $f(k)$ to be analytic in the upper half of k-plane. Since there are no bound states Eq. (5.31) can be written as

$$A_0\left(r+r'\right) = \frac{1}{2\pi}\int_{-\infty}^{\infty}\left[\frac{f(-k)}{f(k)} - 1\right]\exp[ik\left(r+r'\right)]dk$$

$$= \frac{1}{2\pi}\int_{-\infty}^{\infty}\left[\frac{(k-ia)(k+ib)}{(k-ib)(k+ia)} - 1\right]\exp[ik\left(r+r'\right)]dk.$$

(5.43)

By closing the integration contour in the upper half-plane and making use of the residue theorem, we find

$$A_0\left(r+r'\right) = 2b\left(\frac{a-b}{a+b}\right)\exp\left[-b\left(r+r'\right)\right].$$

(5.44)

We can simplify this result if we introduce r_0 by

$$\frac{b-a}{b+a} = e^{2br_0},$$

(5.45)

then the solution to Marchenko's integral equation (5.37) can be obtained as

$$A\left(r,r'\right) = -2b\frac{\exp\left[-b\left(r+r'-2r_0\right)\right]}{1+\exp\left[-2b\left(r-r_0\right)\right]},$$

(5.46)

and the potential in this case found from (5.29) is

$$V(r) = \frac{\hbar^2}{2m}v(r) = -\frac{4b^2\hbar^2}{m}\frac{\exp\left[-2b\left(r-r_0\right)\right]}{\{1+\exp\left[-2b\left(r-r_0\right)\right]\}^2}$$

$$= -\frac{\hbar^2 b^2}{m}\frac{1}{\cosh^2[b(r-r_0)]}.$$

(5.47)

This is a Bargmann potential which we found earlier in Sec. 4.6.

S matrix with Two Poles in the Upper Half of the Complex k-plane —
Suppose that the Jost solution is given by

$$f(\pm k, r) = e^{\pm ikr} \left\{ 1 + 2b \frac{\exp[-2b(r - r_0)]}{(\pm ik - b)[1 + \exp[-2b(r - r_0)]]} \right\}$$

(5.48)

$$f(k) = \frac{k + ia}{k + ib}, \quad b > 0,$$

(5.49)

but now with $a = -\gamma < 0$. In this case we have a single bound state with the energy $E = -\frac{\hbar^2 \gamma^2}{2m}$. For this $f(k)$ the scattering matrix has the form

$$S(k) = \frac{(k + i\gamma)(k + ib)}{(k - i\gamma)(k - ib)}.$$

(5.50)

By substituting this $S(k)$ in (5.31), we find $A_0(r + r')$ to be

$$A_0(r + r') = -\frac{1}{2\pi} \int_{-\infty}^{\infty} \left[1 - \frac{(k + i\gamma)(k + ib)}{(k - i\gamma)(k - ib)} \right] \exp[ik(r + r')]$$
$$- C_n^2 \exp[-\gamma(r + r')],$$

(5.51)

where C_n^2 is the bound state normalization constant. For the contour we choose a semicircle of infinite radius in the upper half of the complex k plane, and from residue theorem we calculate $A_0(r + r')$:

$$A_0(r + r') = -2b \left(\frac{b + \gamma}{b - \gamma} \right) \exp[-b(r + r')] + i \text{Res } S(i\gamma) \exp[-\gamma(r + r')]$$
$$- C_n^2 \exp[-\gamma(r + r')],$$

(5.52)

where Res denotes the residue of the $S(k)$ matrix at $k = i\gamma$. If we choose the normalization C_n^2 such that

$$C_n^2 = i \text{Res } S(k)|_{k=i\gamma} = 2\gamma \frac{b + \gamma}{b - \gamma},$$

(5.53)

then the last two terms in (5.52) cancel each other and we are left with a single term, viz,

$$A_0(r + r') = -2b \left(\frac{b + \gamma}{b - \gamma} \right) \exp[-b(r + r')].$$

(5.54)

This $A_0(r + r')$ is identical with (5.44) and therefore the potential for this case is given by (5.47) and is a Bargmann potential. For this problem the Jost solution is found to be

$$f(i\gamma, r) = e^{-\gamma r} \left\{ \frac{1 - e^{-2br}}{1 + \exp[-2b(r - r_0)]} \right\},$$

(5.55)

where

$$e^{2br_0} = \frac{b + \gamma}{b - \gamma}.$$

(5.56)

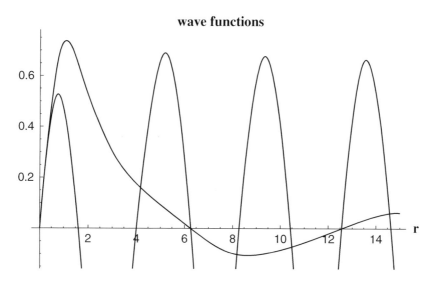

wave functions

Figure 5.2: Bound state in continuum. Here the wave function of the particle for all values of the wave number k is oscillatory, except only for $k = q$ in which case the wave function is localized close to origin. The constant $c = 1$ has been used in this calculation.

5.2   Bound States Embedded in Continuum

Very early in the history of quantum mechanics Neumann and Wigner showed that it is possible to have a particle bounded in continuum with positive energy [8], [9]. Such cases can be easily obtained from Gel'fand–Levitan formulation. Our starting point here will be Eq. (4.82) where $g(r, r')$ has the simple form, viz, the product of two plane waves

$$g(r, r') = c^2 \phi(q, r)\phi(q, r') = \frac{c^2}{q^2}\sin(qr)\sin(qr').\tag{5.57}$$

Using this $g(r, r')$ as the input, we can calculate the Gel'fand–Levitsn kernel $K(r, r')$, Eq. (4.83), from which we can determine the potential $V(r)$, and the corresponding wave function. Let

$$d(r) = 1 + \left(\frac{c}{q}\right)^2 \int_0^r \sin^2(qr')\, dr',\tag{5.58}$$

then according to Gel'fand–Levitan formalism we have

$$V(r) = -\frac{\hbar^2}{m}\frac{d^2}{dr^2}\ln[d(r)],\tag{5.59}$$

and

$$\phi(k, r) = \frac{\sin(kr)}{k} - \left(\frac{c}{q}\right)^2 \frac{\sin(qr)}{kd(r)}\int_0^r \sin(qr')\sin(kr')\, dr'.\tag{5.60}$$

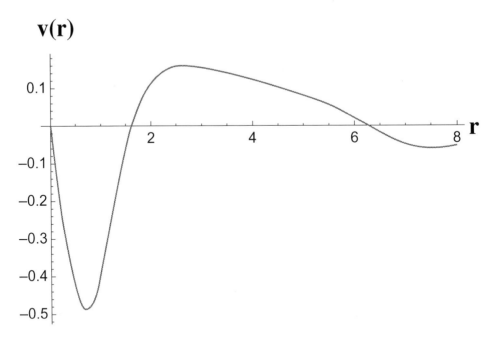

Figure 5.3: The potential that binds the particle in continuum is shown in this figure. Here we have set $\frac{\hbar^2}{2m} = 1$. Other parameters used for this plot are the same as those of the previous figure.

It is important to note that in this example, only for $k = q$, the terms in $\phi(k, r)$ which do not damp out at infinity, cancel each other, with the result that $\phi(q, r)$ goes to zero as r goes to infinity, and becomes square integrable (see Fig. 5.2). For all other k values the wave function is oscillatory, and so is the potential $V(r)$. This special wave number $k = q$, makes the potential attractive and with a damped oscillating tail (Fig. 5.3).

5.3 More Solvable Potentials Found from Inverse Scattering

In our discussion of the Gel'fand–Levitan formulation we observed that if $g(r, r') = \phi_0(r)\phi_0(r')$, then the Gel'fand–Levitan integral equation can be solved exactly (see Eq. (4.84)). This is the simplest case, but a degenerate kernel of $g(r, r')$ having the form

$$g\left(r, r'\right) = \sum_{j=1}^{N} \phi_j(r)\phi_j\left(r'\right), \qquad (5.61)$$

also gives us a solvable integral equation for $K\left(r, r'\right)$, reducing it to a matrix equation. Degenerate kernels of this type can be used in both the Gel'fand–Levitan and also in Marchenko's method by assuming that the Jost function is rational function of k and has the form

$$f(k) = \prod_{j=1}^{N} \frac{(k + ia_j)}{(k + ib_j)}. \tag{5.62}$$

From this expression for $f(k)$ we find the corresponding S matrix and the spectral densities $\rho(k)$ to be

$$S(k) = \prod_{j=1}^{N} \frac{(k - ia_j)(k + ib_j)}{(k - ib_j)(k + ia_j)}, \tag{5.63}$$

and also Eq. (4.75) the spectral density takes the form which we write as

$$d\rho(k) = \frac{2k^2 dk}{\pi f(k) f(-k)} = \frac{2k^2 dk}{\pi} \prod_{j=1}^{N} \frac{k^2 + b_j^2}{k^2 + a_j^2}. \tag{5.64}$$

It should be emphasized that with the condition of integrability of the potential Eq. (4.6);

$$\int_0^{\infty} r|v(r)|dr < \infty, \tag{5.65}$$

we have the same number of factors in the denominator as in the numerator of $f(k)$ and this property guarantees that $\lim_{k \to \infty} f(k) \to 1$, when $\text{Im } k > 0$, a requirement that we have mentioned before. When the potential satisfies the integrability condition then $\text{Re } b_j > 0$, and the zeros of $f(k)$ at the points $k = -ia_j$ lie on the imaginary axis in the upper half of the complex k-plane. In addition the zeros $k = -ia_j$ and the poles $k = -ib_j$ of the function $f(k)$ must be located symmetrically with respect to the imaginary axis (see Fig. 5.4). Now we want to know how it is possible to approximate a given set of scattering data by a combination of Bargmann potentials by increasing the number of factors N in (5.62). There are different ways of constructing the potential:
(a) We can start with an exactly solvable potential $v_1(r)$ and then using a separable kernel $g(r, r')$, Eq. (4.82) we can construct $\Delta v(r)$ such that $v(r) = \bar{v}(r) + \Delta v(r)$.
(b) We may also start with $\bar{v}(r) \equiv 0$ and therefore the initial set in this case are

$$\bar{\phi}(k, r) = \frac{\sin(kr)}{k}, \quad \text{and} \quad f(\pm k) = e^{\pm ikr}. \tag{5.66}$$

By substituting these in the Gel'fand–Levitan equation we obtain [7]

$$\Delta v(r) = 4 \frac{\left[\gamma \left(\frac{r}{2} - \frac{\gamma^2}{c^2} \right) - \sinh^2(\gamma r) \right]}{\left[\frac{\gamma^2}{c^2} + \left(\frac{\sinh(2\gamma r)}{2\gamma} - r \right) \right]^2}, \tag{5.67}$$

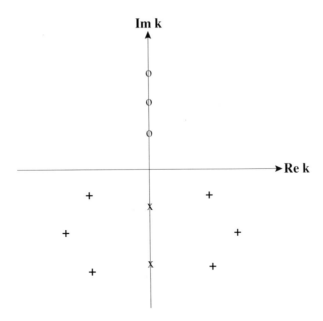

Figure 5.4: The poles of $S(k)$ matrix for a short range potential satisfying the integrability condition. **o** s denotes the poles corresponding to bound states, + s are poles appearing by pairs in the lower half of k-plane, and × s are poles on the negative imaginary axis.

where c^2 is the normalization constant (see Eq. (4.88)). The wave function obtained from the solution of the Gel'fand–Levitan method Eq. (4.85) give us

$$\phi_1(k,r) = \phi_0(k,r) - c^2 \left(\frac{\phi_0(i\gamma,r) \int_0^r \phi_0(i\gamma,r') \phi_0(k,r') \, dr'}{1 + c^2 \int_0^r \phi_0^2(i\gamma,r') \, dr'} \right)$$

(5.68)

$$= \frac{\sin(kr)}{k} - c^2 \left(\frac{\sinh(\gamma r)(\gamma \cosh(\gamma r) \sin(kr) - k \sinh(\gamma r) \cosh(kr))}{k(\gamma^2 + k^2)\left[\gamma^2 + \frac{c^2}{2}\left(\frac{\sinh(2\gamma r)}{2\gamma} - r\right)\right]} \right),$$

(5.69)

which corresponds to the first factor;

$$f_1(\pm k) = \frac{k \mp i\gamma}{k \pm i\gamma}, \quad \text{and} \quad S(k) = \left(\frac{k + i\gamma}{k - i\gamma} \right)^2.$$

(5.70)

Next we choose $v_0(r)$ as the initial potential and we include the second factor in $f(k)$ and proceed as before. Thus we find

$$v(r) = v_0(r) + \Delta v_1(r) + \cdots + \Delta v_N(r),$$

(5.71)

and in this way we find an approximate potential $v(r)$ which gives us the scattering matrix.

(c) Suppose that the system has N_b bound states,

$$ia_j = -i\gamma_j, \quad \gamma_j > 0, \quad j = 1, \ 2, \cdots N_b, \tag{5.72}$$

corresponding to the Jost function

$$f(k) = \prod_{j=1}^{N_b} \frac{(k + ia_j)}{(k + ib_j)}. \tag{5.73}$$

The other singularities of the S matrix must satisfy the following conditions:

$$\text{Re } a_j > 0, \ j = N + 1, \ N + 2, \cdots 2N, \tag{5.74}$$

$$\text{Re } b_j > 0, \ j = 1, \ 2, \cdots N. \tag{5.75}$$

As it was mentioned earlier, for real potentials the points ia_j and ib_j must be located symmetrically relative to the imaginary k-axis. The S matrix which in this case is given by (5.63) can be substituted into Marchenko's equation (5.29), and then choosing the contour for the integration a semi-circle of radius K in the upper half k-plane plus a line, $-K < k < K$, one finds

$$A_0\left(r + r'\right) = \sum_{n=1}^{N+N_b} B_n \exp\left[-\beta_n\left(r + r'\right)\right]. \tag{5.76}$$

Here we have included the contributions of the bound states and poles of the S matrix at the points $k = ib_j$. Thus β_n refers to all the poles of $S(k)$ inside the contour of integration, $\beta_n = \gamma_n$ for $n = N + 1, \ N + 2, \cdots N_b$, and $\beta_n = b_n$ for $n = 1, \ 2, \cdots N$ and B_n s stand for the products

$$B_{n+N} = \frac{-2\gamma_n(\gamma_n + b_n)}{(\gamma_n - b_n)} \prod_{m=1, m \neq n}^{N_b} \frac{(\gamma_n + b_m)(\gamma_n - a_m)}{(\gamma_n + a_m)(\gamma_n - b_m)}, \tag{5.77}$$

for $n = 1, \ 2 \cdots N_b$, and

$$B_n = \frac{-2b_n(b_n - a_n)}{(b_n + a_n)} \prod_{m=1, m \neq n}^{N} \frac{(b_m + b_n)(b_n - a_m)}{(b_n + a_m)(b_n - b_m)}, \tag{5.78}$$

for $n = 1, \ 2 \cdots N$.

Substituting these into Marchenko's equation

$$A\left(r, r'\right) = A_0\left(r + r'\right) + \int_r^\infty A\left(r, r''\right) A_0\left(r'' + r'\right) dr'', \tag{5.79}$$

and simplifying the result we get

$$A\left(r, r'\right) + \sum_{n=1}^{N+N_b} B_n \left\{ e^{-\beta_n r} + \int_r^\infty A\left(r, r''\right) e^{-\beta_n r''} dr'' \right\} e^{-\beta_n r'} = 0. \tag{5.80}$$

As this equation shows $A(r, r')$ is the sum of separable terms, and we can write it as

$$A(r, r') = - \sum_{n}^{N+N_b} B_n f(i\beta_n, r) e^{-\beta_n r'}, \tag{5.81}$$

where $f(i\beta_n, r)$ denotes the part in the curly bracket in (5.80). Hence we get a set of algebraic equations for $f(i\beta_n, r)$;

$$\sum_{n}^{N+N_b} f(i\beta_n, r) \left[\delta_{nm} + B_m \int_{r}^{\infty} e^{-(\beta_n + \beta'_n)r'} dr' \right] = e^{-\beta_n r}. \tag{5.82}$$

Now if we define the matrix $(I + \tilde{A})_{mn}(r)$ by

$$M_{mn} = (I + \tilde{A}_{nm}(r)) = \delta_{mn} + \frac{B_m \exp[-(\beta_m + \beta_n)r]}{(\beta_m + \beta_n)}, \tag{5.83}$$

and denote the inverse of $(I + \tilde{A})_{mn}(r)$ by $\left[(I + \tilde{A})^{-1} \right]_{nm}$, then we can find the Jost function for a given k, calculate the potential and solve the problem completely. The Jost solution for a fixed value of k, determined from the matrix $M^{-1} = (I + \tilde{A})^{-1}$ is given by [7]

$$f(\pm k, r) = e^{\pm kr} + \sum_{n,m}^{N+N_b} \frac{B_n \left(M^{-1} \right)_{mn} \exp[-(\beta_n + \beta_m \mp ik)r]}{\beta_n \mp ik}, \tag{5.84}$$

and the potential is found in terms of the matrix $(I + \tilde{A})$;

$$V(r) = -\frac{\hbar^2}{m} \frac{d^2}{dr^2} \ln[\det(I + \tilde{A})]. \tag{5.85}$$

If we set $r = 0$ in Eq. (5.84), we find the Jost function to be

$$f(k) = \prod_{j=1}^{N} \frac{k + ia_j}{k + ib_j}. \tag{5.86}$$

5.4   The Inverse Problem for Reflection and Transmission from a Barrier

The mathematical methods needed to solve this problem are, in many respects, similar to those that we have encountered for the case of the S wave scattering. Here a complete set of states on the entire axis will be composed of wave functions of continuous spectrum incident on the potential from the right and from the left. We denote these by $\psi_1(k, x)$ and by $\psi_2(k, x)$ respectively (see also Sec. 12.2). In addition if there are bound states with wave functions $\psi(i\gamma_n, x)$ we include these as well. Thus the completeness is expressed by

$$\frac{1}{\pi} \int_0^\infty [\psi_1(k,x)\psi_1^*(k,y) + \psi_2(k,x)\psi_2^*(k,y)]\, dk$$

$$+ \sum_{j=1}^N \psi(i\gamma_n, x)\psi(i\gamma_n, y) = \delta(x-y). \tag{5.87}$$

Now let us consider the scattering solution for the one-dimensional Schrödinger equation

$$\frac{d^2\psi(x)}{dx^2} + k^2\psi(x) = v(x)\psi(x), \quad -\infty < x < +\infty, \tag{5.88}$$

with the following asymptotic solutions

$$\psi_1(k,x) = \begin{cases} S_{11}(k)e^{-ikx} & x \to -\infty \\[2mm] e^{-ikx} + S_{12}(k)e^{ikx} & x \to +\infty \end{cases}, \tag{5.89}$$

and

$$\psi_2(k,x) = \begin{cases} e^{ikx} + S_{21}(k)e^{-ikx} & x \to -\infty \\[2mm] S_{22}(k)e^{ikx} & x \to +\infty \end{cases}. \tag{5.90}$$

We observe that the scattering matrix is a 2×2 matrix of the form

$$S(k) = \begin{bmatrix} S_{11} & S_{21} \\ S_{12} & S_{22} \end{bmatrix} = \begin{bmatrix} T_1(k) & R^+(k) \\ R^-(k) & T_2(k) \end{bmatrix}, \tag{5.91}$$

where $T_{1,2}(k)$ and $R_\pm(k)$ are the transmission and reflection coefficients respectively. Thus it seems that we need four functions of k to determine the potential. However for a real potential the conservation of probability (or the number of particles) implies that the S matrix must be unitary and satisfy the condition

$$S_{11}(k) = S_{22}(k), \quad \text{or} \quad T_1 = T_2, \tag{5.92}$$

and this reduces one of the unknowns. Also one may regard the potential $v(x)$, $-\infty < x < \infty$ as two real functions of x in the interval $0 \le x < \infty$, and this potential completely determines the S matrix. Thus there exists one more relation among the three independent elements of S, all assumed to be real functions of k. This last one is related to the analyticity of the transmission coefficient in the upper half of the k-plane, (Im $k > 0$). Thus the whole of the S matrix is determined if we know $S_{12}(k)$ or $S_{21}(k)$. Choosing either $S_{12}(k) = R^-(k)$ or $S_{21}(k) = R^+(k)$ for the input, we can write the Gel'fand–Levitan integral equation as

$$K^+(x,y) + R_0^+(x+y) + \int_x^\infty K_+(x,z)R_0^+(z+y)dz = 0, \quad y > x, \tag{5.93}$$

or

$$K^-(x,y) + R_0^-(x+y) + \int_{-\infty}^{x} K_-(x,z)R_0^-(z+y)dz = 0, \quad y < x, \qquad (5.94)$$

where in the absence of the bound state $R_0^{\pm}(x)$ is defined by its Fourier transform $r(k)$;

$$R_0^{\pm}(x) = \frac{1}{2\pi} \int_{-\infty}^{\infty} r(k)e^{\pm ikx}dk. \qquad (5.95)$$

Since this one-dimensional problem is over-determined, the two equations (5.94) and (5.95) yield the same barrier profile. In the following account for the solution of the problem we will consider just Eq. (5.94), and we will suppress the superscripts in this equation.

5.5 A Special Problem in Electromagnetic Inverse Scattering

As a model of inverse one-dimensional problem that we discussed in the preceding section, we will study the scattering of electromagnetic waves from a stratified ionized region, a model which has been used to study ionospheric radio wave propagation [10]–[14].

We consider the case where the relative permittivity of the inhomogeneous region is given by

$$\frac{\epsilon(k,x)}{\epsilon_0} = 1 - \frac{v(x)}{k}, \quad x > 0, \qquad (5.96)$$

where ϵ_0 is the permittivity of the free space and k is the wave number of the wave in free space. The profile function $v(x)$, in this model, measured in units of $(\text{length})^{-1}$, and should not be confused with the potential $v(x)$ of potential scattering, is proportional to the electron density. Now we want to replace the amplitude of the electromagnetic field $E_y(x, ct)$ by its time-harmonic amplitude $u(\kappa, x)$ using the Fourier transform. Let κ denote $k \cos \alpha$, where $\cos \alpha$, $\cos \beta$ and $\cos \gamma$ are the direction cosines of the propagating plane wave, then since $E_y(x, ct)$ satisfies the wave equation, $u(x, \kappa)$ will be a solution of a Schrödinger like equation

$$\frac{d^2}{dx^2}u(x,\kappa) + \left[\kappa^2 - v(x)\right]u(x,\kappa) = 0. \qquad (5.97)$$

We also assume that $R(x)$ is zero for $x < 0$. For the common model that is used to study the ionospheric density profile, a possible choice is

$$R(x) = R_1(x)\theta(x), \qquad (5.98)$$

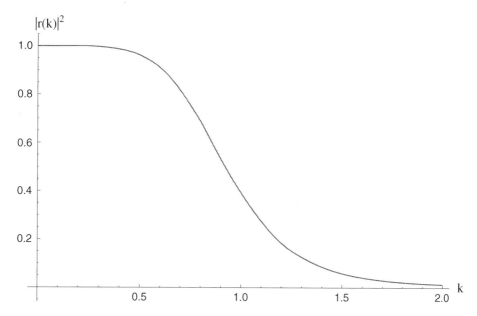

Figure 5.5: The reflection coefficient $r^2(k)$ when $r(k)$ is assumed to have three poles in the lower k-plane [13]–[16].

where $\theta(x)$ is the step function, $\theta(x) = 1$ for $x > 0$ and is zero otherwise. Now by substituting (5.98) in (5.94) we get

$$K_1(x,y) + R_1(x+y) + \int_{-y}^{x} K_1(x,z)R_1(z+y)dz = 0, \quad x+y \geq 0. \quad (5.99)$$

To make the problem solvable, we choose $r(k)$ to be a rational function of k, i.e.

$$r(k) = \frac{N(k)}{D(k)}, \quad (5.100)$$

where both $N(k)$ and $D(k)$ are polynomials in k, and $D(k)$ can be written as a product

$$D(k) = \prod_{j=1}^{N}(k - k_j). \quad (5.101)$$

In this relation k_j s are distinct complex numbers, and are points in the lower half of k-plane. By substituting (5.101) in (5.98) and using contour integration we find

$$R(x) = -i\sum_{j=1}^{N} e^{-ik_j x}C_j\theta(x), \quad (5.102)$$

and

$$R_1(x) = -i\sum_{j=1}^{N} e^{-ik_j x}C_j, \quad x > 0, \quad (5.103)$$

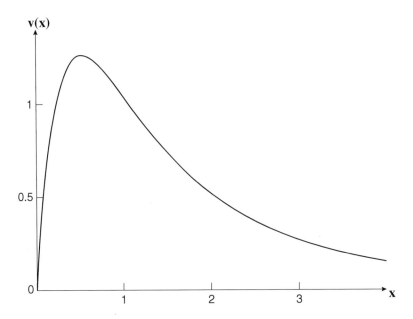

Figure 5.6: The barrier profile found from the inversion of the reflection coefficient. For this case three poles were included in $r(k)$. For other profiles with 5 and 6 poles, see references [13], [16].

where

$$C_j = N(k_j) \prod_{i=1, j \neq i}^{N} \left(\frac{1}{k_j - k_i} \right). \tag{5.104}$$

Now let us consider the trial solution for the Gel'fand–Levitan equation (5.99) of the form

$$K_1(x, y) = \sum_{\alpha} f_\alpha(x) \exp(a_\alpha y), \tag{5.105}$$

where the summation is over a finite number of values of α and $f_\alpha(x)$ s are unknown functions of x to be determined later. Substituting (5.103) and (5.105) in (5.99) we obtain an equation which we can write as

$$A + B = 0, \tag{5.106}$$

where

$$A \equiv \sum_{\alpha} f_\alpha \exp(a_\alpha t) + i \sum_{\alpha} \sum_{j=1}^{N} C_j f_\alpha(x) \frac{1}{a_\alpha - ik_j} \exp(-a_\alpha y), \tag{5.107}$$

and

$$B \equiv -i \left[\sum_{j=1}^{N} f_\alpha \exp[-i(k_j(x+y))]C_j + \sum_\alpha \sum_{j=1}^{N} C_j f_\alpha(x) \right.$$

$$\left. \times \left(\frac{1}{a_\alpha - ik_j} \right) \exp[a_\alpha x - ik_j(x+y)] \right]. \tag{5.108}$$

We will set $A = 0$ and $B = 0$ simultaneously and thus we find the solution to the problem. If we choose

$$a_{-\alpha} = -a_\alpha, \quad \text{for all} \quad \alpha, \tag{5.109}$$

then the equation $A = 0$ may be written as

$$\sum_\alpha \left\{ \left[f_\alpha(x) + i \sum_{j=1}^{N} C_j f_{-\alpha}(x) \left(\frac{1}{-a_\alpha - ik_j} \right) \right] \exp(a_\alpha y) \right\} = 0. \tag{5.110}$$

We now set the coefficient of $\exp(a_\alpha y)$ in (5.110) for each α equal to zero

$$f_\alpha(x) + i \sum_{j=1}^{N} C_j f_{-\alpha}(x) \left(\frac{1}{-a_\alpha - ik_j} \right) = 0, \quad \alpha = 1, 2, \cdots. \tag{5.111}$$

We get a second set of equations by replacing α by $-\alpha$ and we obtain

$$f_{-\alpha}(x) + i \sum_{j=1}^{N} C_j f_\alpha(x) \left(\frac{1}{a_\alpha - ik_j} \right) = 0. \tag{5.112}$$

Thus we have two sets of linear equations (5.111) and (5.112) for $f_\alpha(x)$ and $f_{-\alpha}(x)$. A non-trivial solutions for these equations exists if the determinant of the coefficients for the unknowns f_α and $f_{-\alpha}$ vanish, i.e.

$$\begin{vmatrix} 1 & -i \sum_{j=1}^{N} \left(C_j \frac{1}{a_\alpha + ik_j} \right) \\ i \sum_{l=1}^{N} \left(C_l \frac{1}{a_\alpha - ik_l} \right) & 1 \end{vmatrix} = 0, \tag{5.113}$$

or

$$\left[\sum_{j=1}^{N} \left(C_j \frac{1}{a_\alpha - ik_j} \right) \right] \left[\sum_{l=1}^{N} \left(C_l \frac{1}{a_\alpha + ik_l} \right) \right] = 1. \tag{5.114}$$

This last equation can be solved for the allowed values of a_α. Noting that k is a complex variable, we find that C_j is the residue of $r(k)$ at $k = k_j$ and that $\frac{C_j}{a_\alpha - ik_j}$ is the residue of $\frac{r(k)}{a_\alpha - ik}$ at $k = k_j$. Now $\sum_{j=1}^{N} \frac{C_j}{a_\alpha - ik_j}$ is the sum of all the residues of $\frac{r(k)}{a_\alpha - ik}$ at $k = -ia_\alpha$ except for the one at $k = -ia$. Similarly $\sum_{l=1}^{N} \frac{C_l}{a_\alpha + ik_l}$ is the sum of the residues of $\frac{r(k)}{a_\alpha + ik}$ at $k = ia_\alpha$. However in each case since the integral of

$\frac{r(k)}{a_\alpha}$ around a closed contour (as the radius of the circle tends to infinity) is zero, therefore

$$\sum_{j=1}^{N} \frac{C_j}{a_\alpha - ik_j} = -\text{Res} \left[\frac{r(k)}{a_\alpha - ik} \right]_{k=-ia_\alpha}, \tag{5.115}$$

and

$$\sum_{l=1}^{N} \frac{C_l}{a_\alpha + ik_l} = -\text{Res} \left[\frac{r(k)}{a_\alpha + ik} \right]_{k=ia_\alpha}. \tag{5.116}$$

From these results we conclude that

$$\sum_{j=1}^{N} \frac{C_j}{a_\alpha - k_j} = -ir(-ia_\alpha), \tag{5.117}$$

and

$$\sum_{l=1}^{N} \frac{C_l}{a_\alpha - k_l} = ir(ia_\alpha). \tag{5.118}$$

Thus we find the important relations;

$$r(ia_\alpha)r(-ia_\alpha) = 1, \tag{5.119}$$

and

$$r(-k) = [r(k)]^*. \tag{5.120}$$

These relations must be satisfied by the input $r(k)$. For instance when we choose $n = 3$, then a possible form for $r(k)$ is

$$r(k) = \frac{k_1 k_2 k_3}{(k - k_1)(k - k_2)(k - k_3)}, \tag{5.121}$$

where k_1, k_2 and k_3 are complex numbers and $r(k)$ satisfies both of the conditions (5.119) and (5.120) for real values of k. But as we discussed earlier, in the complex k-plane these three points k_1, k_2 and k_3 are positioned symmetrically about the imaginary k-axis. If we choose $r(k)$ as given by (5.121) and where the poles are located at

$$k_1 = 0.8 - 0.499i, \quad k_2 = -0.8 - 0.499i \quad \text{and} \quad k_3 = -i, \tag{5.122}$$

then the resulting profile $|r(k)|^2$ as a function of k is shown in Fig. 5.5. The conservation of flux in this case implies that $|r(k)|^2 \le 1$ for all values of k.

We assume that $r(k)$ is such that the N possible values of a_α are all distinct, and we allow α s to run from 1 to N. We also notice that from Eq. (5.112) it is clear that $f(x)$ satisfies the equation

$$f_\alpha(x) = \left(-i \sum_{j=1}^{N} \frac{C_j}{a_\alpha - ik_j} \right) f_\alpha(x). \tag{5.123}$$

Now let us examine the solution of $B = 0$. These equations reduce to a set of n equations one for each value of j:

$$\sum_\alpha \frac{1}{a_\alpha - ik_j} e^{a_\alpha x} f_\alpha(x) + 1 = 0, \tag{5.124}$$

where the summation is over both positive and negative values of α. By substituting $f_\alpha(x)$ from (5.123) in (5.124) we find the following relation;

$$\sum_{\alpha=1}^{N} \left(F_{j\alpha} e^{a_\alpha x} + G_{j\alpha} e^{-a_\alpha x} \right) f_\alpha(x) + 1 = 0, \qquad j = 1, 2, \cdots N, \tag{5.125}$$

where

$$F_{j\alpha} = \frac{1}{a_\alpha - ik_j}, \tag{5.126}$$

and

$$G_{j\alpha} = \left(\frac{1}{a_\alpha + ik_j} \right) \sum_{l=1}^{N} \frac{C_l}{a_\alpha - ik_\ell} = \frac{r(-ia_\alpha)}{a_\alpha + ik_\ell}. \tag{5.127}$$

We now solve the set of N equations (5.125) numerically for the unknown functions $f_\alpha(x)$. Once these $f_\alpha(x)$ s are known we obtain the kernels $K_1(x,y)$, Eq. (5.99) and $K(x,y)$:

$$K(x,y) = \sum_{\text{all } \alpha} f_\alpha(x) e^{a_\alpha y} \theta(x+y), \tag{5.128}$$

and thus $K(x,x)$ is given by

$$K(x,x) = \left\{ \sum_{\alpha=1}^{N} \left[e^{a_\alpha x} - r(-ia_\alpha) e^{-a_\alpha x} \right] f_\alpha(x) \right\} \theta(x). \tag{5.129}$$

In arriving at the last equation we have replaced $f_{-\alpha}(x)$ in terms of $f_\alpha(x)$ using equation (5.123). Finally we find the profile of the barrier (or the potential) by taking the derivative of $K(x,x)$

$$v(x) = 2 \frac{dK(x,x)}{dx}. \tag{5.130}$$

Since we have used the reflection coefficient $R_-(x)$ in our formulation, and the integral equation for $K(x,y)$ has an integral from $-\infty$ to y, rather than from y to ∞, the plus sign in (5.130) is opposite to the usual minus sign in the original Gel'fand–Levitan formulation. For the input profile given by $r(k)$, Eq. (5.121) the result of inversion is shown in Fig. 5.6.

5.6   Construction of Reflectionless Potentials

In this and in the following section we want to consider two problems, one which is related to the physics of dielectrics and the other to the particle physics. In the first problem we ask the question as to whether it is possible to find a stratified dielectric medium with the property that at a fixed frequency and polarization a plane wave at any angle of incidence will be transmitted without reflection by the medium [17], [20]. In the second case we ask whether the knowledge of the mass spectrum of the quarkonia can be used to determine the interquark potential [22]–[24].

Reflectionless Transmission through Dielectrics — We want to inquire whether it is possible to find a dielectric medium which is transparent to an incoming plane wave. To simplify the problem we make the following assumptions for such a medium:
(a) We assume that we have a plane stratified medium of variable index of refraction $n(x)$ extending from $x = -\infty$ to $+\infty$.
(b) The index of refraction of the medium vary continuously throughout, but it has the asymptotic value of unity at $x \to \pm\infty$.
(c) For a given frequency and polarization, a plane wave incident from $-\infty$ will be transmitted without reflection, viz,

$$\psi(x, y, z, k) = u(k, x) \exp[ik(y \cos \beta + z \cos \gamma)]. \tag{5.131}$$

Let us write the index of refraction $n(x)$ in terms of another function $\mu(x)$ which is defined by

$$n^2(x) = 1 + \mu^2(x), \tag{5.132}$$

then we can write the amplitude of the wave $\psi(x, y, z, k)$ satisfying the classical equation of wave motion

$$\nabla^2 \psi(x, y, z, k) + k^2 n^2(x)\psi(x, y, z, k) = 0. \tag{5.133}$$

By substituting (5.132) and (5.131) in the equation of wave motion we obtain a one-dimensional Schrödinger type equation;

$$\frac{d^2}{dx^2} u(x, E) + [E - v(x)]u(x, E) = 0. \tag{5.134}$$

In this equation E and $v(x)$ are given by the relations

$$E = k^2 \cos^2 \alpha, \quad \text{and} \quad v(x) = -k^2 \mu^2(x), \tag{5.135}$$

where $\cos \alpha$, $\cos \beta$ and $\cos \gamma$ are the direction cosines for the propagation direction. In this Schrödinger like equation the potential depends on the wave number k and $v(x)$ is negative or zero for all x values.

A different physical interpretation can be given to the potential $v(x)$. That is, $v(x)$ which is negative or zero for all x values, is a potential with the property that it will allow an incident particle to pass through it with probability one, no matter what the particle's initial energy may be [17]–[20]. Now we ask what are the boundary conditions that we have to impose on the solution of (5.134)? Since for the fixed frequency k, we do not want reflected waves at any angle of incidence α, the solution of (5.134) must satisfy

$$u(x, E) \to e^{i\sqrt{E}x}, \quad \text{as} \quad x \to -\infty, \tag{5.136}$$

and

$$u(x, E) \to T(E)e^{i\sqrt{E}x}, \quad \text{as} \quad x \to +\infty, \tag{5.137}$$

where in the last equation $T(E)$ is the transmission amplitude.

To determine $v(x)$, we start with a trial solution for the Schrödinger equation (5.134) which is compatible with the two boundary conditions (5.136) and (5.137);

$$u(x, E) = \left\{ 1 + \sum_{n=1}^{N} \left[\frac{f_n(x)}{\gamma_n + i\sqrt{E}} \right] e^{\gamma_n x} \right\} e^{i\sqrt{E}x}. \tag{5.138}$$

In this relation $f_n(x)$ s are defined as solutions of N linear algebraic equations

$$A_n \exp(\gamma_n(x)) \sum_{\nu=1}^{N} \left[\frac{\exp(\gamma_\nu(x))}{\gamma_n + \gamma_\nu} \right] f_\nu(x) + f_n(x)$$
$$+ A_n \exp(\gamma_n(x)) = 0, \quad n = 1, 2, \cdots N, \tag{5.139}$$

with all A_n s being real positive numbers.

The coefficient of the column matrix $[f_n]$ has the form $I + A$, where I is the unit matrix and A is a square matrix with the elements

$$A = \left[A_n \frac{\exp[(\gamma_n + \gamma_\nu)x]}{\gamma_n + \gamma_\nu} \right]. \tag{5.140}$$

The matrix A is not symmetric, so we apply a similarity transformation generated by the matrix D with the elements

$$D_{n\nu} = \frac{\delta_{n\nu}}{\sqrt{A_n}}, \tag{5.141}$$

to Eq. (5.140) to make the resulting matrix symmetric. In this way we find the following equation:

$$\sum_{\nu=1}^{N} \left\{ \sqrt{A_n A_\nu} \frac{\exp[(\gamma_n + \gamma_\nu)x]}{\gamma_n + \gamma_\nu} g_\nu(x) \right\}$$
$$+ g_n(x) + \sqrt{A_n} \exp(\gamma_n x) = 0, \tag{5.142}$$

where

$$g_n(x) = \frac{1}{\sqrt{A_n}} f_n(x). \tag{5.143}$$

For the matrix (5.140) we calculate $\tilde{A} = DAD^{-1}$ to find the symmetric matrix

$$(\tilde{A})_{n\nu} = (DAD^{-1})_{n\nu} = \frac{\sqrt{A_n A_\nu}}{\gamma_n + \gamma_\nu} \exp[(\gamma_n + \gamma_\nu)x]. \tag{5.144}$$

Since DAD^{-1} is a similarity transformation, therefore [21]

$$\det A = \det \tilde{A}, \tag{5.145}$$

and also

$$\det(I + A) = \det(I + \tilde{A}). \tag{5.146}$$

Thus we can work with the symmetric matrix $I + \tilde{A}$ formed from coefficients. Now we will show that $\tilde{A}$ is a positive-definite matrix, i.e. for any set of N numbers y_n, not all of them zero, we have

$$\sum_{n,\nu=1}^{N} \left\{ y_n y_\nu^* \frac{\sqrt{A_n A_\nu}}{\gamma_n + \gamma_\nu} \exp[(\gamma_n + \gamma_\nu)x] \right\} > 0. \tag{5.147}$$

To show this we write the left-hand side of (5.147) as

$$\int_{-\infty}^{x} \sum_{n,\nu=1}^{N} \sqrt{A_n A_\nu} \, y_n y_\nu^* \exp[(\gamma_n + \gamma_\nu)z] \, dz$$

$$= \int_{-\infty}^{x} \left| \sum_{n=1}^{N} \sqrt{A_n} y_n \exp(\gamma_n z) \right|^2 dz. \tag{5.148}$$

Therefore we conclude that unless the integrand in the right-hand side of (5.148) is zero for all z between $-\infty$ and x, the integral must be greater than zero. Consequently both $\det \tilde{A}$ and $\det(I + \tilde{A})$ are greater than zero, and the latter has a well-defined logarithm.

Next we want to relate the potential $v(x)$ in the Schrödinger equation to the $\det(I + \tilde{A})$. To this end we define two differential operators L and M_n by the relations [17]

$$L = \frac{d^2}{dx^2} + E - v(x), \tag{5.149}$$

and

$$M_n = \frac{d^2}{dx^2} - [\gamma_n^2 + v(x)]. \tag{5.150}$$

Let us first consider the action of L on the wave function $u(x, E)$, Eq. (5.138). By operating L on $u(x, E)$ we find

$$Lu(x, E) = \left\{ \sum_{n=1}^{N} \frac{[M_n f_n(x)] \exp(\gamma_n x)}{\gamma_n + i\sqrt{E}} \right.$$

$$\left. + 2\frac{d}{dx} \left(\sum_{n=1}^{N} f_n(x) e^{\gamma_n x} \right) - v(x) \right\} e^{i\sqrt{E}x}. \tag{5.151}$$

We observe that $Lu(x, E)$ is zero provided that

$$M_n f_n(x) = 0, \qquad (5.152)$$

and at the same time

$$v(x) = 2\frac{d}{dx}\left(\sum_{n=1}^{N} f_n(x)e^{\gamma_n x}\right). \qquad (5.153)$$

Next we apply the operator M_n to the nth equation of (5.139), and for $v(x)$ we substitute from Eq. (5.153). In this way we obtain

$$M_n f_n(x) + A_n \exp(\gamma_n x)\sum_{\nu=1}^{N}\left[\frac{M_\nu f_\nu(x)}{\gamma_n + \gamma_\nu}\exp(\gamma_\nu x)\right] = 0, \quad n = 1, 2 \cdots N. \qquad (5.154)$$

This is a set of N homogeneous linear equations in $M_n f_n(x)$ with coefficients being the matrix $I + A$. As we discussed earlier this matrix has an inverse, therefore $M_n f_n(x) = 0$ and hence $v(x)$ must satisfy (5.153). To find an expression for $v(x)$ in terms of $\det(I + A)$, we get the solution for f_n from (5.139):

$$\sum_{n=1}^{N} f_n e^{\gamma_n x} = -\frac{d}{dx}\left[\ln \det(I + A)\right] = -\frac{d}{dx}\left[\ln \det\left(I + \tilde{A}\right)\right]. \qquad (5.155)$$

Substituting this solution in (5.153) we obtain

$$v(x) = -2\frac{d^2}{dx^2}\left[\ln \det\left(I + \tilde{A}\right)\right]. \qquad (5.156)$$

An Example — The special case of having one other constant A_1 and one constant γ_1 will give us a simple potential. For such a special problem (5.139) reduces to

$$Ae^{\gamma_1 x} + f_1(x) + \frac{A_1}{2\gamma_1}e^{2\gamma_1 x}f_1(x) = 0, \qquad (5.157)$$

and therefore

$$f_1(x) = -\frac{A_1 e^{\gamma_1 x}}{\left[1 + \frac{A_1 e^{2\gamma_1 x}}{2\gamma_1}\right]}, \qquad (5.158)$$

and

$$u(x, E) = \left\{1 - \frac{A_1 \exp(2\gamma_1 x)}{\left[1 + \frac{A_1}{2\gamma_1}\exp(2\gamma_1 x)\right]}(\gamma_1 + i\sqrt{E})\right\}e^{i\sqrt{E}x}. \qquad (5.159)$$

Since we have

$$\det\left(I + \tilde{A}\right) = 1 + \frac{A_1}{2\gamma_1}\exp(2\gamma_1 x), \qquad (5.160)$$

therefore the potential will be

$$v(x) = \frac{-4\gamma_1 A_1 e^{2\gamma_1 x}}{\left(1 + \frac{A_1}{2\gamma_1} \exp(2\gamma_1 x)\right)^2}. \tag{5.161}$$

This potential which depends on two parameters γ_1 and A_1 can be written in a simpler form if we introduce x_0 by

$$A_1 = 2\gamma_1 \exp(-2\gamma_1 x_0), \tag{5.162}$$

and write the reflectionless potential $v(x)$ as

$$v(x) = \frac{-2\gamma_1^2}{\cosh^2[2\gamma_1(x - x_0)]}. \tag{5.163}$$

Then the wave function (5.159) takes the following form:

$$u(x, E) = \left\{ \frac{i\sqrt{E} - \gamma_1 \tanh[\gamma_1(x - x_0)]}{\gamma_1 + i\sqrt{E}} \right\} e^{i\sqrt{E}x}. \tag{5.164}$$

If we are looking for a potential which is symmetric about the origin, i.e. $v(x) = v(-x)$, the we choose $A_1 = 2\gamma_1$ in (5.162), or $x_0 = 0$.

5.7   Symmetric Reflectionless Potentials Supporting a Given Set of Bound States

Let us start by studying the form of the Marchenko equation for the one-dimensional scattering of a plane wave. As in the case of S-wave scattering, Eq. (5.31), the input is

$$A_0(x + y) = \frac{1}{2\pi} \int_{-\infty}^{\infty} [S(k) - 1] e^{ik(x+y)} dk + \sum_{n=1}^{N} C_n^2 e^{-\gamma_n(x+y)},$$

$$-\infty \leq x \leq \infty, \quad -\infty \leq y \leq \infty. \tag{5.165}$$

If the reflection coefficient is zero for all wave numbers then $S(k) \equiv 1$ and the integral in (5.165) will not contribute to $A_0(x + y)$. Therefore we try to solve Marchenko's integral equation with only the bound state terms in Eq. (5.165).

We have already seen the result of solving Marchenko's integral equation for just bound states. For example, for the case of S-wave bound states, we found the potential to be given by (5.85) and the Jost solution for wave function given by (5.84).

In the case of the one-dimensional problem, we obtained (5.156) and (5.138) for the potential and the wave function respectively, where in the latter $f_n(x)$ s are given by the solution of a set of linear equations (5.139). We can write the wave function in the compact form of

$$u_n(x) = \frac{1}{\sqrt{A_n}e^{\gamma_n(x)}} \frac{\det\left(I + \tilde{A}\right)^{(n)}}{\det\left(I + \tilde{A}\right)}, \qquad (5.166)$$

where $\left(I + \tilde{A}\right)^{(n)}$ denotes the matrix $\left(I + \tilde{A}\right)$ with its nth column differentiated once with respect to x. In general the potential found from (5.156) is not symmetric about the origin, as we can see from the example (5.162). In order to get a symmetric potential, $v(x) = v(-x)$, it is necessary to choose A_n such that [22], [23]

$$\frac{A_n}{2\gamma_n} = \prod_{m \neq n} \left|\frac{\gamma_m + \gamma_n}{\gamma_m - \gamma_n}\right|, \quad m = 1, 2. \cdots N. \qquad (5.167)$$

If we consider the case where all γ_m s are zero except γ_1, then $A_1 = 2\gamma_1$, which is the same result that we found in the previous section, Eq. (5.162).

Reflectionless Potential is Negative Everywhere — The potential $V(x)$ is negative for all x and it goes to zero exponentially as $x \to \pm\infty$.

We have already shown that the matrix $\tilde{A}$ which we write as

$$\tilde{A}_{n\nu} = \frac{\sqrt{A_n A_\nu}}{\gamma_n + \gamma_\nu} \exp[(\gamma_n + \gamma_\nu)x], \qquad (5.168)$$

has a positive definite determinant [17]. Similarly the determinant of $\left|\frac{1}{\gamma_n+\gamma_\nu}\right|$ which has the same form as (5.147) is also positive definite. Therefore $\det \tilde{A}$ has the form of $\alpha e^{-\gamma x}$, where α and γ are positive constants [17]. We write this determinant as

$$\Delta = \det(I + \tilde{A}) = 1 + \sum_n \alpha_n e^{\gamma_n x}, \qquad (5.169)$$

which has the correct asymptotic property as $x \to \infty$. Now we will show that $v(x)$ is negative for all values of x. To prove this point we consider the determinant Δ defined by (5.169). Then according (5.156) we have

$$v(x) = -\frac{2\left(\Delta\Delta'' - (\Delta')^2\right)}{\Delta^2}, \qquad (5.170)$$

where primes denote derivatives with respect to x. Next we show that

$$\Delta\Delta'' - (\Delta')^2 > 0. \qquad (5.171)$$

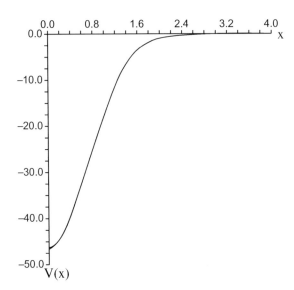

Figure 5.7: The Gaussian well $V(x) = -46\exp(-x^2)$ has five energy levels, -39.595, -27.592, -17.289, -8.901 and -2.809. Using these values, from Eq. (5.156) a symmetric reflectionless potential is constructed, which in this figure overlaps with the original Gaussian potential [24].

For this we substitute for Δ and its derivatives from (5.169) to get

$$\Delta\Delta'' - (\Delta')^2 = \left[1 + \sum_n \alpha_n e^{\gamma_n x}\right]\left[\sum_n \gamma_n^2 \alpha_n e^{\gamma_n x}\right] - \left[\sum_n \gamma_n \alpha_n e^{\gamma_n x}\right]^2$$

$$= \sum_n \gamma_n^2 \alpha_n e^{\gamma_n x} + \frac{1}{2}\sum_{n,\nu}\left\{\alpha_n\alpha_\nu(\gamma_n - \gamma_\nu)^2 \exp[(\gamma_n + \gamma_\nu)x] \geq 0\right\}. \quad (5.172)$$

This relation and Eq. (5.170) show that $v(x) \leq 0$ for all values of x.

Asymptotic Limits of $v(x)$ — Let us first write down the potential found from Eq. (5.170) with the Δ given by (5.169)

$$v(x) = -2\frac{\sum_n \gamma_n^2 \alpha_n \exp(\gamma_n x) + \frac{1}{2}\sum_{n,\nu}\left\{\alpha_n\alpha_\nu(\gamma_n - \gamma_\nu)^2 \exp[(\gamma_n + \gamma_\nu)]\right\}}{(1 + \sum_n \alpha_n \exp(\gamma_n x))^2}. \quad (5.173)$$

We first examine the denominator of Eq. (5.173). If γ_μ is the largest member of the set of γ_n s then as $x \to \infty$ the denominator grows as $\exp(2\gamma_\mu x)$ but then the second term in the numerator for $n = \nu$ becomes zero. Therefore in this limit $v(x)$ goes to zero. When $x \to -\infty$, it is obvious that $v(x)$ vanishes exponentially.

S-wave Bound States and One-dimensional Problems — In the case of one-dimensional problem there is a redundancy of information compared with the three-dimensional problem with $\ell = 0$. In the former case only the odd parity wave functions vanish at the origin, whereas in the three-dimensional case, in order that

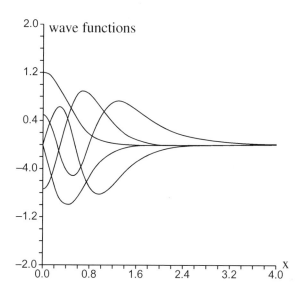

Figure 5.8: The lowest five bound state wave functions for the Gaussian potential of the previous figure [24].

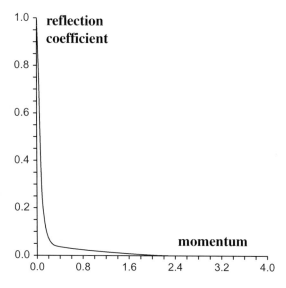

Figure 5.9: The reflection coefficient for the "reflectionless" approximation to the Gaussian potential [24].

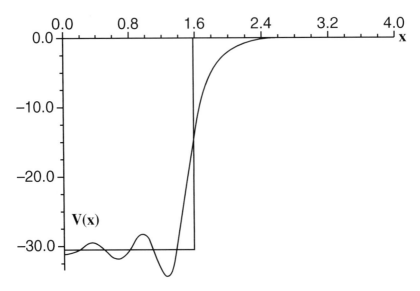

Figure 5.10: Approximate reconstruction of a finite square well of depth -30.5 and range 1.6 using $N = 6$ eigenvalues.

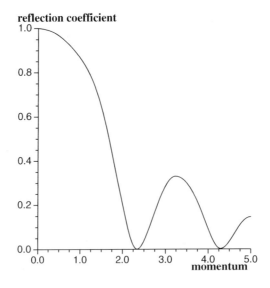

Figure 5.11: The reflection coefficient for the square well potential. The finite square well because of the sharp boundaries is a highly reflective potential [24].

the wave function $\psi(\mathbf{r})$ remains finite at $\mathbf{r} = 0$, $u_\ell(r = 0)$ must vanish. Thus the physical states can only be related to the odd-parity states of the one-dimensional case.

Reconstruction of Two Symmetric Potential Wells — Let us now consider short range potential wells which are nonsingular and negative on a finite interval $(-\infty < x < \infty)$. These potentials are not reflectionless, but we want to approximate them with reflectionless potentials. This is done by calculating the exact bound state energies and using them to determine the reflectionless potentials which support exactly these energies, and compare the reconstructed potentials with the original ones. In addition we want to calculate the reflection coefficient in each case as a function of the wave number k. We will examine a smooth Gaussian potential and a discontinuous square well potential.

(a) **Reflectionless Approximation to a Gaussian Potential** — Let us examine a Gaussian potential given by $V(x) = -46 \exp\left(-x^2\right)$ (arbitrary units). This well admits five bound states [24], $E_1 = -\gamma_1^2 = -39.595$, $E_2 = -\gamma_2^2 = -27.59$, $E_3 = -\gamma_3^2 = -17.289$, $E_4 = -\gamma_4^2 = -8/901$, and $E_5 = -\gamma_5^2 = -2.809$. This well is a smooth Gaussian potential shown in Fig. 5.6. Using these γ_n s we can construct a reflectionless potential from Eq. (5.156), where matrices $\tilde{A}$ and A_n s are given by Eqs. (5.144) and (5.167) respectively. The constructed potential, to a good approximation, overlaps with the Gaussian well Fig. 5.7. The wave functions for the reconstructed reflectionless potential are shown in Fig. 5.8. Due to the approximate nature of the reconstructed potential, the reflection coefficient is not zero but is very small Fig. 5.9.

(b) **Approximating Square Well by Reflectionless Potentials** — What happens when the original potential is not smooth, e.g. when it has a discontinuity? In this case even when we include a larger number of eigenvalues, the resulting potential that we find from (5.156) is noticeably different from a square well. For instance if we take an attractive well given by $V(x) = -30.5\theta(x - 1.6)$ ($\theta(x)$ is a step function), this well supports 6 bound states. From the bound state energies we obtain the potential shown in Fig. 5.10, together with the square well. Obviously when the original and the reconstructed potentials are not similar, the reflection coefficient will not be zero or even small, as can be seen in Fig. 5.11.

For the underlying physics of the reflectionless potentials and their applications in particle physics see the detailed discussion in the papers of Quigg *et al.* [22]–[26].

References

[1] V.A. Marchenko, Dokl. Akad. Nauk SSSR, 75 457 (1950).

[2] V.A. Marchenko, Dokl. Akad. Nauk SSSR, 77, 557 (1951).

[3] Z.S. Agranovich and V.A. Marchenko, *The Inverse Problem of Scattering Theory*, (Gordon and Breach, New York, 1963).

[4] E.C. Titchmarch, *Eigenfunction Expansion* Part 1, (Oxford University Press, Oxford, 1962) p. 189.

[5] G.M.L. Gladwell, *Inverse Problems in Scattering: An Introduction*, (Kluwer Academic Publishers, 1993) Chapter 10.

[6] R.G. Newton, *Scattering Theory of Waves and Prticles*, Second Edition (Springer New York, 1982) p. 336.

[7] B.N. Zakhariev and A.A. Suzko, *Direct and Inverse Problems: Potentials in Quantum Scattering* (Springer-Verlag, Berlin 1990) pp. 58-69.

[8] J. von Neumann and E. Wigner, Phys. Z. 30, 465 (1929).

[9] F.H. Stillinger and D.R. Herrick, Phys. Rev. A 11, 446 (1975).

[10] K.L. Hopcraft and P.R. Smith, *An Introducrion to Electromagnetic Inverse Scattering*, (Springer, Dordrecht, 1992).

[11] A.K. Jordan and H.N. Kritikos, IEEE Trans. Antennas Prop. AP-21, 245 (1981).

[12] M.H. Reilly and A.K. Jordan, IEEE Trans. Antennas Prop. AP-29, 910 (1973).

[13] S. Ahn and A.K. Jordan, IEEE. Trans. Antennas Prop. 24, 879 (1976).

[14] Y. Zhang, J.A. Kong and A.K. Jordan, Microwave Optical Tech. Lett. 15, 277 (1997).

[15] K.R. Pechenick and J. M. Cohen, Phys. Lett. 82A, 156 (1981).

[16] K.R. Pechenick and J. M. Cohen, J. Math. Phys. 22, 1513 (1981).

[17] I. Kay and H.E. Moses, J. Appl. Phys. 21, 1503 (1956).

[18] I. Kay and H.E. Moses, Nouvo Cimento, 2, 917 (1955).

[19] I. Kay and H.E. Moses, Nouvo Cimento, 3, 66 (1956).

[20] I. Kay and H.E. Moses, Supp. Nouvo Cimento, 5, 22 (1957).

[21] G. Arfken, *Mathematical Methods for Physicists*, Third Edition, (Academic Press, New York, 1985) p. 201.

[22] C. Quigg and J.L. Rosner, Phys. Lett. 71B, 153 (1977).

[23] C. Quigg and J.L. Rosner, Phys. Rev. D17, 2364 (1978).

[24] P. Asthana and A.N. Kamal, Z. Phys. C. 19, 37 (1983).

[25] J. Lekner, Am. J. Phys. 75, 1151 (2007).

[26] J.F. Shoenfeld, W. Kwong, J.L. Rosner, C. Quigg, and H.B. Thacker, Ann. Phys. 128,1 (1980).

Chapter 6

Newton–Sabatier Approach to the Inverse Problem at Fixed Energy

There are a number of problems in quantum theory related to the scattering of a particle where the incident energy of the particle is fixed but a large number of partial waves contribute to the cross section. This fixed energy must be sufficiently high, so that the number of partial waves making a sizeable contribution is large. To simplify the notation in the case of fixed energy scattering, we choose this energy so that in the system of units that we are working with, the wave number is $k = 1L^{-1}$, where L is the unit of length [1].

6.1 Construction of the Potential at Fixed Energy

We begin our study by considering the linear differential operator defined by

$$\bar{D}(r) = r^2 \left[\frac{d^2}{dr^2} + 1 - \bar{v}(r) \right],$$ (6.1)

where we have assumed that $\bar{v}(r)$ is a well-behaved "reference" potential. The regular solution of the radial Schrödinger equation for the ℓth partial wave, $\bar{\phi}_\ell(r)$ is the solution of the differential equation

$$\bar{D}(r)\bar{\phi}_\ell(r) = \ell(\ell+1)\bar{\phi}_\ell(r),$$ (6.2)

subject to the boundary condition

$$\lim_{r\to 0} r^{-\ell-1}(2\ell+1)!!\,\bar{\phi}_\ell(r) = 1. \tag{6.3}$$

Using these regular wave functions we define

$$g(r,r') = \sum_{\ell=0}^{\infty} c_\ell \bar{\phi}_\ell(r)\bar{\phi}_\ell(r'), \tag{6.4}$$

where c_ℓ s are real coefficients to be determined later. We note that $g(r,r')$ satisfies the partial differential equation

$$\bar{D}(r)g(r,r') = \bar{D}(r')g(r,r'), \tag{6.5}$$

and the boundary conditions

$$g(0,r') = g(r,0) = 0. \tag{6.6}$$

Just like the case of Gel'fand–Levitan equation for a single partial wave (e.g. S-wave), Eq. (4.56), let us consider the integral equation

$$K(r,r') = g(r,r') - \int_0^r \frac{1}{r''^2} K(r,r'')g(r'',r')\,dr'', \tag{6.7}$$

and assume that $K(r,r')$ is the unique solution of (6.7). Next we define $\xi(r,r')$ by

$$\xi(r,r') = D(r)K(r,r') - \bar{D}(r')K(r,r'), \tag{6.8}$$

where the differential operator $D(r)$ is defined by

$$D(r) = \bar{D}(r) - \Delta v(r), \tag{6.9}$$

and

$$\Delta v(r) = -\frac{2}{r}\frac{d}{dr}\left[\frac{1}{r}K(r,r')\right]. \tag{6.10}$$

Now we can show, by differentiation, integration by parts, and using the differential equation (6.5) and the boundary conditions (6.6) that $\xi(r,r')$ defined by (6.8) satisfies the homogeneous differential equation

$$\xi(r,r') = -\int_0^r \frac{1}{r''^2} K(r,r'')\xi(r'',r')\,dr''. \tag{6.11}$$

Our assumption that (6.7) has a unique solution implies that $\xi(r,r') \equiv 0$. Therefore $K(r,r')$ is a solution of the differential equation

$$D(r)K(r,r') = \bar{D}(r')K(r,r'). \tag{6.12}$$

Moreover from (6.7) it follows that

$$K(0,r') = K(r,0) = 0. \tag{6.13}$$

Now we can define the set of functions $\phi_\ell(r)$ by the relation

$$\phi_\ell(r) = \bar{\phi}_\ell(r) \quad \int_0^r \frac{1}{r'^2} K\left(r, r'\right) \bar{\phi}_\ell\left(r'\right) dr', \tag{6.14}$$

and by applying the operator $D(r)$ to (6.14) and using (6.12) and (6.2) we find that $\phi_\ell(r)$ is a solution of the differential equation

$$D(r)\phi_\ell(r) = \ell(\ell+1)\phi_\ell(r). \tag{6.15}$$

The boundary condition for this equation can be found from (6.13):

$$\phi_\ell(0) = 0. \tag{6.16}$$

In this way from the solution of the Schrödinger equation for the ℓth partial wave, when the potential is $\bar{v}(r)$, we have found the solution for

$$v(r) = \bar{v}(r) + \Delta v(r). \tag{6.17}$$

Next we try to express $K\left(r, r'\right)$ in terms of the new set of wave functions, $\bar{\phi}_\ell(r)$. From the expressions for $K\left(r, r'\right)$, Eq. (6.7) and for $g\left(r, r'\right)$, Eq. (6.4) we find that $K\left(r, r'\right)$ can be written as

$$K\left(r, r'\right) = \sum_{\ell=0}^{\infty} c_\ell K_\ell(r) \bar{\phi}_\ell\left(r'\right), \tag{6.18}$$

where $K_\ell(r)$ is defined by

$$K\left(r\right) = \bar{\phi}_\ell(r) - \int_0^r \frac{1}{r'^2} K\left(r, r'\right) \bar{\phi}_\ell\left(r'\right) dr'. \tag{6.19}$$

If we compare (6.19) with (6.14) we conclude that $K_\ell(r) = \phi_\ell(r)$, therefore

$$K\left(r, r'\right) = \sum_{\ell=0}^{\infty} c_\ell \phi_\ell(r) \bar{\phi}_\ell\left(r'\right). \tag{6.20}$$

Now we want to know how to relate, the so far undetermined constants c_ℓ, to the empirically determined phase shifts $\delta_\ell(k = 1)$, $\ell = 0, 1, \cdots$. We first observe that in the case of fixed ℓ problem in the Gel'fand–Levitan formulation we used the orthogonality of different $\bar{\phi}(E, r)$ s and $\phi(E, r)$ s, Eqs. (4.73) and (4.74), to relate the bound state energies and the scattering amplitudes to the weight function $h(E)$. But here we do not have the orthogonality condition among different $\phi_\ell(r)$ s.

To find the relationship between c_ℓ s and the phase shifts we substitute (6.18) in (6.14) and interchange the order of summation and integration, and obtain the following result

$$\phi_\ell(r) = \bar{\phi}_\ell(r) - \sum_{\ell'} L_{\ell\ell'}(r) c_{\ell'} \phi_{\ell'}(r), \tag{6.21}$$

where

$$L_{\ell\ell'}(r) = \int_0^r \frac{1}{r'^2} \bar{\phi}_\ell(r') \bar{\phi}_{\ell'}(r') \, dr'. \tag{6.22}$$

To simplify this expression for the matrix $L_{\ell\ell'}$ we multiply (6.15) by $\phi_{\ell'}(r)$, and then write (6.15) for $\phi_{\ell'}(r)$, multiply it by $\phi_\ell(r)$ subtract the resulting equations and integrate what has been found over r. Then using the boundary condition (6.13) we get the following relation for $L_{\ell\ell'}(r)$:

$$L_{\ell\ell'}(r) = \frac{1}{(\ell' - \ell)(\ell' + \ell + 1)} \left[\bar{\phi}_\ell(r) \frac{d}{dr} \bar{\phi}_{\ell'(r)} - \bar{\phi}_{\ell'}(r) \frac{d}{dr} \bar{\phi}_\ell(r) \right], \quad \ell' \neq \ell, \tag{6.23}$$

and

$$L_{\ell\ell}(r) = \int_0^r \frac{1}{r'^2} [\bar{\phi}_\ell(r')]^2 dr'. \tag{6.24}$$

Now let us consider the limit of $L_{\ell\ell'}$ as r tends to infinity. Remembering that we have set $k = 1$, in this limit we have

$$\lim_{r \to \infty} \to A \sin\left(r - \frac{1}{2}\ell\pi + \delta_\ell\right) \equiv \phi_\ell^\infty(r), \tag{6.25}$$

$$\lim_{r \to \infty} \to \bar{A}_\ell \sin\left(r - \frac{1}{2}\ell\pi + \bar{\delta}_\ell\right) \equiv\equiv \bar{\phi}_\ell^\infty(r), \tag{6.26}$$

$$\lim_{r \to \infty} L_{\ell\ell'}(r) \to L_{\ell\ell'}^\infty(r) \tag{6.27}$$

$$\lim_{r \to \infty} L_{\ell\ell}^\infty(r) \to \left(\bar{A}_\ell\right)^2 \left(\frac{\frac{\pi}{2} + \frac{d\bar{\delta}_\ell}{d\ell}}{2\ell + 1}\right), \tag{6.28}$$

and

$$\lim_{r \to \infty} L_{\ell\ell'}^\infty(r) \to \frac{\left(\bar{A}_\ell\right)\left(\bar{A}_{\ell'}\right) \sin\left[\frac{\pi}{2}(\ell - \ell') + \bar{\delta}_\ell - \delta_{\ell'}\right]}{(\ell' - \ell)(\ell' + \ell + 1)}. \tag{6.29}$$

To simplify these sets of equations, let us consider the case where the "reference" potential, $v(r)$, is equal to zero. Then we have

$$\bar{A}_\ell = 1, \quad \bar{\delta}_\ell = 0, \tag{6.30}$$

and

$$L_{\ell\ell'}^\infty = \begin{cases} i^{\ell'-\ell-1} M_{\ell\ell'} & \ell' \neq \ell \\ \frac{\pi}{2(2\ell+1)} & \ell' = \ell \end{cases} \tag{6.31}$$

where the matrix $M_{\ell\ell'}$ is defined by

$$M_{\ell\ell'} = \begin{cases} \frac{1}{(\ell'-\ell)(\ell'+\ell+1)} & \ell' - \ell \quad \text{odd} \\ 0 & \ell' - \ell \quad \text{even} \end{cases} \tag{6.32}$$

By substituting (6.25)–(6.27) in (6.21) we find an equation for the asymptotic form of $\phi_\ell^\infty(r)$

$$\phi_\ell^\infty(r) = \bar{\phi}_\ell^\infty(r) - \sum_{\ell'} L_{\ell\ell'}^\infty c_{\ell'} \phi_{\ell'}^\infty(r). \tag{6.33}$$

Now we write the sine functions in terms of their exponential forms, and equate the coefficients of e^{ir} and e^{-ir} separately to obtain two equations;

$$e^{i\delta_\ell} A_\ell = 1 - \frac{\pi}{2} \frac{c_\ell A_\ell e^{i\delta_\ell}}{2\ell+1} - \sum_{\ell'\neq\ell} L_{\ell\ell'}^\infty i^{\ell-\ell'} c_{\ell'} e^{i\delta_{\ell'}} A_{\ell'}, \tag{6.34}$$

and the complex conjugate of this equation. Next we multiply (6.34) by $e^{-i\delta_\ell}$ and introduce b_ℓ by

$$b_\ell = c_\ell A_\ell, \tag{6.35}$$

then we find

$$A_\ell = e^{-i\delta_\ell} - \frac{\pi}{2}\left(\frac{b_\ell}{2\ell+1}\right) + i\sum_{\ell'\neq\ell} M_{\ell\ell'} b_{\ell'} e^{i(\delta_{\ell'}-\delta_\ell)}. \tag{6.36}$$

Noting that A_ℓ s are real, we separate the real and imaginary parts of (6.36) to get

$$\sin\delta_\ell = \sum_{\ell'} M_{\ell\ell'} b_{\ell'} \cos(\delta_{\ell'} - \delta_\ell), \tag{6.37}$$

and

$$A_\ell = \cos\delta_\ell - \frac{\pi}{2}\frac{b_\ell}{2\ell+1} - \sum_{\ell'\neq\ell} M_{\ell\ell'} b_{\ell'} \sin(\delta_{\ell'} - \delta_\ell). \tag{6.38}$$

We can write these equations in a compact matrix form. Thus we consider a diagonal matrix whose diagonal elements are $\tan\delta_\ell$, and denote this matrix by $\tan\Delta$. Also let M be a square matrix whose elements are $M_{\ell\ell'}$, and let $\mathbf{a}$ denote the column matrix with elements $b_\ell \cos\delta_\ell$. In addition let $\mathbf{e}$ represent a column matrix with unity for all its matrix elements. Using these matrices Eq. (6.37) can be written as [1]

$$(M + \tan\Delta M \tan\Delta)\mathbf{a} = \tan\Delta\mathbf{e}. \tag{6.39}$$

Now if we introduce R as the matrix

$$R = M^{-1} \tan \Delta M \tan \Delta, \tag{6.40}$$

then (6.39) becomes

$$(1 + R)\mathbf{a} = M^{-1} \tan \Delta \mathbf{e}, \tag{6.41}$$

or

$$\mathbf{a} = \left[(1 + R)^{-1} M^{-1} \tan \Delta \right] \mathbf{e}. \tag{6.42}$$

Once the matrix M which is independent of the phase shifts is determined, then both of the matrices M^{-1} and $(1 + R)^{-1}$ can be found. From Eq. (6.42) the matrix elements $a_\ell = b_\ell \cos \delta_\ell$, formally can be calculated. The unknown c_ℓ s are now related to a_ℓ s by

$$\frac{1}{c_\ell} = \frac{\sum_{\ell'} M_{\ell\ell'} a_{\ell'}}{a_\ell \tan \delta_\ell} - \frac{\pi}{2(2\ell + 1)}. \tag{6.43}$$

Knowing c_ℓ s enable us to find $g(r, r')$ from (6.4) and by solving the integral equation (6.7) calculate $K(r, r')$. Finally by substituting $K(r, r')$ in Eq. (6.10) we find the potential $\Delta v(r)$.

An Example — Let us assume that a single c_ℓ for $\ell = \Lambda$ is given as a real number and all other c_ℓ s are zero. Then from (6.33) we find

$$\phi_\Lambda(r) = \frac{\bar{\phi}_\Lambda(r)}{1 + c_\ell L_{\Lambda\Lambda}(r)}, \tag{6.44}$$

and

$$\phi_\ell(r) = \bar{\phi}_\ell(r) - \frac{c_\Lambda L_{\ell\Lambda}(r) \bar{\phi}_\Lambda(r)}{1 + c_\Lambda L_{\Lambda\Lambda}(r)}. \tag{6.45}$$

These wave functions solve the radial Schrödinger equation with the potential

$$\Delta v(r) = -\frac{2}{r} c_\Lambda \frac{d}{dr} \left\{ \frac{1}{r} \frac{[\bar{\phi}_\Lambda(r)]^2}{1 + c_\Lambda L_{\Lambda\Lambda}(r)} \right\}. \tag{6.46}$$

This $\Delta v(r)$ is an oscillating function of r.

We can see from Eq. (6.36) δ_ℓ is zero when $\ell - \Lambda$ is an even integer. For odd integers $\tan \delta_\ell$ is given by [1]

$$\tan \delta_\ell = \left(\frac{2c_\Lambda(2\Lambda + 1)}{2(2\Lambda + 1) + \pi c_\Lambda} \right) \left(\frac{1}{(\ell - \Lambda)(\ell + \Lambda + 1)} \right). \tag{6.47}$$

A remarkable property of this potential is that we can determine its scattering amplitude $A(\theta)$ analytically

$$A(\theta) = (-1)^{\Lambda+1} \frac{\pi a}{4k} \left[P_b(\cos \theta) - (-1)^\Lambda P_b(- \cos \theta) \right], \tag{6.48}$$

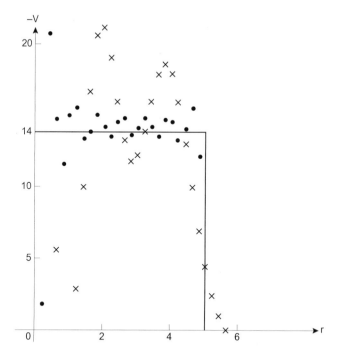

Figure 6.1: Numerical results of inversion using Newton–Sabatier method for a square well of depth 14 MeV and the range of 5 fermis. For the energies of $E = 50$ and $E = 400$ MeV the results of inversion are shown by crosses and dots respectively [7], [8].

where $P_b(\cos\theta)$ is the Legendre polynomial and

$$a = A_\Lambda c_\Lambda,\tag{6.49}$$

$$b = \sqrt{\left(A_\Lambda + \frac{1}{2}\right)^2 + ia} - \frac{1}{2}, \quad \text{Re } b > 0 \tag{6.50}$$

and A_Λ is defined in Eq. (6.36).

6.2 Criticism of the Newton–Sabatier Method of Inversion at a Fixed Energy

Ramm in a series of papers [2]–[5] criticizes the Newton–Sabatier (N-S) method that we discussed in this chapter. He points out that in the N-S method one starts from the phase shift for all partial waves at a fixed energy (or wave number) $\delta_\ell(k)$ $\ell = 0, 1, 2, \cdots$ and then summarizes their technique using the following

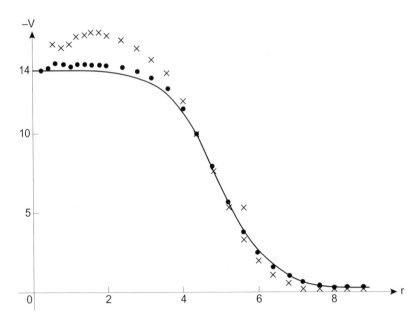

Figure 6.2: Numerical results of inversion using Newton–Sabatier method [7]. Here the input potential is $V(r) = -\dfrac{14}{\left[1+\exp\left(\frac{r-5}{0.65}\right)\right]}$. The output for the energy $E = 50$ MeV is shown by crosses and for $E = 400$ by dots [7], [8].

scheme:

$$\{\delta_\ell(k)\}_{\ell=0,\,1,\,2,\cdots} \Longrightarrow \{c_\ell\} \Longrightarrow K\,(r,r') \Longrightarrow -\frac{2}{r}\frac{d}{dr}\left(\frac{K(r,r)}{r}\right) = \Delta v(r). \qquad (6.51)$$

Here c_ℓ s are defined by (6.4) and $K\,(r,r')$ by (6.7).

Ramm claims that Eq. (6.19) is fundamentally wrong, and in general $K\,(r,r')$ does not solve the inverse problem as suggested in (6.51). He considers the case where the reference potential $v(r)$ is zero, i.e. when $g\,(r,r')$, Eq. (6.4), is defined by

$$g\,(r,r') = \sum_{\ell=0}^{\infty} c_\ell \left(\frac{\pi k}{2}\right) \sqrt{rr'} J_{\ell+\frac{1}{2}}(kr) J_{\ell+\frac{1}{2}}\,(kr'). \qquad (6.52)$$

In this relation $J_{\ell+\frac{1}{2}}$ is the regular Bessel function of the order $\ell+\frac{1}{2}$. In his criticism Ramm makes the following remarks about N-S approach to the inverse problem:
(1) Newton did not prove that (6.19) is solvable for all values of $r > 0$. But if it is not solvable for at least one $r > 0$, then N-S method breaks down.
(2) Newton did not prove the existence of the transformation kernel $K\,(r,r')$ independent of ℓ nor did he study its properties.
(3) If we are given for the input data a set of phase shifts δ_ℓ s derived from a certain potential $\Delta v(r)$, can we find a potential $\Delta v_N(r)$ which generates the same data? Ramm states that Newton has not solved this problem and the proof is missing. The set of potential $\Delta v_N(r)$ obtained by this method with the condition $\sum_{\ell=0}^{\infty} |c_\ell| < \infty$

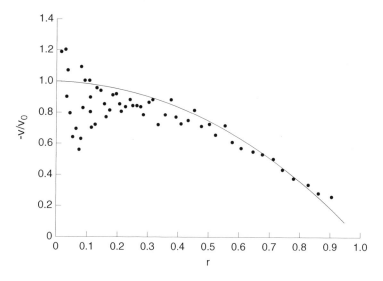

Figure 6.3: Result of inversion proposed by Markushevich *et al.* for forces with a well-defined range. Here the potential is parabolic $v(r) = -v_0 \left(1 - \frac{r^2}{R^2}\right)$, $v(r) = 0$, $R > 0$, shown by a solid line, and the dots are the result of inversion. For this calculation the parameters $R = 1L$, $v_0 = 6000L^{-2}$ and $k = 0$ have been used.

is not dense in $L_{1,1}$ (see [2]). For the debate between Sabatier and Ramm about the very concept and the mathematics involved in the inverse scattering problem at fixed energy and unresolved issues see references [2]–[6].

6.3 On the Results of the Numerical Solution of Inverse Problems

Regarding the examples of inversion using Newton–Sabatier method, Chadan and Sabatier in their book make the following assertion:

When the inverse problem has infinite number of exact solutions, the algorithm that one chooses to do the calculation, determines which solution will be found from the numerical calculation. Thus this algorithm should not be called a reconstruction (of the potential), rather a construction of a solution belonging to a large class of solutions. In other words they suggest that the procedure outlined in Newton–Sabatier method is not a recovery of the actual potential $v_1(r)$, but a determination of a potential $v(r)$ which gives exactly all the phase shifts that $v_1(r)$ produces only at a given energy. Thus at best it is a reliable class of approximation for any potential $v_1(r)$ satisfying reasonable constraints.

For solving the direct problem we can start with local energy independent potential, $v(r)$, and solve the Schrödinger equation for the ℓth partial wave to find

$\delta_\ell(k)$ very accurately. For simplicity assuming that there are no bound states, we can choose $\delta_\ell(k)$ for one partial wave, say $\ell = 0$ and find the potential $v(r)$. If the inverse problem has a unique result, then we will recover $v(r)$, no matter which partial wave we choose. The same definition for the uniqueness can be applied for inversion at a fixed energy. In the latter case if for a different energy we determine the potential we will find a different potential. If we use an algorithm which is stable for the numerical calculation and find the potential, we observe that the results fall between two curves, depending on the algorithm that we use.

Now let us examine the predictions of the inverse method for the determination of the potential at fixed energy. We will consider two simple cases, a square well and an exponentially decaying potential of the form

$$V(r) = -14 \left[1 + \exp \left(\frac{r-5}{0.65} \right) \right]^{-1}, \tag{6.53}$$

where $V(r)$ is given in MeV, and the lengths are measured in fermis. Both of these potentials have been studied by Sabatier and Quyen Van Phu [7], [8]. First let us consider the square well which is interesting because the result of inversion is poorly reproduced. Figure 6.1 shows what one gets from inversion by applying the Newton–Sabatier method to the phase shifts produced by a well of depth 14 MeV and the range of 5 fermis. The result is similar to what Markushevich *et al.* found using their own method (see the following section) [9]. Caudray has done a detailed calculation for the same square well and has shown that the second potential oscillates and that the largest oscillations appear near the edge of the well. These oscillations disappear when a smooth potential such as a Gaussian or $V(r)$ given by (6.53) are used. Therefore the oscillations can be regarded as a Gibb phenomenon. In Fig. 6.2 the solid line shows the potential function which generates the partial wave phase shifts. Again the results for two energies 50 and 400 Mev have been obtained. The points for 50 MeV are shown by crosses and for 400 MeV by dots. In this case the potential varies smoothly with r, and the results are much better than those for the square well.

6.4 Modified Form of the Gel'fand–Levitan for Fixed Energy Problems and the Langer Transform

Is it possible to reformulate the inverse scattering problem of Gel'fand and Levitan in such a way that it can be used to solve the inverse problem but now for the case when the energy is fixed, assuming that we know the phase shift for all partial waves?

Here we want to discuss this possibility in the case of potentials with finite range. Let us suppose that the potential $v(r)$ is of short range R such that for $r > R$, $V(r) = \frac{\hbar^2}{2m} v(r) = 0$, then we use the Langer transformation [10] which we will define shortly, to map the finite range, $0 \leq r \leq R$, into a semi-infinite range, and hence it becomes possible to replace the problem at fixed angular momentum with the corresponding problem at fixed energy. Thus the original problem with the boundary condition given at $r = 0$ and the logarithmic derivative of the wave function given at $r = R$ can be transformed to an equivalent inverse problem in which the boundary condition is at infinity and the logarithmic derivative is known at the origin.

As we we will see in the case of a number of inverse problems, the input can be the $\mathcal{R}$ matrix and the output will be a number of points giving us the approximate shape of the potential. For the present discussion, we prefer to work with the inverse of the $\mathcal{R}$ matrix and assume that this $\mathcal{R}$ is given for a number of partial waves $\ell_0, \ell_1, \cdots \ell_N$. From these $\mathcal{R}$ matrix elements, using Thiele's reciprocal-difference method, we can construct a rational function approximation for the logarithmic derivative of the wave function at the point R (see Appendix B). This rational approximation enables us to find the position of the poles and also the residues of the logarithmic derivative in the complex ℓ-plane. With the analytic representation of the $\mathcal{R}$ matrix we can use a variant of the one-dimensional Gel'fand–Levitan method of inversion which we discussed in Chapter 4 and find the potential, first in the Langer coordinate, and then in the original radial coordinate.

The Langer Transformation and the Zeros of the $\mathcal{R}$ Matrix — We start with the radial Schrödinger equation for the ℓth partial wave

$$\frac{d^2 \phi_\ell(r)}{dr^2} + \left[k^2 - v(r) - \frac{\ell(\ell + 1)}{r^2} \right] \phi_\ell(r) = 0, \tag{6.54}$$

where we have set $\hbar^2 = 2m = 1$, and where $v(r)$ which is the potential function is assumed to be integrable

$$\int_0^R r|v(r)|dr < \infty, \tag{6.55}$$

and vanishes outside a range R;

$$v(r) = 0, \quad r \geq R. \tag{6.56}$$

The differential equation (6.54) is subject to the boundary condition

$$\phi_\ell(r) \to r^\ell, \quad \text{as} \quad r \to \infty. \tag{6.57}$$

In the direct problem we integrate (6.54) with the initial condition (6.57) from zero to R. At this point we find the reaction matrix which contains all the information about the angular momentum dependence of the logarithmic derivative of the wave function;

$$R \frac{d}{dr} [\ln \phi_\ell(r)]_{r=R} = \frac{1}{\mathcal{R}(\ell)} = I(\ell). \tag{6.58}$$

To formulate the inverse problem we need to discuss the Langer transformation briefly.

The Langer Transformation Applied to the ℓth Partial Wave — In this transformation, first studied by Langer, we change the radial distance r to the dimensionless coordinate x were [10]

$$r = Re^{-x}. \tag{6.59}$$

We also change $\phi_\ell(r)$ to $y(\ell, x)$;

$$y(\ell, x) = \exp\left(\frac{x}{2}\right)\phi_\ell\left(Re^{-x}\right). \tag{6.60}$$

Substituting these in the differential equation for $\phi_\ell(r)$ (6.54) we obtain

$$y''(\ell, x) + q(x)y(\ell, x) = \left(\ell + \frac{1}{2}\right)^2 y(\ell, x), \tag{6.61}$$

where primes denote derivative with respect to x and

$$q(x) = k^2 R^2 e^{-2x}\left[1 - k^{-2}v\left(Re^{-x}\right)\right]. \tag{6.62}$$

This equation satisfies the boundary condition

$$y(\ell, x \to \infty) \to 0, \tag{6.63}$$

a result which follows from Eq. (6.57). The reaction matrix in terms of x and y becomes

$$\mathcal{R}^{-1}(\ell) = \frac{1}{2} - \left[\frac{d}{dx}\ln y(\ell, x)\right]_{x=0}. \tag{6.64}$$

To simplify the notation, let us replace $\left(\ell + \frac{1}{2}\right)^2$ by λ^2, and consider $\mathcal{R}$ as a function of λ. Now suppose that the reaction matrix $\mathcal{R}(\ell(\lambda))$ in (6.64) has been given empirically for a number of λ s, say, $\lambda = L_1, L_2, \cdots L_n$, where

$$L_{j+1}^2 = L_j(L_j + 1), \tag{6.65}$$

i.e. $\mathcal{R}(L_j)$ are known for $j = 0, 1, \cdots (N-1)$. From these data we want to construct $q(x)$, Eq. (6.62), and subsequently determine $v(r)$. The reciprocal of $\mathcal{R}$ at these points, i.e. $I(\lambda = L_i)$ will also be known for N points. At this point it is convenient to consider $\mathcal{J}(\lambda)$ which is defined as

$$\mathcal{J}(\lambda) = I(\lambda) + \lambda. \tag{6.66}$$

This function passes through N points $\lambda = L_j$ and at these points it takes the values $\mathcal{J}(L_j)$. We want to find a rational fraction approximation for $\mathcal{J}(\lambda)$ and for this purpose we utilize Thiele's reciprocal-difference method, [11] (see Appendix B). Thus we express $\mathcal{J}(\lambda)$ as a continued fraction

$$\mathcal{J}(\lambda) = \frac{\mathcal{J}(\lambda_1)}{1 + K_N\left[\frac{(\lambda - L_j)a_j}{1+}\right]}, \tag{6.67}$$

where

$$K_N \left[\frac{(\lambda - L_1)a_1}{1+} \right] \equiv \frac{(\lambda - L_1)a_1}{1+} \frac{(\lambda - L_2)a_2}{1+} \cdots \frac{(\lambda - L_N)a_N}{1+\cdots}. \qquad (6.68)$$

From (6.67) we get a set of $N - 1$ equations

$$\mathcal{J}(L_{j+1}) = \frac{\mathcal{J}(L_1)}{1+} \frac{(L_{j+1} - L_j)a_1}{1+} \cdots \frac{(L_{j+1} - L_j)a_j}{1}. \qquad (6.69)$$

Since $\mathcal{J}(L_{j+1}) = 0$, $j = 0, 1, \cdots (N-1)$ are known we solve (6.69) for the coefficients a_j;

$$a_j = (L_j - L_{j+1})^{-1}(1 + K_j). \qquad (6.70)$$

In this relation K_j denotes the following continued fraction:

$$K_j = \frac{(L_{j+1} - L_{j-1})a_{j-1}}{1+} \frac{(L_{j+1} - L_{j-2})a_{j-2}}{1+} \cdots \frac{(L_{j+1} - L_1)}{\frac{1 - \mathcal{J}(L_1)}{\mathcal{J}(L_1+1)}}, \qquad (6.71)$$

and

$$a_1 = \left[\frac{\mathcal{J}(L_1)}{\mathcal{J}(L_2)} \right] \frac{1}{(L_2 - L_1)}. \qquad (6.72)$$

The continued fraction (6.67) is generated by the difference equation [12]

$$s_{k-1} = s_k + (\lambda - L_k)a_k s_{k+1}. \qquad (6.73)$$

That is if we calculate $\frac{s_0}{s_1}$ with the condition that $s_N = s_{N+1}$ we find that

$$\frac{s_0}{s_1} = 1 + K_N \left[(\lambda - L_j) \frac{a_j}{1+} \right]. \qquad (6.74)$$

By comparing (6.74) and (6.67) we notice that the right-hand side of (6.74) is just the denominator of (6.67). Therefore by setting (6.74) equal to zero we find the poles of $\mathcal{J}(\lambda)$. Let λ_k be a root of (6.67), then we introduce $\mathcal{F}(\lambda)$ by

$$\mathcal{J}(\lambda) = (\lambda - \lambda_k)\mathcal{F}(\lambda), \qquad (6.75)$$

i.e. we denote the residue of $\mathcal{J}(\lambda)$ by $\mathcal{F}(\lambda_k)$. For real positive roots of $\mathcal{J}(\lambda)$ we can write

$$\mathcal{F}(\lambda_k) = \frac{c_k}{2\lambda_k}. \qquad (6.76)$$

Thus, the Thiele fit of the reaction matrix gives us two set of quantities λ_k and $\mathcal{F}(\lambda_k)$. The set of λ_k and $\mathcal{F}(\lambda_k)$ is the information that we need to construct the potential.

Inversion of the Sturm–Liouville Equation — We have seen that the Langer transformation transforms the Schrödinger equation (6.54) to a Sturm–Liouville type equation (6.61) subject to the boundary condition $y(x \to \infty) \to 0$. Again we replace $\ell(\ell+1)$ by λ^2 and consider $I(\lambda) = \mathcal{R}^{-1}(\lambda)$ as a function of λ. According to Levitan, the logarithmic derivative of $I(\lambda)$ can be expressed as [13]:

$$I(\lambda) = -\lambda + \sum_{k=1}^{N} \frac{c_k}{\lambda^2 - \lambda_k^2} + \int_{-\infty}^{0} \frac{d\sigma(s)}{\lambda^2 - s}, \tag{6.77}$$

i.e. from the analytic properties of $I(\lambda)$ in the complex λ^2 plane it follows that this function has a finite number of simple poles at the points $\lambda^2 = \lambda_k^2$, $k = 1, 2, \cdots N$ and in addition there is a cut for $-\infty \le \lambda^2 \le 0$ [13], [14]. The constants c_k are positive and real quantities. Because of the boundary condition (6.63) and the exponentially damping potential (6.62), we only include positive λ_k s, and these we arrange according to their magnitudes, $\lambda_1 > \lambda_2 > \cdots > \lambda_N$. We can also approximate the integral in (6.77) by a finite sum of the form [13]

$$\sum_{j=1}^{N} \frac{d_i}{\lambda^2 + \mu_j^2}. \tag{6.78}$$

The contribution of these terms to the determination of the potential $q(x)$ in the Schrödinger like equation (6.61) is small and will be ignored in this work [14].

Noting that (6.61) is like the wave equation for S-wave except for k^2 being replaced by $-\lambda^2$, we now use the Gel'fand–Levitan formulation (see Eq. (4.98)), and write

$$q(x) = 2 \frac{d^2}{dx^2} \left[\ln \det W_{sr}(x) \right], \tag{6.79}$$

where $W_{rs}(x)$ is an $N \times N$ matrix with the elements

$$W_{sr}(x) = \frac{1}{\lambda_s + \lambda_r} \{ 1 - \exp[-2(\lambda_s + \lambda_r)x] \} + \frac{(1 - \delta_{sr})}{\lambda_s - \lambda_r}$$
$$\times [\exp(-2\lambda_s x) - \exp(-2\lambda_r x)] - \delta_{sr} \exp(-2\lambda_r x) \left[2x - \left(\frac{4\lambda_r^2}{c_r} \right) \right]. \tag{6.80}$$

Let us again remind ourselves that λ_s and λ_r are the positive poles of $\mathcal{R}^{-1}\left(\lambda^2\right)$, and $c_r > 0$ is the residue at $\lambda^2 = \lambda_r^2$. Thus by calculating the determinant of the matrix W_{sr} and substituting it in (6.79) we find $q(x)$. This direct method of numerically calculating $q(x)$ and subsequently $v(r)$ is not reliable nor accurate.

Markushevich and Novikova have suggested a better algorithm for this calculation based on the properties of the Cauchy matrix [14]. This lengthy method is described in Refs. [9], [14] and the reader can find a detailed discussion in these references.

Accuracy of the Present Method — For testing the stability of the numerical method used, and also determine the effect of omission of the integral in (6.77) and that of the roundoff errors, we study two simple but exactly solvable potential models. From these models we calculate $\mathcal{R}(\ell)$ and $I(\ell)$ and use these to find λ_k s and c_k s, and then find points for the potential as the output. The first is the parabolic potential $v(r)$ given by

$$v(r) = \begin{cases} -v_0 \left(1 - \frac{r^2}{R^2}\right) & 0 \le r < R \\ 0 & r > R \end{cases}. \tag{6.81}$$

The Schrödinger equation for this potential when $r < R$ takes the form

$$\frac{d^2\phi_\ell(r)}{dr^2} + \left[2\beta\mu - \beta^2 r^2 - \frac{\ell(\ell+1)}{r^2}\right]\phi_\ell(r), \tag{6.82}$$

where

$$\beta = \frac{\sqrt{v_0}}{R} \quad \text{and} \quad \beta\mu = \frac{1}{2}\left(k^2 + v_0\right). \tag{6.83}$$

If we write $\phi_\ell(r)$ as

$$\phi_\ell(r) = r^{\ell+1} \exp\left(-\frac{1}{2}\beta r^2\right) F_\ell(r), \tag{6.84}$$

and substitute this in (6.82) we find that $F_\ell(r)$ satisfies the differential equation for degenerate (or confluent) hypergeometric function [15]:

$$r\frac{d^2 F_\ell(r)}{dr^2} + \left(\ell + \frac{3}{2} - r\right)\frac{dF_\ell(r)}{dr} - \left(\frac{\ell}{2} + \frac{3}{4} - \frac{\mu}{2}\right) F_\ell(r) = 0. \tag{6.85}$$

The regular solution of this equation that we want is

$$F_\ell(r) = {}_1F_1\left[\frac{1}{2}\left(\ell + \frac{3}{2} - \mu\right), \ell + \frac{3}{2}; \beta r^2\right]. \tag{6.86}$$

At $r = R$ we calculate the logarithmic derivative of $\phi_\ell(r)$ and find $I(\ell)$

$$I(\ell) = R\left(\frac{d}{dr}\left[\ln\phi_\ell(r)\right]\right)_{r=R} = \ell + 1 - \beta R^2 + R\left[\frac{d}{dr}\ln({}_1F_1(r))\right]_{r=R}. \tag{6.87}$$

Now the derivative of ${}_1F_1(a, b; z)$ can be expressed as

$$\frac{d({}_1F_1(a, b; z))}{dz} = \left(\frac{a}{b}\right) {}_1F_1(a+1, b+1; z), \tag{6.88}$$

thus by substituting for the derivative of ${}_1F_1$ in (6.87) we find $I(\ell)$:

$$I(\ell) = \ell + 1 - \beta R^2 + \sqrt{v_0} R \left\{\frac{\ell + \frac{3}{2} - \mu}{\ell + \frac{3}{2}}\right\}$$
$$\times \left\{\frac{{}_1F_1\left\{\frac{1}{2}\left[\ell + \frac{7}{2} - \mu\right], \ell + \frac{5}{2}; \sqrt{v_0}R\right\}}{{}_1F_1\left\{\frac{1}{2}\left[\ell + \frac{3}{2} - \mu\right], \ell + \frac{3}{2}; \sqrt{v_0}R\right\}}\right\}. \tag{6.89}$$

As a second example consider the square well potential;

$$v(r) = \begin{cases} -v_0 & 0 \le r \le R \\ 0 & r > R \end{cases}. \tag{6.90}$$

The regular solution of the Schrödinger equation for this potential is

$$\phi_\ell(r) = Aqr j_\ell(qr), \quad q^2 = k^2 + v_0, \tag{6.91}$$

and the logarithmic derivative at R is given by

$$I(\ell) = 1 + R \left\{ \frac{d}{dr} [\ln j_\ell(qr)] \right\}_{r=R}. \tag{6.92}$$

If we test the present method of inversion for the two potentials (6.81) and (6.90) for different strengths, v_0, ranges R and the incident energies k^2, we find that it works reasonably well when the potential is strong and the incident energy of the particle is very small. Only near the origin, the result of inversion becomes unreliable (see Figs. 6.3 and 6.4). Choosing the unit of length to be L, for the case of parabolic potential (6.81) we have chosen $R = 1\ L$, $v_0 = 6000\ L^{-2}$ and $k = 0\ L^{-2}$. For these parameters we have found the points shown in Fig. 6.3. In the case of square well, the parameters that we have used are $R = 2.5\ L$, $v_0 = 252\ L^{-2}$ and $k^2 = 4\ L^{-2}$. Such a strong short range potential occurs in molecular scattering theory such as $He - W$ or $N_2O - Ru$ scattering [16]. For the results shown in Figs. 6.3 and 6.4, we have used 12 λ_j s for parabolic and 18 λ_j s for the square well to find $I(\lambda)$. These results are comparable with the results found by Sabatier and Quyen Van Phu [7] (see Figs. 6.1 and 6.2).

6.5   Lipperheide and Fiedeldey Approach to the Inverse Problem at Fixed Energy

We have seen that the Bargmann potential can be constructed from the assumption that the Jost function is a rational function of the particle's energy or momentum $\hbar k$. Can we have a parallel method where we can find exactly solvable potentials using the inverse scattering theory for fixed energy? To find the answer to this question we start with the assumption that the scattering matrix has the form of a product of a complex rational function of angular momentum multiplied by the known scattering matrix of a "reference" potential.

   Let us begin with the Schrödinger equation for the ℓth partial wave in the absence of any potential. In this case the Jost solution is given by [22]–[25]

$$\left[\frac{d^2}{dr^2} - \frac{\lambda^2 - \frac{1}{4}}{r^2} + k^2 \right] f(\lambda, \pm k, r) = 0, \tag{6.93}$$

where $\lambda = \ell + \frac{1}{2}$ and $\hbar k = \sqrt{2mE}$. This Jost function has the asymptotic form of

$$f(\lambda, \pm k, r) \to e^{\pm ikr}, \quad \text{as} \quad r \to \infty. \tag{6.94}$$

For the case of no interaction $f(\lambda, \pm k, r)$ has the exact Jost solution

$$
f(\lambda, \mp k, r) = \sqrt{\frac{\pi k r}{2}} \frac{(\pm i) \exp\left(\frac{\pm i\pi}{4}\right)}{\sin(\pi\lambda)}
$$
$$
\times \left[\exp\left(\pm\frac{i\lambda\pi}{2}\right) J_\lambda(kr) - \exp\left(\pm\frac{i\lambda\pi}{2}\right) J_{-\lambda}(kr)\right], \qquad (6.95)
$$

where $J_\lambda(kr)$ is the Bessel function. Let us assume that $\lambda > 0$, then we have the following asymptotic form for $f(\lambda, \pm k)$;

$$
f(\lambda, \pm k, r) \to \frac{1}{2\lambda} f(\lambda, \pm k) r^{-\lambda + \frac{1}{2}}, \quad r \to 0, \qquad (6.96)
$$

where the Jost function for this case is

$$
f(\lambda, \pm k, r) = 2^\lambda \sqrt{\frac{2}{\pi}} \Gamma(\lambda + 1)(\mp k)^{(-\lambda + \frac{1}{2})} \exp\left[\frac{i\pi}{2}\left(\lambda - \frac{1}{2}\right)\right]. \qquad (6.97)
$$

To simplify the notation we introduce $f_\lambda(r)$ and $f_\lambda^*(r)$ by

$$
f_\lambda^*(r) \equiv f(\lambda, k, r), \quad f_\lambda(r) \equiv f(\lambda, -k, r), \qquad (6.98)
$$

and

$$
f_\lambda^* \equiv f(\lambda, k), \quad f_\lambda \equiv f(\lambda, -k). \qquad (6.99)
$$

Now we define the function

$$
x_{\lambda\nu} = \frac{1}{\lambda^2 - \mu^2} W\left[f_\lambda^*(r), f_\mu(r)\right] = \int_r^\infty \frac{1}{r'^2} f_\lambda^*(r') f_\mu(r') dr' - \frac{2ik}{\lambda^2 - \mu^2}, \qquad (6.100)
$$

where $W\left[f_\lambda^*(r), f_\mu(r)\right]$ is the Wronskian of the two functions $f_\lambda^*(r)$ and $f_\mu(r)$. We write for $\lambda = \beta_n$ and or $\mu = \alpha_m$, $(n, m = 1, 2, \cdots N)$, and use the subscripts m and n to denote the components of vectors and the matrix elements describing the system.

From (6.100) we can calculate the elements of the matrix x_{mn} and also the set of N functions $K_m(r)$ and these are the solutions of the linear set of equations

$$
\sum_m x_{nm}(r) K_m(r) = f_n(r), \quad n, m = 1, 2 \cdots N. \qquad (6.101)
$$

By solving this set and finding $K_m(r)$, we determine a potential $v(r)$ by

$$
v(r) = \frac{2}{r} \frac{d}{dr}\left[\frac{1}{r} \sum_m K_m^*(r) f_m(r)\right]. \qquad (6.102)
$$

When we add this potential to the Schrödinger equation (6.93) we obtain an outgoing Jost-type solution, having the form

$$
M_\lambda(r) = f_\lambda(r) - \sum_m x_{\lambda m}(r) K_m(r), \qquad (6.103)
$$

where the set of vectors $x_{\lambda m}(r)$ are defined by (6.100). To show that $M_\lambda(r)$ is indeed a solution of the Schrödinger equation with the potential $v(r)$ we start with (6.100) and differentiate $x_{\lambda\mu}(r)$ with respect to r;

$$\frac{d}{dr}x_{\lambda m}(r) = -\frac{1}{r^2}f_\lambda(r)f_m^*(r). \tag{6.104}$$

In addition we find that

$$\left[-\frac{d^2}{dr^2} + \frac{\lambda^2 - \frac{1}{4}}{r^2} + v(r) - k^2\right]M_\lambda(r)$$

$$= \left[\frac{d^2}{dr^2} - \frac{\lambda^2 - \frac{1}{4}}{r^2} - v(r) + k^2\right]\sum_m x_{\lambda m}(r)K_m(r)$$

$$+ \frac{2}{r}\frac{d}{dr}\left(\frac{1}{r}\sum_m K_m(r)f_m^*(r)\right)f_\lambda(r) = -\sum_m x_{\lambda m}R_m(r), \tag{6.105}$$

where

$$R_m(r) \equiv \left[-\frac{d^2}{dr^2} + \frac{\alpha_m^2 - \frac{1}{4}}{r^2} + v(r) - k^2\right]K_m(r). \tag{6.106}$$

Now if we set $\lambda = \beta_n$ in (6.103), then from (6.101) it follows that

$$M_n(r) \equiv 0. \tag{6.107}$$

Therefore from (6.105) we find

$$\sum_m x_{nm}(r)R_m(r) = 0, \tag{6.108}$$

and since in general $\det x_{mn}(r) \neq 0$, we conclude that

$$R_m(r) = 0. \tag{6.109}$$

From (6.105) also we get

$$\left[-\frac{d^2}{dr^2} + \frac{\lambda^2 - \frac{1}{4}}{r^2} + v(r) - k^2\right]M_\lambda(r) = 0. \tag{6.110}$$

Using the relation

$$\frac{dX(r)}{dr} = x\,\text{trace}\left(\frac{1}{X(r)}\frac{dX(r)}{dr}\right), \tag{6.111}$$

where $X(r)$ represents the matrix with the elements $x_{nm}(r)$. we can write the potential in the more familiar form of

$$v(r) = -\frac{2}{r}\frac{d}{dr}\left[r\frac{d}{dr}\ln x(r)\right]. \tag{6.112}$$

Finding the Relation between the Constructed Potential and the Jost Function — In order to simplify the equations leading to $v(r)$ we introduce the dimensionless variable $\rho = kr$, and introduce the set of quantities $y_{\lambda\mu}(r)$ by

$$y_{\lambda\mu}(\rho) = \frac{i}{k}\frac{1}{f_{\lambda(r)}f_\mu^*(r)}x_{\lambda\mu}(r) = \frac{L_\lambda(\rho) + L_\mu^*(\rho)}{\lambda^2 - \mu^2}, \qquad (6.113)$$

where

$$L_\mu(\rho) = -\frac{i}{f_\mu(\rho)}\frac{df_\mu(\rho)}{d\rho}, \qquad (6.114)$$

and

$$L_\lambda^*(\rho) = \frac{i}{f_\lambda^*(R\rho)}\frac{df_\lambda^*(\rho)}{d\rho}. \qquad (6.115)$$

Each of the two functions $L_\mu(\rho)$ and $L_\lambda^*(\rho)$ satisfy the Ricatti differential equation

$$-i\frac{d}{d\rho}L_\lambda^*(\rho) - L_\lambda^{*2} + \left(1 - \frac{\lambda^2 - \frac{1}{4}}{\rho^2}\right) = 0, \qquad (6.116)$$

and a similar equation for $L_\mu(\rho)$. Both $L_\lambda^*(\rho)$ and $L_\mu(\rho)$ approach 1 as $\rho \to \infty$. We can write the potential $v(r)$ in terms of $L_\mu(\rho)$ and $L_\lambda^*(\rho)$:

$$v(r) = -\frac{2k^2}{\rho}\frac{d}{d\rho}\left\{i\rho\left[\sum_n L_n(\rho) - \sum_m L_m^*(\rho)\right] + \rho\frac{d}{d\rho}\ln y(\rho)\right\}, \qquad (6.117)$$

where $L_n^*(\rho)$ and $L_m(\rho)$ are solutions of Ricatti differential equation for $\mu = m$ and $\lambda = n$, and $y(\rho) = \det y_{nm}(\rho)$. If Y is the matrix formed from elements $y_{nm}(\rho)$, then we can write the potential as

$$v(r) = \frac{2ik^2}{\rho}\frac{d}{d\rho}\left[\frac{1}{\rho}\sum_{nm}\left(Y^{-1}(\rho)\right)_{mn}\right], \qquad (6.118)$$

or noting that

$$\frac{d}{d\rho}Y^{-1} = -Y^{-1}\frac{dY}{d\rho}Y, \qquad (6.119)$$

we can write the final form of the potential $v(r)$ as

$$v(r) = -\frac{2ik^2}{\rho^3}\left\{\sum_{nm}\left(Y^{-1}\right)_{mn}[1 - i\rho(L_n - L_m^*)] - \frac{i}{\rho}\left[\sum_{nm}\left(Y^{-1}\right)_{mn}\right]^2\right\}. \qquad (6.120)$$

The Jost Solution for the Outgoing Wave — We have already found the Jost type solution $M_\lambda^*(r)$, Eq. (6.103). With the help of Eqs. (6.101) and (6.113) we can write this solution in the following form:

$$M_\lambda^*(r) = f_\lambda(r)\left[1 - \sum_{nm}y_{\lambda m}\left(Y^{-1}\right)_{mn}\right]. \qquad (6.121)$$

From the algebraic relation

$$\det\left[y_{nm} - y_{\lambda m}\right] = \det[y_{nm}]\left[1 - \sum_{nm} y_{\lambda m}\left(Y^{-1}\right)_{nm}\right], \tag{6.122}$$

we find

$$M_\lambda(r) = f_\lambda(r)\frac{1}{\det[y_{mn}]}\det[y_{nm} - y_{\lambda m}]. \tag{6.123}$$

Also from (6.113) and (6.116) we find the asymptotic form of $M_\lambda(r)$ as $r \to \infty$;

$$M_\lambda(r) \to f_\lambda(r)\frac{1}{\det[y_{mn}]}\det\left[y_{nm}\left(1 - \frac{\beta_n^2 - \alpha_m^2}{\lambda^2 - \alpha_n^2}\right)\right]$$

$$= \frac{\prod\left(\lambda^2 - \beta_n^2\right)}{\prod\left(\lambda^2 - \alpha_m^2\right)}\exp(ikr) \quad \text{as} \quad r \to \infty \tag{6.124}$$

which is the outgoing Jost solution of (6.110). Since $M_\lambda(r)$ is the solution of the Schrödinger equation (6.110) with the asymptotic condition (6.124), therefore the Jost solution $F_\lambda(r)$ is given by

$$F_\lambda(r) = \frac{\prod\left(\lambda^2 - \alpha_m^2\right)}{\prod\left(\lambda^2 - \beta_n^2\right)}M_\lambda(r). \tag{6.125}$$

To find the dependence of $F_\lambda(r)$ when r goes to zero, from Eqs. (6.96), (6.113) and (6.94), (6.116), we have

$$y_{\lambda m}(r) \to \left(\frac{i}{\lambda + \alpha_m}\right)\frac{1}{\rho}, \quad \text{as} \quad r \to 0, \tag{6.126}$$

and

$$\lim_{r \to 0} F_\lambda(r) \to f_\lambda(r)\frac{\prod\left(\lambda^2 - \alpha_m^2\right)}{\prod\left(\lambda^2 - \beta_n^2\right)}\left(\frac{1}{\det[y_{mn}]}\right)\det\left[y_{nm}\left(1 - \frac{\beta_n + \alpha_m}{\lambda + \alpha_m}\right)\right]$$

$$= \frac{1}{2\lambda}f_\lambda\frac{\prod(\lambda - \alpha_m)}{\prod(\lambda + \beta_n)}r^{-\lambda + \frac{1}{2}}. \tag{6.127}$$

From the last equation we find the Jost function to be;

$$F_\lambda = f_\lambda\frac{\prod(\lambda - \alpha_m)}{\prod(\lambda + \beta_n)}. \tag{6.128}$$

Let us consider the scattering of a particle of fixed energy E and assume that the scattering matrix can be represented by a complex rational function of the form

$$S(\lambda) = S_0(\lambda)\prod_{n=1}^{N}\frac{\lambda^2 - \beta_n^2}{\lambda^2 - \alpha_n^2}, \tag{6.129}$$

where $S_0(\lambda)$ is a reference S matrix corresponding to the background potential $v_0(r)$. Now the total scattering potential $v(r) = v_N(r)$ can be obtained by iteration

$$v_n(r) = v_{n-1}(r) + \Delta v_n(r). \tag{6.130}$$

In this equation $\Delta v_N(r)$ is the potential resulting from each addition of a pole and a zero pair of the N numbers which defines the S matrix. Each $\Delta v_n(r)$ is given in terms of the Jost solutions from the preceding iterate of the potential, i.e. from $v_{n-1}(r)$.

Solution for a Single Term — In the Lipperheide–Fiedeldey inversion method one approximates the empirical S matrix by an S matrix which is a rational function of the angular momentum ℓ or $\lambda = \ell + \frac{1}{2}$. Another way of tackling this problem is to start with a rational approximation to the Jost function for λ, i.e.

$$f(k,\lambda) = f_0(k,\lambda) \sum_{j=1}^{N} \frac{\lambda - \alpha_j}{\lambda - \beta_j}, \tag{6.131}$$

where $f_0(\lambda, k)$ is the Jost function for the background or reference potential. Then one follows the Gel'fand–Levitan method to construct the potential. The function $f(k, \lambda, k)$ here must satisfy the asymptotic condition

$$\lim_{k \to \infty} \frac{f(k,\lambda)}{f_0(k,\lambda)} = 1, \quad |\arg \lambda| < \frac{\pi}{2}. \tag{6.132}$$

For the potential $v(r)$ to be real, the condition $\beta_j = -\alpha_j^*$ must also be satisfied [10],

$$f(k,\lambda) = f_0(k,\lambda) \sum_{j=1}^{N} \frac{\lambda - \alpha_j}{\lambda + \alpha_j^*}. \tag{6.133}$$

The requirement that $f(k, \lambda)$ must be analytic in the right-hand complex λ half-plane and the continuity for $\mathrm{Re}\,\lambda \geq 0$, it is necessary to choose $\mathrm{Re}\,\alpha_j > 0$.

Asymptotic Forms of the Potential — If the scattering matrix S is unitary, then $\beta_n = \alpha_n^*$, $f_{\beta_n} = [f_{\alpha_n}]^*$, and $x_{nm}(r)$ is a Hermitian matrix and its determinant is real, resulting in a real potential. Otherwise the potential is complex. Now let us consider the asymptotic form of the potential as $r \to \infty$. In this limit we have

$$\lim L_n(\rho) \to 1 - \frac{\beta_n^2 - \frac{1}{4}}{2\rho^2} + \cdots, \tag{6.134}$$

$$\lim L_m^*(\rho) \to 1 - \frac{\alpha_m^2 - \frac{1}{4}}{2\rho^2} + \cdots, \tag{6.135}$$

and

$$\lim y_{nm} \to \frac{1}{\beta_n^2 - \alpha_m^2} \left(1 - \frac{\beta_n^2 + \alpha_m^2}{4\rho^2} + \cdots \right), \quad \text{as} \quad \rho \to \infty. \tag{6.136}$$

For large r (or ρ), the second term in (6.136) can be ignored compared to the first term, and therefore

$$v(r) \to -\frac{i}{k} \sum_n \frac{\beta_n^2 - \alpha_n^2}{r^3} + \cdots. \tag{6.137}$$

Now let us determine the behaviour of the potential near the origin. To this end we first calculate the following limits

$$\left.\begin{array}{c} f_\lambda^*(r) \\ f_\lambda(r) \end{array}\right\} \rightarrow \left(\frac{1}{\rho}\right) \rho^{-\lambda+\frac{1}{2}} - \frac{1}{4}\left(\frac{1}{1-\lambda}\right)\rho^{-\lambda+\frac{1}{2}}, \tag{6.138}$$

and therefore using Eq. (6.116) we get

$$y_{nm} \rightarrow \frac{1}{\beta_n + \alpha_m}\left[1 + \frac{1}{2(\beta_n-1)(\alpha_n-1)}\rho^2 + \cdots\right], \quad \text{as } \rho \rightarrow 0. \tag{6.139}$$

Therefore

$$\frac{1}{\rho}\sum_{nm}\left[Y^{-1}\right]_{mn} \rightarrow a + b\rho^2 + \cdots, \tag{6.140}$$

and from (6.118) we find the limit of $v(r)$ as r tends to zero,

$$v(r) \rightarrow 4k^2 b = \text{finite}. \tag{6.141}$$

Other properties of the potential $v(r)$ have been studied by Lipperheide and Fiedeldey [22].

6.6   Completeness of the Set of Jost Solutions $f(\lambda, k, r)$

In the Gel'fand–Levitan and also Abraham–Moses formulations of the inverse problem, the completeness relations, Eqs. (4.73), (4.74) and (4.93) played an essential role in finding the integral equation for the kernel $K(r, r')$. There, we were interested in the completeness relation for a given partial wave, i.e. $\ell = $ constant, and for all wave numbers k (or energies E). The question that we want to discuss now is whether we can find the corresponding completeness of the set of wave functions when they are given at a fixed energy. In the Newton–Sabatier method we bypassed this question with the introduction of $L_{\ell\ell'}$ and $M_{\ell\ell'}$, Eqs. (6.31) and (6.32). Here we want to show that a completeness relation similar to what we found for the Gel'fand–Levitan formulation can also be obtained for the Jost function at fixed energy and variable $\lambda = \ell + \frac{1}{2}$, if the latter is assumed to be complex. Again we assume that both the first and the second absolute moments of the potential $v(r)$ are finite, i.e.

$$\int_0^\infty r|v(r)|dr = \text{finite}, \quad \text{and} \quad \int_0^\infty r^2|v(r)|dr = \text{finite}. \tag{6.142}$$

Following the work of Burdet *et al.* we consider the integral [27], [28]

$$I(k, r) = \int_\Gamma \lambda d\lambda \int_0^\infty h_1(r')\frac{-G(\lambda, k, r, r')}{rr'}dr', \tag{6.143}$$

where $G(\lambda, k, r, r')$ is the complete Green function, and $h_1(r')$ is an arbitrary square integrable function. Since $G(\lambda, k, r, r')$ is the complete Green function we can write the exact solution of the Schrödinger equation in terms of $\bar{\phi}(\lambda, k, r)$ which is the solution of the wave equation in the absence of any potential. The solution in this case is given by

$$\phi(\lambda, k, r) = \bar{\phi}(\lambda, k, r) + \int_0^\infty G(\lambda, k, r, r') v(r') \bar{\phi}(\lambda, k, r') dr'. \tag{6.144}$$

The Green function can be expressed in terms of the regular solution obtained with the boundary condition

$$\lim_{r \to 0} \phi(\lambda, k, r) r^{-\lambda - \frac{1}{2}} = 1, \tag{6.145}$$

and by the Jost solution which is defined by

$$\lim_{r \to \infty} f(\lambda, k, r) \exp(-ikr) = 1. \tag{6.146}$$

Thus

$$G(\lambda, k, r, r') = \frac{1}{f(\lambda, k)} \begin{cases} -\phi(\lambda, -k, r) f(\lambda, k, r') & r < r' \\ \\ -\phi(\lambda, -k, r') f(\lambda, k, r) & r' < r \end{cases}, \tag{6.147}$$

where the Jost function is given by the Wronskian

$$f(\lambda, k) = W\left[f(\lambda, k, r),\ \phi(\lambda, -k, r)\right]. \tag{6.148}$$

When the potential $v(r)$ satisfies the conditions given by (6.142), $\phi(\lambda, k, r)$ is a holomorphic function in the right half-plane $\mathrm{Re}\, \lambda \geq 0$ and $f(\lambda, k, r)$ is a holomorphic function in $\mathrm{Re}\, \lambda \geq 0$. We choose the contour Γ to be composed of the imaginary λ axis plus a semicircle C and radius R $(R \to \infty)$ in the half-plane $\mathrm{Re}\, \lambda \geq 0$ as shown in Fig. 6.5. We first evaluate the integral in (6.143) with the help of Cauchy's theorem, and we express it in terms of individual contributions. Since $\phi(\lambda, k, r)$ and $f(\lambda, k, r)$ are analytic functions regular in the right half-plane $\mathrm{Re}\, \lambda \geq 0$, the only poles in the integrand are simple zeros of the Jost function $f(\lambda, k)$ at $\lambda = \alpha_j$. Thus the integral will be

$$I(k, r) = 2\pi i \sum_j \int_0^\infty \frac{h_1(r')}{rr'} \frac{dr'}{\left(\frac{\partial f(\lambda, k)}{\partial \lambda}\right)_{\lambda = \alpha_j}}$$

$$\times \begin{cases} \phi(\alpha_j, -k, r) f(\alpha_j, k, r') & r < r' \\ \phi(\alpha_j, -k, r') f(\alpha_j, k, r) & r' < r \end{cases}. \tag{6.149}$$

The partial derivative of $f(\lambda, k)$ with respect to λ at $\lambda = \alpha_j$ can be expressed as

$$\left(\frac{\partial f(\lambda, k)}{\partial \lambda}\right)_\lambda = \frac{-i\alpha_j}{k} f(\alpha_j, -k) C_j^{-2}(\alpha_j, -k), \tag{6.150}$$

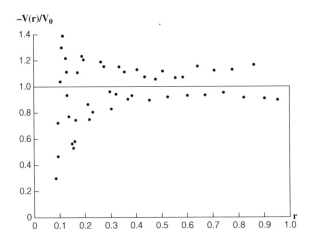

Figure 6.4: The attractive potential assumed for the direct calculation is a square well shown here by solid line. The $\mathcal{R}$ matrix obtained in this way have been used in the method of Markushevich *et al.* for inversion, with the result shown by dots. The parameters used for this calculation are $R = 2.5L$, $v_0 = 252L^{-2}$ and $k^2 = 4L^{-2}$.

where

$$\frac{1}{C_j^2(\alpha_j, -k)} = \int_0^\infty \frac{1}{r^2} f^2(\alpha_j, k, r) dr. \tag{6.151}$$

Since α_j is a root of the Jost function we have

$$\phi(\alpha_j, k, r) = \frac{1}{2ik} f(\alpha_j, -k) f(\alpha_j, k, r). \tag{6.152}$$

In addition we have the following relations for the solutions of the Schrödinger equation:

$$\frac{d}{dr} \left(W\left[\psi\left(\lambda'\right), \psi(\lambda)\right] \right) = \frac{\lambda^2 - \lambda'^2}{r^2} \psi\left(\lambda'\right) \psi(\lambda), \tag{6.153}$$

and

$$\frac{d}{dr} \left(W\left[\left(\frac{\partial \psi}{\partial \lambda}\right)_{\lambda = \alpha_j}, \psi\left(\alpha_j\right) \right] \right) = -\frac{2\alpha_j}{r^2} \psi^2(\alpha_j), \tag{6.154}$$

where W denotes the Wronskian. From these results we conclude that

$$I(k, r) = i\pi \sum_j \frac{1}{C_j^2(\alpha_j, -k)} \frac{h_1\left(r'\right)}{rr'} f(\alpha_j, k, r) f\left(\alpha_j, k, r'\right) dr'. \tag{6.155}$$

We also note that $f(\lambda, k, r)$ is an even function of λ and that

$$f(\lambda, k, r) = \frac{1}{2\lambda} \left[f(\lambda, k)\phi(-\lambda, -k, r) - f(-\lambda, k)\phi(\lambda, -k, r) \right]. \tag{6.156}$$

Now we can evaluate (6.143) along the contour Γ. First we calculate the contribution to the integral coming from the integration along the imaginary axis:

$$\int_{\text{Im }\lambda} = -2 \int_0^{i\infty} \lambda^2 d\lambda \int_0^\infty \frac{h_1(r') f(\lambda, k, r) f(\lambda, k, r')}{rr' f(\lambda, k) f(-\lambda, k)} dr'. \tag{6.157}$$

Second, using the asymptotic expansion for large λ, we can evaluate the contribution to (6.143) coming from the large semicircle. We note that the asymptotic values of $\phi(\lambda, k, r)$ and $f(\lambda, k)$, are given by

$$\phi(\lambda, k, r) \sim \bar{\phi}(\lambda, k, r) \sim r^{\lambda + \frac{1}{2}}, \quad \text{as } |\lambda| \to \infty, \tag{6.158}$$

and

$$f(\lambda, k) \sim f_0(\lambda, k) \sim 2\sqrt{-\lambda k} \left(\frac{2\lambda}{ek}\right)^\lambda \exp\left(\frac{i\pi\lambda}{2} - \frac{i\pi}{4}\right), |\lambda| \to \infty, \tag{6.159}$$

and $|\arg \lambda| < \pi$. This expression can also be written as [29]

$$f(\lambda, k) = \sqrt{\frac{-kr}{\lambda}} \exp\left(\frac{i\pi\lambda}{2}\right) \left[\left(\frac{2\lambda}{-ek}\right)^\lambda \frac{\exp\left(\frac{-i\pi}{4}\right)}{r^\lambda} - \exp\left(\frac{i\pi}{4}\right) r^\lambda \left(\frac{-ek}{2\lambda}\right)^\lambda\right],$$
$$\text{as } |\lambda| \to \infty, \quad |\arg \lambda| < \pi. \tag{6.160}$$

Therefore for the integral along the semicircle we get

$$\lim_{R \to \infty} \int_C \to \frac{1}{2} \int_0^\infty \frac{h(r') dr'}{\sqrt{rr'}} \left\{\int_C \left[\left(\frac{r'}{r}\right)^\lambda \theta(r - r') + \left(\frac{r}{r'}\right)^\lambda \theta(r' - r)\right] d\lambda\right\}. \tag{6.161}$$

If we denote $\ln\left(\frac{r'}{r}\right)$ by Λ, then

$$\lim_{R \to \infty} \frac{\sin(R\Lambda)}{\Lambda} = \pi\delta(\Lambda). \tag{6.162}$$

Thus in the limit of $R \to \infty$, the integral taken along the semicircle can be evaluated with the result that

$$\int_C = i\pi h(r). \tag{6.163}$$

From Eqs. (6.155), (6.157) and (6.163) we get [27], [30]

$$h_1(r) = \int_0^\infty h_1(r') dr' \left[\sum_j \frac{f(\alpha_j, k, r)}{r} \frac{f(\alpha_j, k, r')}{r'} C^2(\alpha_j, -k)\right.$$
$$\left. + \frac{2i}{\pi} \int_0^{i\infty} \frac{f(\lambda, k, r)}{r} \frac{f(\lambda, k, r')}{r'} \frac{\lambda^2 d\lambda}{f(\lambda, k) f(-\lambda, k)}\right]. \tag{6.164}$$

This relation will hold for any square integrable function $h_1(r)$. Thus we conclude that

$$
\sum_j \frac{f(\alpha_j, k, r)}{r} \frac{f(\alpha_j, k, r')}{r'} C^2(\alpha_j, -k)
$$

$$
+ \frac{2i}{\pi} \int_0^{i\infty} \frac{f(\lambda, -k, r)}{r} \frac{f(\lambda, k, r')}{r'} \frac{\lambda^2 d\lambda}{f(\lambda, k) f(-\lambda, k)}
$$

$$
= \delta(r - r'). \tag{6.165}
$$

Now we introduce the spectral function $\rho(\lambda)$ by

$$
\frac{d\rho(\lambda)}{d\lambda} = \begin{cases} \frac{2i}{\pi} \frac{\lambda^2}{f(\lambda, k) f(-\lambda, k)} & \text{for } 0 \le \lambda \le i\infty \\ \\ \sum_j C_j^2(\alpha_j, -k) \delta(\lambda - \alpha_j), & \text{where } f(\lambda = \alpha_j, k) = 0 \end{cases}. \tag{6.166}
$$

This has exactly the form of the spectral function for the constant energy case, Eq. (4.75). The Jost function, in addition to satisfying the boundary condition at infinity, behaves as $r^{\lambda + \frac{1}{2}}$ near the origin. Following the same method that we discussed for the constant energy problem, here we also find the Gel'fand–Levitan type equation of the form [31]

$$
K(r, r') + g(r, r') + \int_r^\infty K(r, r'') g(r'', r') dr'' = 0, \tag{6.167}
$$

where in the present case $g(r, r')$ is defined by

$$
g(r, r') = \int_0^{i\infty} \frac{\bar{f}(\lambda, k, r)}{r} \frac{\bar{f}(\lambda, k, r')}{r'} \left[\frac{1}{f(\lambda, k) f(-\lambda, k)} - \frac{1}{\bar{f}(\lambda, k) \bar{f}(-\lambda, k)} \right]
$$

$$
\times \lambda^2 d\lambda + \sum \frac{C^2(\alpha_j, -k)}{rr'} \bar{f}(\alpha_j, k, r) \bar{f}(\alpha_j, k, r) -
$$

$$
\sum_j \frac{\bar{C}^2(\alpha_j^{(0)}, k)}{rr'} \bar{f}(\alpha_j^{(0)}, k, r) \bar{f}\left(\alpha_j^{(0)}, k, r'\right). \tag{6.168}
$$

In this equation $\bar{f}(\lambda, k, r)$, $\bar{f}(\lambda, k)$ and $\bar{C}(\alpha_j^{(0)}, -k)$ denote the Jost solution, the Jost function and the normalization constant for the bound states of the reference potential $\bar{v}(r)$. We also use the symbols $f(\lambda, k, r)$, $f(\lambda, k)$ and $C(\alpha_j^{(0)}, -k)$ as the corresponding quantities for the unknown potential $v(r) = \bar{v}(r) + \Delta v(r)$.

Bargmann Type Potential — If $\bar{f}(\lambda, k)$ is the known Jost function for the reference or background potential $\bar{v}(r)$, let us construct the new Jost function by multiplying the old one, $\bar{f}(\lambda, k)$, by the sum of rational functions of λ, and define $f(\lambda, k)$ by

$$
f(\lambda, k) = \bar{f}(\lambda, k) \sum_{j=1}^N \frac{\lambda - \alpha_j}{\lambda - \beta_j}. \tag{6.169}
$$

The new $f(\lambda, k)$ has been chosen in such a way that the condition (6.132) is satisfied and that $f(\lambda, k)$ is given by (6.133), with Re $\lambda \geq 0$ and Re $\alpha_j > 0$.

We can find the exact potential for the case of $N = 1$ in (6.169). To this end we calculate $g(r, r')$ from (6.168):

$$g(r, r') = \frac{C_1^2}{rr'} \bar{f}(\alpha_1, k, r) \bar{f}^*(\alpha_1, k, r') \tag{6.170}$$

$$+ \frac{\alpha_1^{*2} - \alpha_1^2}{rr' \bar{f}(\alpha_1, k)} \begin{cases} \bar{\phi}(\alpha_1, k, r) \bar{f}(\alpha_1, k, r') & \text{for } r < r' \\ \bar{\phi}(\alpha_1, k, r') \bar{f}(\alpha_1, k, r) & \text{for } r > r' \end{cases}, \tag{6.171}$$

where $\bar{\phi}(\alpha_1, r, k)$ is defined as in Eq. (4.15)

$$\bar{\phi}(\lambda, k, r) = \frac{1}{2ik} \left[\bar{f}(\lambda, k) \bar{f}(\lambda, -k, r) - \bar{f}(\lambda, k, r) \bar{f}(\lambda, -k) \right]. \tag{6.172}$$

By substituting (6.171) in (6.168) we find $K(r, r')$. From $K(r, r')$ we can calculate the potential $v(r)$ [10];

$$v(r) = \bar{v} - \frac{2}{r} \frac{d}{dr} [r K(r, r)]. \tag{6.173}$$

For real potential $v(r)$, we need to choose the normalization constant $C_1^2(\alpha_1, k)$ from the relation

$$C_1^2(\alpha_1, -k) = \frac{(\alpha_1^{*2} - \alpha_1^2) \bar{f}(-\alpha, k)}{2ik \bar{f}(\alpha, -k)}, \tag{6.174}$$

and when this condition is satisfied, then the potential is given by

$$v(r) = \frac{-2 (\alpha_1^{*2} - \alpha_1^2) |f(\alpha_1^*, -k, r)|^2}{W [\bar{f}(\alpha_1^*, k, r), \bar{f}(\alpha_1, k, r)]}, \tag{6.175}$$

where again W is the Wronskian.

6.7 Generalized Gel'fand–Levitan Approach to Inversion

In the standard Gel'fand–Levitan or Marchenko formulations of the inverse problem, we selected a given partial wave, (in this book $\ell = 0$), and with the input $\delta_0(k) \equiv \delta(k)$ we showed how it is possible to construct a unique local potential $v(r)$. We also discussed the Newton–Sabatier method where at a fixed energy $E = \frac{\hbar^2 k^2}{2m}$ from the angular dependence of the phase shift $\delta_\ell(k)$, we can determine the shape of the potential. Let us consider the general form of the Schrödinger equation which we write as

$$\left[-\frac{d^2}{dr^2} + \bar{w}(r) \right] \bar{\phi}(\gamma, r) = \gamma h(r) \bar{\phi}(\gamma, r), \tag{6.176}$$

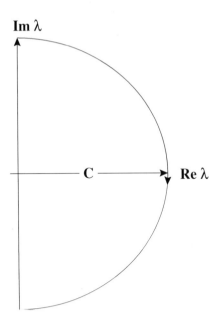

Figure 6.5: The Γ contour in the λ-plane.

where

$$\bar{w}(r) = \bar{v}(r) + \frac{\ell_0(\ell_0 + 1)}{r^2} - E_0. \tag{6.177}$$

In these relations ℓ_0 and E_0 are constants, γ is a parameter and $h(r) = \frac{a + br^2}{r^2}$. The boundary condition for the differential equation (6.176) is

$$\bar{\phi}(\gamma, r = 0) = 0. \tag{6.178}$$

Thus depending on the choice of a or b in $h(r)$ we can have different types of inverse problems. To see the connection between the generalized inverse problem that we want to study and the ones that we have already discussed let us consider these problems in the following way:

In the $E - \lambda^2$ plane, $(\lambda = \ell + \frac{1}{2})$ a straight line parallel to the vertical axis shows the information needed to solve the Gel'fand–Levitan or Marchenko inverse problems (fixed λ), whereas for the Newton–Sabatier method the information is needed along parallel to the horizontal axis (fixed E),(see Fig. 6.6). We can inquire whether we can solve the inverse problem when the input is given along a line with inclination, e.g. along the line $\alpha E + \beta \lambda^2$ in $E - \lambda^2$ plane [32]. To be more specific, let us consider the simple generalization of the Gel'fand–Levitan formulation by changing the integral equation (4.56) to the following

$$K(r, r') = g(r, r') - \int_0^r K(r, r'') h(r'') g(r'', r') \, dr'', \tag{6.179}$$

with the symmetrical input

$$g\left(r, r'\right) = \int \phi_0(\gamma, r)\phi_0\left(\gamma, r'\right) d\,\Gamma(\gamma). \tag{6.180}$$

The functions $\phi_0(\gamma, r)$ are the solutions of (6.176). As we have seen in Chapter 4, $K(r, r')$ in (6.179) is the kernel of the transformation from the known functions $\phi_0(\gamma, r)$ to the functions $\phi(\gamma, r)$;

$$\phi(\gamma, r) = \bar\phi(\gamma, r) - \int_0^r K\left(r, r'\right) h\left(r'\right) \bar\phi\left(\gamma, r'\right) dr'. \tag{6.181}$$

These set of $\phi(\gamma, r)$ s, in turn, are the solutions of the Schrödinger equation

$$\left[-\frac{d^2}{dr^2} + w(r)\right]\phi(\gamma, r) = \gamma h(r)\phi(\gamma, r), \tag{6.182}$$

with the boundary condition

$$\phi(\gamma, 0) = 0, \tag{6.183}$$

and with $w(r)$ given by

$$w(r) = v(r) + \frac{\ell_0(\ell_0 + 1)}{r^2} - E_0, \tag{6.184}$$

and

$$v(r) = \bar v(r) - 2\sqrt{h(r)}\frac{d}{dr}\left[\sqrt{h(r)}K(r, r)\right]. \tag{6.185}$$

The integral equation (6.179) reduces to the ordinary Gel'fand–Levitan equation if $h(r) = 1$. It also reduces to Newton–Sabatier equation, (6.7), if we choose $h(r) = \frac{1}{r^2}$.

Let us assume that

$$h(r) = \frac{a + br^2}{r^2}, \tag{6.186}$$

then the Schrödinger equation for this $h(r)$ is

$$\left[-\frac{d^2}{dr^2} + v(r) + \frac{\lambda_0^2 - \frac{1}{4}}{r^2} - E_0\right]\phi(\gamma, r) = \gamma\left(\frac{a + br^2}{r^2}\right)\phi(\gamma, r), \tag{6.187}$$

where

$$v(r) = \bar v(r) - \frac{2}{r}\sqrt{a + br^2}\frac{d}{dr}\left[\frac{1}{r}\sqrt{a + br^2}\,K(r, r)\right]. \tag{6.188}$$

The solution of the Schrödinger equation with this potential is obtained from

$$\phi(\gamma, r) = \bar\phi(\gamma, r) - \int_0^r \left[K\left(r, r'\right) + \frac{a + br'^2}{r'^2}\right]\bar\phi\left(\gamma, r'\right) dr'. \tag{6.189}$$

Bargmann Type Potentials — We can solve the Gel'fand–Levitan equation if the input $g\left(r, r'\right)$ is separable. Let us consider the simplest case where $g\left(r, r'\right)$ has a single term;

$$g\left(r, r'\right) = c_1\bar\phi(\gamma_1, r)\bar\phi\left(\gamma_1, r'\right). \tag{6.190}$$

Since for the separable $g(r, r')$, $K(r, r')$ also becomes separable, we have

$$K(r, r') = c_1 \phi(\gamma_1, r) \bar{\phi}(\gamma_1, r').$$ (6.191)

By substituting $g(r, r')$ and $K(r, r')$ in Eq. (6.185) we find $\phi(\gamma_1, r)$:

$$\phi(\gamma_1, r) = \frac{\bar{\phi}(\gamma_1, r)}{1 + c_1 \int_0^r \bar{\phi}^2(\gamma_1, r') h(r') dr'}.$$ (6.192)

This function $\phi(\gamma_1, r)$ is the bound state wave function for those values of γ_1 which corresponds to a negative energy.

To determine the potential, we first substitute (6.192) in (6.179) and find $K(r, r')$;

$$K(r, r') = \frac{c_1 \bar{\phi}(\gamma_1, r) \bar{\phi}(\gamma_1, r')}{1 + c_1 \int_0^r \bar{\phi}^2(\gamma_1, r'') h(r'') dr''},$$ (6.193)

and then from (6.188) we obtain $v(r)$;

$$v(r) = \bar{v}(r) - 2c_1 \sqrt{h(r)} \frac{d}{dr} \left[\frac{\sqrt{h(r)} \bar{\phi}^2(\gamma_1, r)}{1 + c_1 \int_0^r \bar{\phi}^2(\gamma_1, r') h(r') dr'} \right].$$ (6.194)

In addition, from Eq. (6.181) we find the wave function $\phi(\gamma, r)$ to be

$$\phi(\gamma, r) = \bar{\phi}(\gamma, r) - \frac{c_1 \phi_0(\gamma_1, r) \int_0^r \bar{\phi}(\gamma_1, r') h(r') \bar{\phi}(\gamma, r') dr'}{1 + c_1 \int_0^r \bar{\phi}^2(\gamma_1, r') h(r') dr'},$$ (6.195)

where γ is related to γ_1 by

$$(\gamma - \gamma_1) = \frac{W[\phi_0(\gamma, r), \phi(\gamma_1, r)]}{\int_0^r \bar{\phi}(\gamma_1, r') h(r') \bar{\phi}(\gamma, r') dr'}.$$ (6.196)

Here W is the Wronskian of $\phi_0(\gamma, r) \phi(\gamma_1, r)$. This relation can be obtained directly from the Schrödinger equation.

Now let us consider the special case where

$$\alpha E + \beta \lambda^2 = \text{constant}.$$ (6.197)

The potential in this case is found from (6.195) with $h(r) = \frac{\alpha + \beta r^2}{r^2}$

$$v(r) = v_0(r) - 2c_1 \frac{\sqrt{\alpha + \beta r^2}}{r} \frac{d}{dr} \left[\frac{\frac{\sqrt{\alpha + \beta r^2}}{r} \bar{\phi}^2(\gamma_1, r)}{1 + c_1 \int_0^r \bar{\phi}^2(\gamma_1, r') (\alpha + \beta r'^2) \frac{dr'}{r'^2}} \right].$$ (6.198)

For this example we can easily find the Jost function for the potential $v(r)$. For this we substitute $h(r) = \frac{\alpha + \beta r^2}{r^2}$ in Eq. (6.196) and take the limit as $\to \infty$ (see Eq. (4.15))

$$\phi(\gamma, r) \to \frac{1}{2ik} \left[f(\gamma) e^{ikr} - f(-\gamma) e^{-ikr} \right] \quad \text{as} \quad r \to \infty.$$ (6.199)

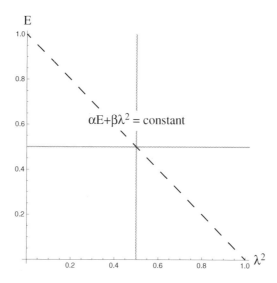

Figure 6.6: The (E, λ^2) values at which the knowledge of the S matrix is required in different inversion problems. In this figure the horizontal solid line corresponds to the inversion at fixed energy, E, and the vertical line corresponds to inversion at fixed $\lambda^2 = (\ell + \frac{1}{2})^2$, where ℓ is the angular momentum of the particle. The dashed line shows the case of general inversion scheme with linear dependence on λ and E values.

Thus we find the Jost function $f(\gamma)$ is related to the original Jost function $\bar{f}(\gamma)$ by [32]

$$f(-\gamma) = \bar{f}(-\gamma)\frac{k - i\kappa}{k + i\kappa}, \qquad (6.200)$$

where

$$-\kappa_1^2 = E_0 + \gamma_1\beta, \quad \text{and} \quad k^2 = E_0 + \gamma\beta. \qquad (6.201)$$

6.8   The Method of Schnizer and Leeb

Schnizer and Leeb have solved this problem by directly using the Jost solution for the Schrödinger equation [33]–[38], but we find the same results using the formulation of Zakhariev and Rudyak [31], [32].

Let us consider two specific solutions of the Schrödinger equation

$$\left[\frac{d^2}{dr^2} - \frac{\lambda_0^2 - \frac{1}{4}}{r^2} + k_0^2 - v_0(r)\right]\psi_0(\nu, r) = \nu^2 h(r)\psi_0(\nu, r), \qquad (6.202)$$

where ν, k_0 and λ_0 can be complex numbers. Now we define two solutions of (6.202) for $\nu = \gamma$ and $\nu = \alpha$ and we denote these solutions by η_0 and ζ_0, where

$$\psi_0(\gamma, r) = \eta_0(\gamma, r), \quad \psi_0(\alpha, r) = \zeta_0(\alpha, r). \qquad (6.203)$$

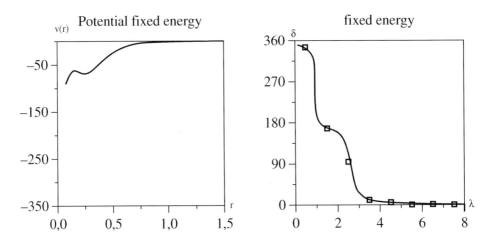

Figure 6.7: Results of the numerical calculation of Schnizer and Leeb for the potential function obtained from the generalized Bargmann method. For this calculation the parameters $\alpha_1 = \beta_1^* = 8.0 + 0.7i$ and $\alpha_2 = \beta_2^* = 3.0 + 0.1i$ have been used and the value of $\frac{b}{a} = 0$ is assumed [33]. This is the case of fixed energy. With no background potential, the figure to the left shows the radial dependence of the potential (measured in MeV) as a function of r (measured in fm). The figure on the right shows the dependence of the phase shift (measured in degrees) versus the parameter λ. The symbol ■ shows the half integer values of angular momentum [33].

From $\eta_0(\gamma, r)$ and $\zeta_0(\alpha, r)$, we can define a new function $\eta_1(\gamma, r)$ [34]:

$$\eta_1(\gamma, r) = C \frac{W[\eta_0(\gamma, r); \ \zeta_0(\alpha, r)]}{\sqrt{h(r)}\zeta_0(\alpha r)}, \tag{6.204}$$

where C is an arbitrary constant, and $W[\eta_0(\gamma, r); \ \zeta_0(\alpha, r)]$ is the Wronskian for these two functions. A straightforward calculation shows that $\eta_1(\gamma, r)$ is also a solution of (6.203), but now the potential $v_1(r)$ is given by

$$v_1(r) = v_0(r) - 2\sqrt{h(r)} \frac{d}{dr} \left[\frac{1}{\sqrt{h(r)}} \frac{d}{dr} \ln \left((h(r))^{\frac{1}{4}} \zeta_0(\alpha, r) \right) \right]. \tag{6.205}$$

Here we observe that the change in the potential function depends only on $\zeta_0(\alpha, r)$. If we choose $\eta_0(\gamma, r)$ and $\xi_0(\gamma, r)$, as two linearly independent solutions of (6.202), these two will yield two linearly independent solutions $\eta_1(\gamma, r)$ and $\xi_1(\gamma, r)$, of the Schrödinger equation with the potential $v_1(r)$. We can go one step further and construct a new function $\eta_2(\gamma, r)$ from these three known solutions, i.e. $\eta_0(\gamma, r)$, $\zeta_0(\alpha, r)$ and $\xi_0(\beta, r)$. For this we need the following extension of (6.204) [33]–[38].

$$\eta_2(\gamma, r) = \frac{1}{(h(r))^{\frac{1}{4}}} \frac{\tilde{\mathcal{B}}(r)}{W[\xi_0(\beta, r); \ \zeta_0(\alpha, r)]}, \tag{6.206}$$

where

$$\tilde{\mathcal{B}} = \tilde{C}(h(r))^{-\frac{3}{4}} \det \begin{bmatrix} \eta_0(\gamma,r) & \xi_0(\beta,r) & \zeta_0(\alpha,r) \\ \eta_0'(\gamma,r) & \xi_0'(\beta,r) & \zeta_0'(\alpha,r) \\ \eta_0''(\gamma,r) & \xi_0''(\beta,r) & \zeta_0''(\alpha,r) \end{bmatrix} \tag{6.207}$$

where primes denote derivatives with respect to r. The function $\eta_2(\gamma,r)$ is the solution of Eq. (6.202) with the potential

$$v_2(r) = v_0(r) - 2\sqrt{h(r)}\frac{d}{dr}\left\{\frac{1}{\sqrt{h(r)}}\frac{d}{dr}\ln W[\xi_0(\beta,r);\ \zeta_0(\alpha,r)]\right\}. \tag{6.208}$$

Schnizer and Leeb how shown how this approach can be extended further (see [33]–[38]). The particular choice of $\tilde{C} = \frac{1}{\gamma^2-\beta^2}$ is of special interest. For this choice one has

$$\eta_2(\gamma,r) = \eta_0(\gamma,r) - \left(\frac{\alpha^2-\beta^2}{\gamma^2-\beta^2}\right)\frac{W[\eta_0(\gamma,r);\ \xi_0(\beta,r)]}{W[\zeta_0(\alpha,r);\ \xi_0(\beta,r)]}\zeta_0(\alpha,r). \tag{6.209}$$

Next, let us examine the solution of this problem for the case where $h(r) = (a + br^2)/r^2$ and find the behaviour of the generalized Bargmann potential between the scattering problem at fixed energy and fixed angular momentum. Schnizer and Leeb considered the numerical solution of the iterated potentials with two complex conjugate parameter pairs [35]

$$\alpha_1 = \beta_1^* = 8.0 + 0.7i, \tag{6.210}$$

$$\alpha_2 = \beta_2^* = 3.0 + 0.1i. \tag{6.211}$$

This choice leads to a real potential and to a unitary S matrix. In Figs. 6.7 and 6.8 the radial dependence of two potentials for two different values of a and b are shown. Since the ratio $\frac{b}{a}$ is the quantity which enters in the calculation of the potentials, by varying them we can go from the fixed energy case $b = 0$ to $a = 0$ i.e. fixed angular momentum.

6.9 Analysis of Atom-Atom Scattering Using Complex Angular Momentum Formulation

We have seen that the S matrix which is parametrized as rational fraction of λ^2 ($\lambda = \ell+\frac{1}{2}$) is helpful in finding solvable potentials at fixed energy. Such a parametrization is also useful in describing the atom-atom scattering where it greatly facilitates the determination of phase shift from scattering data [39]. Thus we start with the construction of the phase shift δ_ℓ from experimental differential cross section $\sigma(\theta)$, and in order to have a classical picture, we calculate the corresponding deflection

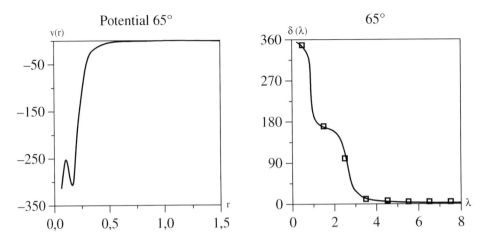

Figure 6.8: Same as in previous figure, except that the potential and phase shifts are shown for the value of $\frac{b}{a} = \tan\theta$ with $\theta = 65$ degrees [33].

angle, $\theta(\ell)$. Once the deflection angle is found as we have seen in Sec. 2.2 we can use the inverse scattering technique and determine the potential between two atoms. Rich *et al.* have used this method to determine the intermolecular potential from the low energy (less than 14 eV) differential scattering data. They have applied this method to calculate the atom-atom scattering for HeH^4, NeH^4 and KrH^4 [39].

In the direct method, one calculates δ_ℓ by the WKB method and substitutes these in the well-known expression

$$\sigma(\theta) = \frac{1}{4k^2}\left|\sum_\ell (2\ell + 1)P_\ell(\cos\theta)\left(e^{2i\delta_\ell} - 1\right)\right|^2 , \qquad (6.212)$$

and determines the angular dependence of the differential cross section. However there are two major issues in this direct method. First, it may be necessary to use a very large number of partial waves to get a reasonable result, and second, many important features in the differential cross section cannot be simply connected with the form of summand in (6.212). A better alternative to this procedure is to replace the sum in (6.212) by an integral and use the method of stationary phase to evaluate the integral. As we have seen earlier WKB approximation with the stationary phase method connects the phase shift to the classical deflection function $\theta(\ell)$ [20]–[21]:

$$\theta(\ell) = 2\frac{\partial\delta_\ell}{\partial\ell}. \qquad (6.213)$$

A more interesting way is to follow the work of Reimer–Regge and calculate the differential cross section in the following way which avoids the calculation of many terms in the partial wave sum and also without using stationary phase approximation [17]–[19].

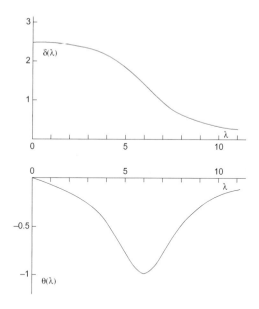

Figure 6.9: The phase shift $\delta(\lambda)$ and the deflection function $\theta(\lambda)$ for the problem where the S matrix has a single pole and a single zero Eqs. (6.221) with $\alpha = 6 + 2i$ and $\beta + \alpha^* = 6 - 2i$ [22].

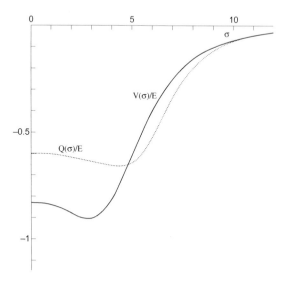

Figure 6.10: The potential $V(\sigma)$ which is obtained from semiclassical inversion of the λ dependence of the deflection function. The parameters used in this calculation are the same as those given in Fig. 6.9 and since $\beta = \alpha^*$ the potential is real. The dotted line represents the quasi-potential $Q(\sigma)$, Eqs. (2.25)–(2.27) [22].

Regge observed that by multiplying the summand in (6.212) by $\frac{1}{\sin(\ell\pi)}$ for integer ℓ and integrating the product over a contour along the positive real ℓ axis one finds a result identical to (6.212). Now if the S matrix contains a finite number N of singularities which are in the first quadrant of the complex ℓ-plane, the contour of integration may be deformed to include this finite set of poles, and $f(\theta)$ may be written as a finite sum. As we have seen in the work of Lipperheide and Fiedeldey, we can write the S matrix such that it is unitary, symmetric and contain only first order poles [22]

$$S_N = \sum_{n=1}^{N} S_n = \sum_{n+1}^{N} \frac{\lambda^2 - \alpha_n^{*\,2}}{\lambda^2 - \alpha_n^2}, \quad \lambda = \ell + \frac{1}{2}, \tag{6.214}$$

so that

$$S_n(\lambda) = S_n(-\lambda), \quad \text{and} \quad S_n(\lambda) = [S_n^*(\lambda)]^{-1}. \tag{6.215}$$

Noting that

$$S_n = e^{2i\delta_n}, \quad \text{or} \quad \delta_n = \frac{1}{2i}\ln S_n, \tag{6.216}$$

then from (6.213) we find the deflection angle arising from the nth pole to be

$$\theta_n = 2\frac{\partial \delta_n}{\partial \lambda} = -\frac{4\lambda\,\text{Im}\,\alpha_n^2}{|\lambda^2 - \alpha_n^2|^2}. \tag{6.217}$$

Now the total phase δ is the sum of contributions, one from each pole

$$\delta(\lambda) = \sum_{n=1}^{N} \delta_n(\lambda), \tag{6.218}$$

where

$$\delta_n(\lambda) = -\arg(\lambda - \alpha_n) - \arg(\lambda + \alpha_n). \tag{6.219}$$

Thus the total deflection function is given by

$$\theta(\lambda) = \sum_{n} \theta_n(\lambda), \tag{6.220}$$

with

$$\theta_n(\lambda) = -\frac{4\lambda\,\text{Im}\,\alpha_n^2}{|\lambda^2 - \alpha_n^2|^2}. \tag{6.221}$$

Once $\theta(\lambda)$ is known then we can find the potential using semiclassical approximation discussed in Sec. 2.2, Eqs. (2.25)–(2.27).

References

[1] R.G. Newton *Scattering Theory of Waves and Particles*, Second Edition (Springer, New York, 1982).

[2] A.G. Ramm, Comm. Math. Phys. 207, 231 (1999).

[3] A.G. Ramm, App. Analysis, 81, 965 (2002).

[4] A.G. Ramm, Comments on the Letter of P.C. Sabatier, Google Scholar, 2003.

[5] A.G. Ramm, Mod. Phys. Lett. B 22, 2217 (2008).

[6] P.C. Sabatier, Mod. Phys. Lett. B 23, 2069 (2009).

[7] P.C. Sabatier and F. Quyen Van Phu, Phys. Rev. D4, 127 (1971).

[8] K. Chadan and P.C. Sabatier, *Inverse Problems in Quantum Scattering Theory*, (Spring-Verlag, New York, 1989).

[9] V.M. Markushevich, N.N. Novikova and M. Razavy, Can. J. Phys. 68, 1429 (1990).

[10] R.E. Langer, Phys. Rev. 51, 669 (1937).

[11] G.A. Baker, Jr., *Essentials of Padé Approximants*, (Academic Press, New York, 1975).

[12] M. Hron and M. Razavy, J. Stat. Phys., 38, 655 (1985).

[13] B.M. Levitan and I.S. Sargsjan, *Introduction to the Spectral Theory*, (American Mathematical Society, Providence, 1975) Chapter 2.

[14] V.M. Markushevich and N.N. Novikova, Comp. Seismology, 18, 173 (1985).

[15] I.S. Gradshteyn and I.M. Ryzhik, *Tables of Integrals, Series, and the Products*, (Academic Press, New York, 1965) p. 1057.

[16] Z.W. Gortel and A. Wierzbicki, Surface Sc. Lett. 239, L565 (1990).

[17] E.A. Reimer, Phys. Rev. A 3, 1949 (1971).

[18] T. Regge, Nuovo Cimento, 24, 139 (1979).

[19] V. D'Alfaro and T.E. Regge, *Potentia Scattering*, (North-Holland, Amsterdam, 1965).

[20] R. Bellman and R. Vasudevan in *Mathematics and Applications*, vol. 385, (Springer, New York, 1977).

[21] M. Razavy, *Quantum Theory of Tunneling*, Second Edition, (World Scientific, Singapore, 2014) Chapter 2.

[22] R. Lipperheide and H. Fiedeldey, Z. Phys. A286, 45 (1978).

[23] R. Lipperheide and H. Fiedeldey, Z. Phys. A301, 81 (1981).

[24] R. Lipperheide, S.A. Sofianos and H. Fiedeldey, Phys. Rev. C 26, 770 (1982),

[25] R. Lipperheide and H. Fiedeldey, Nucl. Phys. A 419, 13 (1984).

[26] B.N. Zakhariev and A.A. Suzko, *Direct and Inverse Problems: Potentials in Quantum Scattering* (Springer-Verlag, Berlin 1990) p. 84.

[27] G. Burdet, M. Giffon and E. Predazzi, Nuovo Cimento, 36, 1337 (1965).

[28] G. Burdet, M. Giffon and E. Predazzi, Nuovo Cimento, A44, 138 (1966).

[29] A. Bottino, A. Longoni and T. Regge, Nuovo Cimento, 23, 954 (1961).

[30] K. Chadan and P.C. Sabatier, *Inverse Problems in Quantum Scattering Theory*, (Spring-Verlag, New York, 1989) p. 222.

[31] B.N. Zakhariev and A.A. Suzko, *Direct and Inverse Problems: Potentials in Quantum Scattering* (Springer-Verlag, Berlin 1990) p. 88.

[32] B.V. Rudyak and B.V. Zakhariev, Inverse Problems, 3, 125 (1987).

[33] W.A. Schnizer and H. Leeb, *Inverse Methods in Action*, Edited by P.C. Sabatier, (Springer-Verlag, New York 1990) p. 456.

[34] G. Darboux, C.R. Acad. Sci. (Paris) 94, 1456 (1882).

[35] W.A. Schnizer and H. Leeb, J. Phys. A 26, 5145 (1993).

[36] H. Leeb in *Quantum Inversion Theory and Application*, Edited by H.V. von Germab, (Springer-Verlag, Berlin 1994) p. 241.

[37] H. Leeb, W.A. Schnizer, H. Fiedeldey, S.A. Sofianos, and R. Lipperheide, Inverse Problems, 5, 817 (1989).

[38] W.A. Schnizer and H. Leeb, J. Phys. A 27, 2606 (1994).

[39] W.G. Rich, S.M. Bobbio, R.L. Champion and L.D. Doverspike, Phys. Rev. A 4, 2253 (1971).

Chapter 7

Discrete Forms of the Schrödinger Equation and the Inverse Problem

We know that in quantum theory there are few exactly solvable two-body problems with local potentials, where the phase shift for each partial wave can be written exactly for all energies. Even less are the examples where at fixed energy we know the phase shift for all partial waves [1], [2]. For almost all other problems we have to obtain numerical solutions of quantities of interest from the appropriate wave equation, or resort to one of the large number of approximate techniques. For the numerical solution of Schrödinger equation, or wave equation, in general, we replace the given differential equation by a difference equation, the latter is chosen in such a way that as the step size tends to zero, the differential equation is recovered. Similar procedure can be applied to the solution of wave equation in two or three dimensions when the system is not separable. In the direct problem, we can replace a linear differential equation by a nonlinear equation if we can get a more accurate result or if it is easier way to calculate certain properties of the system. For instance we can use the variable phase method, a nonlinear first order differential equation to calculate the phase shift for a given partial wave [3].

While in most of the physical systems that we encounter the direct problem is given in terms of differential equation(s), there are instances where the direct problem can be expressed as a difference equation. For example the problem of light reflection from a stack of glass plates was formulated and solved by Stokes. This problem is expressed as a difference equation [4]. A very important case of such a physical system which is the reflection of elastic waves from a layered medium where each layer has a constant travel time for the propagating wave has been discussed by Goupillaud (see Sec. 13.4).

153

Linear Difference Equation Approximating the Wave Equation — A straight forward replacement of derivatives by finite differences allows us to solve the direct and in many cases the inverse problem. Zakhariev and Suzko in their book show the remarkable similarity between solving the inverse problem for the Schrödinger equation and the Gel'fand–Levitan's and Marchenko's methods [5].

An extensive study of the connection between the standard methods of inversion in quantum mechanics and a particular discrete form of the Schrödinger equation has been made by Case and Kac, and in this section we present their approach which is very interesting both for its mathematical elegance and its physical utility. There are many ways that we can construct discrete forms of the Schrödinger equation and in this chapter we will consider some of these forms which can be used to solve inverse problems. Let us mention a few of these difference equations which can be used for solving inverse problems (see references [6]–[8]). For instance an asymmetric discrete form of the Scrödinger equation which is suitable for calculation of the Jost function in both direct and inverse problems [8], or the method of equal travel time mentioned before. As we will see in later chapters, we can also use difference equations to find the potential from the partial phase shifts all given at a fixed energy.

7.1   Zakhariev's Method

A simple approach for the derivation of the discrete version of the Gel'fand–Levitan equation has been formulated by Zakhariev, and will be studied in this section [10]. In order to simplify the formulae in what follows we set $\frac{\hbar^2}{m}$ equal to unity and write the Schrödinger equation as

$$\frac{1}{2}\frac{d^2\phi(E,x)}{dx^2} - v(x)\phi(E,x) = -E\phi(E,x) \tag{7.1}$$

with the boundary conditions

$$\phi(0) = 0, \quad \text{and} \quad \phi(1) = \Delta. \tag{7.2}$$

In the absence of potential we can replace (7.1) by the difference equation

$$\bar{\phi}(E,n+1) - 2\bar{\phi}(E,n) + \bar{\phi}(E,n-1) = 2E\bar{\phi}(E,n), \tag{7.3}$$

and the solution of this equation is given by (7.22), where λ is given by $\lambda = 1 - E\Delta^2$, (see Eq. (7.12) below), then for $-1 < \lambda < 1$, as we will see shortly, Eq. (7.22) shows that $\bar{\phi}(E,n)$ s are polynomials of the $(n-1)$th order in the energy E. Since

$$\phi(E,0) = \bar{\phi}(E,0) = 0, \quad \text{and} \quad \phi(E,1) = \bar{\phi}(E,1) = \Delta, \tag{7.4}$$

the first two members of these sets of polynomials are independent of E.

Starting with (7.4), we observe that at the next step, i.e. for $\phi(E, 2)$ and, $\bar{\phi}(E, 2)$ the factor E will increase the order of the energy dependence of the polynomials by one, however their coefficients will be different. We can construct $\phi(E, n)$ by writing it as a linear combination of the known polynomials $\bar{\phi}(E, m)$ of the orders, $m - 1 = n - 1$, $n - 2, \cdots 0$, with the coefficients $K(n.m)$ containing all the information about the interaction.

$$\phi(E, 1) = -\Delta K(1, 1)\bar{\phi}(E, 1), \tag{7.5}$$

$$\phi(E, 2) = -\Delta K(2, 2)\bar{\phi}(E, 2) - \Delta K(2, 1)\bar{\phi}(E, 1), \tag{7.6}$$

$$\cdots\cdots\cdots\cdots\cdots\cdots\cdots\cdots\cdots\cdots\cdots\cdots\cdots$$

$$\phi(E, n) = -\Delta K(n, n)\bar{\phi}(E, n) - \Delta K(n, n - 1)\bar{\phi}(E, n - 1) - \cdots - \Delta K(n, 1)\bar{\phi}(E, 1). \tag{7.7}$$

Now let us consider the limit of (7.7) as $\Delta \to dx$, and $n \to x$. In this limit the summation over n is replaced by integration, i.e.

$$\phi(E, x) = \bar{\phi}(E, x) - \int_0^x K(x, y)\bar{\phi}(E, y)dy. \tag{7.8}$$

This equation is the same as the Gel'fand–Levitan equation (4.62).

7.2 The Method of Case and Kac for Discrete Form of Inverse Scattering Problem

We assume that the boundary conditions are given at the origin for the physical solution of the wave equation

$$\frac{1}{2}\frac{d^2\psi(E, x)}{dx^2} - v(x)\psi(E, x) = -E\psi(E, x), \tag{7.9}$$

with the initial conditions given at $x = 0$;

$$\psi(E, 0) = 0, \quad \left(\frac{d\psi}{dx}\right)_{x=0} = 1. \tag{7.10}$$

The simplest form of difference equation which can replace (7.9) is of the form

$$\frac{1}{2\Delta^2}\left[\psi(E, (n + 1)\Delta) - 2\psi(E, n\Delta) + \psi(E, (n - 1)\Delta)\right] + [E - v(n\Delta)]\psi(E, n\Delta) = 0. \tag{7.11}$$

However to solve the inverse problem in such a way that we can establish a connection between the discrete version of the Schrödinger equation and the discrete form of the Gel'fand–Levitan integral equation, we follow the work of Case and Kac and start with the following difference equation [11]–[13]:

$$\frac{1}{2}\psi(E, (n+1)\Delta) + \frac{1}{2}\psi(E, (n-1)\Delta) = \left(1 - E\Delta^2\right)\exp\left[v(n\Delta)\Delta^2\right]\psi(E, n\Delta). \tag{7.12}$$

In this equation we replace E by λ where

$$\lambda = 1 - E\Delta^2, \tag{7.13}$$

and then $\psi(E, n)$ can be written as $\psi(\lambda, n)$. In the limit of $\Delta \to 0$, the equation like (7.11) goes over into Schrödinger equation (7.9) provided that in (7.12) we choose

$$q(n) = \Delta^2 v(n\Delta). \tag{7.14}$$

By setting

$$\phi(\lambda, n) = \exp\left[\frac{1}{2}\Delta^2 v(n\Delta)\right] \psi(E, n) \tag{7.15}$$

we can change (7.12) into the following difference equation:

$$\frac{1}{2}\exp\left[-\frac{1}{2}(q(n) + q(n+1))\right]\phi(\lambda, n+1)$$
$$+ \frac{1}{2}\exp\left[-\frac{1}{2}(q(n-1) + q(n))\right]\phi(\lambda, n-1) = \lambda\phi(\lambda, n), \tag{7.16}$$

which we can write in the matrix form as,

$$A\phi = \lambda\phi, \tag{7.17}$$

where A is a symmetric tri-diagonal matrix with elements $a_{m,n}$ $(m, n \geq 1)$ and these elements are given by

$$a_{m,n} = \frac{1}{2}\exp\left\{-\left[\frac{1}{2}(q(m) + q(n))\right]\right\}\delta_{m,n+1}. \tag{7.18}$$

In order to simplify the algebra we set

$$q(0) = 0. \tag{7.19}$$

The boundary condition $\phi(\lambda, 0)$ is satisfied automatically, and for the other boundary condition rather than choosing $\phi(\lambda, 1) = \Delta$, we set

$$\phi(\lambda, 1) = 1, \tag{7.20}$$

and remember that with this choice of the boundary condition we cannot take the limit of $\Delta \to 0$.

Next, let us consider the case where $q(n) = 0$ for all n s, and let us denote the matrix A for this choice of $q(n)$ by $\bar{A}$ and its elements by $\bar{a}_{m.n}$, then we have

$$\bar{a}_{m.n} = \frac{1}{2}\left(\delta_{m,n-1} + \delta_{m,n+1}\right). \tag{7.21}$$

For this case $\bar{\phi}(\lambda, n)$ can be found exactly;

$$\bar{\phi}(\lambda, n) = \frac{\left[\lambda + \sqrt{\lambda^2 - 1}\right]^n - \left[\lambda - \sqrt{\lambda^2 - 1}\right]^n}{2\sqrt{\lambda^2 - 1}}, \tag{7.22}$$

and it satisfies the recursion relation

$$\frac{1}{2}\left[\bar{\phi}(\lambda, n+1) + \bar{\phi}(\lambda, n-1)\right] = \lambda\bar{\phi}(\lambda, n). \tag{7.23}$$

The spectral distribution $\bar{\rho}(\lambda)$ for the $\bar{\phi}(\lambda, n)$ is given by [11]

$$\bar{\rho}(\lambda) = \begin{cases} 0 & \lambda < -1 \\ \frac{2}{\pi}\int_{-1}^{\lambda}\sqrt{1-\mu^2}\, d\mu & -1 < \lambda < 1 \\ 1 & \lambda > 1. \end{cases} \tag{7.24}$$

If the spectral distribution of A is $\rho(\lambda)$ then

$$\frac{\int \bar{\phi}(\lambda, m)\bar{\phi}(\lambda, n)\, d\rho(\lambda)}{d\lambda d\lambda} = \delta_{m,n}, \tag{7.25}$$

and

$$\int \lambda\phi(\lambda, m)\phi(\lambda, n)\, d\rho(\lambda) = a_{m,n}. \tag{7.26}$$

Thus the set $\phi(\lambda, n)$ form a set of orthogonal polynomials, properly normalized with the weight function $\rho(\lambda)$.

The Inverse Problem — Given $\rho(\lambda)$ it is rather simple to determine the orthogonal polynomials with the weight function $\rho(\lambda)$ and then calculate $a_{n,n+1}, \cdots$. Let us note two important points in this regard:
(1) We observe that

$$a_{n,n+1} = \frac{1}{2}\exp\left[-\frac{1}{2}[q(n) + q(n+1)]\right], \tag{7.27}$$

is a positive quantity, and we have the condition that the orthogonal polynomials satisfy the relation

$$\int \lambda\phi(\lambda, n)\phi(\lambda, n+1)\, d\rho(\lambda) > 0. \tag{7.28}$$

(2) The weight function $\rho(\lambda)$ cannot be prescribed quite arbitrarily, since we must have

$$a_{n,n} = \int \lambda\phi^2(\lambda, n)\, d\rho(\lambda) = 0. \tag{7.29}$$

Therefore to insure that (7.29) is satisfied we choose $\rho(\lambda)$ such that

$$\rho(-\lambda) = -\rho(\lambda), \tag{7.30}$$

and this guarantees that the orthogonal polynomials are either even or odd [11].

Gel'fand–Levitan Type Equation for Discrete Schröfinger Equation —
Here, just as in the formulation of the Gel'fand–Levitan approach we assume that

the spectral function $\rho(\lambda)$ is known and we want to determine the set of orthogonal polynomials of the form

$$\phi(\lambda, n) = K(n.n)\bar{\phi}(\lambda, n) + \sum_{m=1}^{n-1} K(n, m)\bar{\phi}(\lambda, m). \tag{7.31}$$

The polynomial $\phi(\lambda, n)$ which is of degree $n - 1$ is orthogonal to every polynomial of degree lower that $n - 1$ and hence to every $\bar{\phi}(\lambda, n)$ for $m < n$. Thus we have the orthogonality relation

$$\int \phi(\lambda, n)\bar{\phi}(\lambda, m)\, d\rho(\lambda) = 0, \quad m < n. \tag{7.32}$$

Now following Gel'fand and Levitan let us introduce the matrix $q(m, j)$ by

$$q(m, j) = \int \bar{\phi}(\lambda, m)\bar{\phi}(\lambda, j)\, d[\rho(\lambda) - \bar{\rho}(\lambda)], \tag{7.33}$$

and write (7.31) in the form of

$$K(n, n)q(n, m) + K(n, m) + \sum_{j=1}^{n-1} K(n, j)q(j, m), \quad n > m, \tag{7.34}$$

which resembles the Gel'fand–Levitan equation. The normalization condition

$$\int \phi^2(\lambda, n)\, d\rho(\lambda) = 1, \tag{7.35}$$

is equivalent to the condition

$$K(n, n) \int \phi(\lambda, n)\bar{\phi}(\lambda, n)\, d\rho(\lambda) = 1. \tag{7.36}$$

Using Eqs. (7.31) and (7.33) we can write this condition as

$$\frac{1}{K(n, n)} = K(n, n)[1 + q(n, n)] + \sum_{j=1}^{n-1} K(n, j)q(j, n). \tag{7.37}$$

Defining $\kappa(n, m)$ by

$$\kappa(n, m) = \frac{K(n, m)}{K(n, n)}, \quad m < n, \tag{7.38}$$

we can write (7.34) as

$$q(n, m) + \kappa(n, m) + \sum_{j=1}^{n-1} \kappa(n, j)q(j, m) = 0, \tag{7.39}$$

and these form a system of $(n-1)$ linear equations from which $\kappa(n, m)$ s, $1 \leq m \leq n-1$ are uniquely determined. Also we can write the normalization condition (7.37) as

$$\frac{1}{K^2(n, n)} = [1 + q(n, n)] + \sum_{j=1}^{n-1} K(n, j) q(j, n). \qquad (7.40)$$

This last relation gives us $K^2(n, n)$, but $K(n, n)$ can have positive or negative sign. We can determine this sign by observing that the recursion relation (7.16) implies that the leading coefficient of $\phi(\lambda, n)$ which is $K(n, n)$ has to be positive, therefore we choose the positive sign.

Having found $\kappa(n, n)$, we note that

$$a_{n,n+1} = \int \lambda \phi(\lambda, n) \phi(\lambda, n+1) \, d\rho(\lambda)$$

$$= K(n, n) \int \lambda \bar{\phi}(\lambda, n) \phi(\lambda, n+1) \, d\rho(\lambda). \qquad (7.41)$$

Since

$$\lambda \bar{\phi}(\lambda, n) = \frac{1}{2} \bar{\phi}(\lambda, n+1) + \frac{1}{2} \bar{\phi}(\lambda, n-1), \qquad (7.42)$$

therefore by using Eq. (7.36), but with n being replaced by $n+1$, we get

$$a_{n.n+1} = \frac{1}{2} K(n, n) \int \bar{\phi}(\lambda, n+1) \phi(\lambda, n+1) \, d\rho(\lambda) = \frac{1}{2} \frac{K(n, n)}{K(n+1, n+1)}, \qquad (7.43)$$

or equivalently

$$\frac{1}{2} [q(n) + q(n+1)] = \ln K(n+1, n+1) - \ln K(n, n). \qquad (7.44)$$

Equation (7.44) does not in general determine $q(n) = \Delta^2 v(\Delta n)$ but if we know that

$$\lim_{n \to \infty} q(n) = 0, \qquad (7.45)$$

then we can find $q(n)$ uniquely. Thus we can express $q(n+1)$ in terms of $\frac{K(m+1,m+1)}{K(m,m)}$ in the following way;

$$\frac{1}{2} [q(1) \pm q(n+1)] = \sum_{m=1}^{n} (-1)^{m+1} \ln \frac{K(m+1, m+1)}{K(m, m)}, \quad q(n) = v(\Delta n)\Delta^2. \qquad (7.46)$$

Thus we have the solution of the inverse problem, i.e. if we are given $\rho(\lambda)$, we can find $v(r)$.

Gel'fand–Levitan Equation as the Limit of the Case and Kac Formulation
— By taking the limit of $n \to \infty$ and noting that $q(n \to \infty) \to 0$ we find

$$\frac{1}{2} q(1) = \sum_{n=1}^{\infty} (-1)^{r+1} \ln \frac{K(r+1, r+1)}{K(r, r)}, \qquad (7.47)$$

where the convergence of (7.47) is a consequence of the convergence of Eq. (7.45).

From Eq. (7.13) it is evident that for $-1 < \lambda < 1$, the energy E must be positive, and since we have set $\frac{\hbar^2}{m} = 1$, we have

$$E = \frac{1}{2}k^2. \tag{7.48}$$

Now we define θ by

$$\lambda = \cos\theta, \quad \text{or} \quad \theta = \cos^{-1}\left(1 - 2\Delta k^2\right), \tag{7.49}$$

and then from (7.22) we find

$$\bar{\phi}(\lambda, m) = \frac{\sin(m\theta)}{\sin\theta}. \tag{7.50}$$

Also for small Δ and fixed E we have

$$\theta = k\Delta + \mathcal{O}\left(\Delta^2\right). \tag{7.51}$$

Therefore we can write Eq. (7.31) as

$$\Delta\phi(k, n\Delta) \sim K(n, n)\Delta\frac{\sin(nk\Delta)}{\sin(k\Delta)} + \Delta\sum_{m=1}^{n-1} K(n, m)\frac{\sin(mk\Delta)}{\sin(k\Delta)}. \tag{7.52}$$

Noting that the assumption that the series $\sum_{n=1}^{\infty} q(n)$ converges, implies that the term $K(n, n)$ will also tend to a finite limit as $n \to \infty$. Therefore in the limit

$$\Delta \to 0, \quad n\Delta \to x, \tag{7.53}$$

Eq. (7.52) will go over into

$$\psi(k, x) = a\frac{\sin(kx)}{k} + \int_0^{\pi} K(x, \xi)\frac{\sin(k\xi)}{k}\,d\xi, \tag{7.54}$$

where $\psi(k, x)$ is related to the physical wave function $\phi(k, x)$ by Eq. (7.15). In addition to satisfy the boundary condition $\left(\frac{d\psi(k, x)}{dx}\right)_{x=0} = 1$, we have to set $\alpha = 1$. With this choice of α (7.54) becomes similar to the Gel'fand–Levitan equation (4.62).

When bound states are present one can use the same approach, details of this generalization can be found in the paper of Case and Kac [11].

The Algebraic Version of the Gel'fand-Levitan Equation — For discrete set of states $\bar{\phi}(E, m)$ and $\phi(E, n)$ we have the completeness relations

$$\sum_{\nu=1}^{N} \bar{C}_{\nu}^2\bar{\phi}(\bar{E}_{\nu}, m)\bar{\phi}(\bar{E}_{\nu}, n) = \frac{1}{\Delta}\delta_{mn}, \tag{7.55}$$

$$\sum_{\nu=1}^{N} C_\nu^2 \phi(E_\nu, m)\phi(E_\nu, n) = \frac{1}{\Delta}\delta_{mn}. \tag{7.56}$$

Here E_ν and $\bar{E}_\nu$ are the eigenvalues of (7.11) with and without $v(n\Delta)$ respectively, and C_ν and $\bar{C}_\nu$ are the normalization constants,

$$\sum_{n=1}^{N} C_\nu{}^2 \phi^2(E_\nu, n)\Delta = 1, \tag{7.57}$$

and

$$\sum_{n=1}^{N} \bar{C}_\nu^2 \bar{\phi}^2(E_\nu, n)\Delta = 1. \tag{7.58}$$

In addition we have the orthogonality of the eigenfunctions:

$$\sum_{\nu=1}^{N} \psi(E_\nu, n)\psi(E_\nu, m) = \frac{1}{\Delta}\delta_{nm} = \sum_{\nu=1}^{N} C_\nu{}^2 \phi(E_\nu, n)\phi(E_\nu, m). \tag{7.59}$$

The two relations (7.59) and (7.56) are the orthogonality conditions with respect to the variable E_ν of the sets of vectors ϕ_n and $\bar{\phi}_n$ with two different weight factors C_ν^2 and $\bar{C}_\nu^2$ respectively. Thus if the set $\bar{\phi}$ is known and it spans the N dimensional space, we want to choose this set and find the coefficients of the expansion of the desired set, i.e. ϕ in terms of $\bar{\phi}$. Using the condition for the orthonormality, we can determine the coefficients of the expansion $K(m, n)$.

Let us choose the first vector $\phi(E, 1) = \bar{\phi}(E, 1)$ according to (7.4). We find the coefficient $K(2, 1)$ for the second vector $\phi(E, 2)$ from the orthogonality relation, viz, $\phi(E, 2) \perp \phi(E, 1)$ which is equivalent to $\phi(E, 2) \perp \bar{\phi}(E, 1)$. Next we determine $\phi(E, 3)$ by requiring that $\phi(E, 3) \perp \phi(E, 1)$ and $\phi(E, 3) \perp \bar{\phi}(E, 1)$. But $\bar{\phi}(E, 2)$ is a combination of $\bar{\phi}(E, 1)$ and $\bar{\phi}(E, 2)$, and from $\phi(E, 3) \perp \phi(E, 2)$ it follows that $\phi(E, 3) \perp \bar{\phi}(E, 2)$. The two conditions $\phi(E, 3) \perp \bar{\phi}(E, 1)$ and $\phi(E, 3) \perp \bar{\phi}(E, 2)$ determine $K(3, 1)$ and $K(3, 2)$. By continuing this process we can find the coefficients $K(n, m)$ from Eq. (7.8), i.e. by orthogonalizing $\phi(E, n)$ with respect to all $\bar{\phi}(E, m)$ with $m < n$. Thus we have

$$\sum_{\nu=1}^{N} C_\nu^2 \phi(E_\nu, n)\bar{\phi}(E_\nu, m) = 0, \quad m < n. \tag{7.60}$$

Now if we substitute from (7.7) in (7.60) we find a linear algebraic equation for $K(n, m)$:

$$K(n, m) = Q(n, m) - \sum_{j=1}^{N} \Delta K(n, j)Q(j, m) = 0, \quad m < n \tag{7.61}$$

which is the algebraic analogue of the Gel'fand–Levitan equation. In this relation $Q(n, m)$ is given by

$$Q(n, m) = \sum_{j=1}^{N} C_\nu^2 \phi(E_\nu, n)\phi(E_\nu) - \frac{1}{\Delta}\delta_{nm}. \tag{7.62}$$

Now if we replace $\frac{1}{\Delta}\delta_{nm}$ by the sum in (7.60) we find

$$Q(n,m) = \sum_{j=1}^{N} \left(C_\nu^2 - \bar{C}_\nu^2\right) \bar{\phi}(E_\nu, n)\bar{\phi}(E_\nu). \tag{7.63}$$

We can also derive (7.61) in a way which is more suitable for taking the limit of $\Delta \to 0$. For this, we multiply (7.7) at $E = E_\nu$ by $\bar{\phi}(E_\nu, j)\bar{C}_\nu^{\,2}$ and sum over ν making use of (7.60), and for the coefficients $K(n,m)$ we obtain

$$K(n,j) = \sum_{\nu=1}^{N} \bar{C}_\nu^2 \phi(E_\nu)\bar{\phi}(\bar{E}_\nu). \tag{7.64}$$

Next we subtract (7.60) from this equation to find

$$K(n,m) = -\sum_{\nu=1}^{N} C_\nu^2 \phi(E_\nu, n)\bar{\phi}(\bar{E}_\nu, m) + \bar{C}_n^2 \phi(\bar{E}_\nu)\bar{\phi}(\bar{E}_\nu). \tag{7.65}$$

Defining the spectral function by

$$\begin{cases} d\rho(E) = \sum_\nu C_\nu^2 \delta(E - E_\nu)dE \\[2mm] d\bar{\rho}(E) = \sum_\nu \bar{C}_\nu^2 \delta\left(E - \bar{E}_\nu\right) dE \end{cases}, \tag{7.66}$$

we can write (7.65) as

$$K(n,m) = -\int_{-\infty}^{\infty} \phi(E, n)\bar{\phi}(E, m)d\left[\rho(E) - \bar{\rho}(E)\right]. \tag{7.67}$$

This expression should be compared with Eq. (4.77) of the Gel'fand–Levitan formulation.

To complete our solution of the inverse problem, we want to show how the potential can be obtained if we know $K(n,m)$. For this we substitute $\phi(E_\nu, n)$, Eq. (7.7) with $E = E_\nu$ into the finite difference form of the Schrödinger equation (7.9), multiply both sides of the equation by $\bar{C}_\nu^2 \bar{\phi}^2(\bar{E}, \nu)$, sum over E_ν making use of Eq. (7.60), and the fact that $\bar{\phi}$ satisfies (7.9) with $v(n\Delta) \equiv 0$, we find

$$v(n\Delta) = \frac{1}{2\Delta}\left[K(n+1) - K(n) + K(n) - K(n, n-1)\right]. \tag{7.68}$$

In the limit of $\Delta \to 0$, $v(r) \to -2\frac{dK(r,r)}{dr}$ (or $V(r) = -\frac{\hbar^2}{m}\frac{dK(r,r)}{dr}$ with the factors $\hbar$ and m included). This expression should be compared with Eq. (4.77) of the Gel'fand–Levitan formulation.

7.3 Discrete Form of the Spectral Density for Solving the Inverse Problem on Semi-axis $0 \leq r < \infty$

While for the Gel'fand–Levitan formulation the spectral density had a simple relation to the Jost function, as we will see for discrete form of the Schrödinger equation it is not so simple.

In this section we try to outline the steps needed to find $d\rho(\lambda)$ without giving all the details, but we will show the basic arguments leading to the final result. First let us remind ourselves of the form of the spectral density obtained earlier (Chapter 4) by considering the completeness relation. Setting $\frac{\hbar^2}{m} = 1$, we can write the completeness relation as

$$\int_0^\infty \frac{1}{\sqrt{E}} \psi(E,r)\psi^*(E,r')\, dE + \sum_\nu \psi(E_\nu,r)\psi(E_\nu,r') = \delta(r-r'), \qquad (7.69)$$

where the normalized wave function $\psi(E,r)$ and $\phi(E,r)$ differ from each other by the normalization constant $C(E)$;

$$\psi(E,r) = C(E)\phi(E,r), \qquad (7.70)$$

and where $\phi(E,r)$ is related to the Jost solution by Eq. (4.15). The completeness relation can be written as

$$\int_{-\infty}^\infty \phi(E,r)\phi(E,r')\, d\rho(E) = \delta(r-r'), \qquad (7.71)$$

with

$$\frac{d\rho(E)}{dE} = \begin{cases} \frac{\sqrt{E}}{\pi|f(E)|^2} & E \geq 0 \\[2mm] \sum_\nu C_\nu^2 \delta(E - E_\nu) & E < 0. \end{cases} \qquad (7.72)$$

For the finite difference problem this spectral density will take a special form which we will be discussing now.

We observe that for large enough n, $q(n) = 0$, therefore in this limit we can write

$$\phi(\lambda, n) = \alpha(\lambda)\left[\lambda + \sqrt{\lambda^2 - 1}\right]^n + \beta(\lambda)\left[\lambda - \sqrt{\lambda^2 - 1}\right]^n, \qquad (7.73)$$

where $\alpha(\lambda)$ and $\beta(\lambda)$ are two functions to be determined. Now for $-1 < \lambda < 1$, we can replace λ by $\cos\theta$, then (7.73) takes the form

$$\phi(\theta, n) = A(\theta)e^{in\theta} + B(\theta)e^{-in\theta}. \qquad (7.74)$$

From the normalization condition

$$\int \phi^2(\lambda, n)\, d\rho(\lambda) = 1, \qquad (7.75)$$

it follows that

$$\int_1^\infty \phi^2(\lambda, n)\, d\rho(\lambda) \le 1. \tag{7.76}$$

By dividing Eq. (7.76) by λ^{2n} and letting $n \to \infty$ we find

$$\int_1^\infty \alpha^2(\lambda, n)\, d\rho(\lambda) = 0. \tag{7.77}$$

This result shows that at every point of increase of ρ for $\lambda > 1$, $\alpha(\lambda) = 0$. Similarly $\beta(\lambda) = 0$ at every point of increase of $\rho(\lambda)$ for $\lambda < -1$. That is if $\lambda > 1$ $(\lambda < -1)$ is a point in the spectrum, then $\alpha(\lambda) = 0$, $(\beta(\lambda) = 0)$.

Equation (7.12) for large n has the solution

$$\psi_\pm(\theta, n) = \exp(\pm in\theta), \tag{7.78}$$

and that for large n, $\psi(\lambda, n)$ and $\phi(\lambda, n)$ are identical since $(q(n \to \infty) \to 0)$. Let us write $\psi(\lambda, n)$ as

$$\psi(\lambda, n) = \beta(\lambda) \left[\lambda - \sqrt{\lambda^2 - 1} \right]^n. \tag{7.79}$$

Now by using the recursion relation

$$\psi(\lambda, n - 1) = -\psi(\lambda, n + 1) + 2\lambda e^{q(n)} \psi(\lambda, n), \tag{7.80}$$

repeatedly, we arrive at an expansion for $\psi(\lambda, 0)$):

$$\psi(\lambda, 0) = \beta(\lambda) P\left(\lambda - \sqrt{\lambda^2 - 1}\right), \tag{7.81}$$

where P is a polynomial of degree $2N$ if $q(N) \ne 0$. In deriving (7.81) we have used the following identity repeatedly

$$2\lambda = \lambda - \sqrt{\lambda^2 - 1} + \frac{1}{\lambda - \sqrt{\lambda^2 - 1}}. \tag{7.82}$$

In the same way for $\lambda < -1$, we have

$$\psi(\lambda, 0) = \alpha(\lambda) P\left(\lambda - \sqrt{\lambda^2 - 1}\right). \tag{7.83}$$

Thus to each root of $P(z)$ which lies inside the unit circle, there corresponds a bound state and vice versa [11]. Since $P(z)$ is of the form $Q\left(z^2\right)$, these states come in positive-negative pairs and the real roots are all simple (for proof see [11]).

Now let us study the solution of the difference equation for values of λ between -1 and $+1$. Again we set $\lambda = \cos\theta$ and define $\psi_+(\theta, n)$ and $\psi_-(\theta, n)$ as solutions of the difference equation

$$\frac{1}{2}\psi(\lambda, n - 1) + \frac{1}{2}\psi(\lambda, n + 1) = \lambda e^{q(n)}\psi(\lambda, n). \tag{7.84}$$

Here we find that for sufficiently large n, $\psi_\pm(\theta, n)$ s are given by

$$\psi_\pm(\theta, n) = e^{\pm in\theta}. \tag{7.85}$$

From the definition of the polynomial $P(z)$ it follows that

$$\psi_+(\theta, n) = P\left(e^{\pm i\theta}\right),\tag{7.86}$$

and since

$$\psi(\theta, n) = \exp\left(-\frac{1}{2}q(n)\right)\phi(\theta, n),\tag{7.87}$$

and this equation satisfy the boundary conditions (see (7.2))

$$\psi(\theta, 0) = 0, \quad \psi(\theta, 1) = \exp\left(-\frac{1}{2}q(1)\right),\tag{7.88}$$

which results from the fact that $\phi(\lambda, 1) = 1$. The function $\psi(\theta, n)$ satisfies a second order recursion relation and therefore it is a linear combination of any two independent solutions, i.e.

$$\psi(n, \theta) = A(\theta)\psi_+(\theta, n) + B(\theta)\psi_-(\theta, n).\tag{7.89}$$

By setting $n = 0$ in (7.84), (7.88) and (7.89) we arrive at the result that

$$A(\theta)P\left(e^{i\theta}\right) + B(\theta)P\left(e^{-i\theta}\right) = 0.\tag{7.90}$$

Now if $\psi_1(\lambda, n)$ and $\psi_2(\lambda, n)$ are two solutions of (7.84), i.e. if

$$\frac{1}{2}\psi_1(\lambda, n-1) + \frac{1}{2}\psi_1(\lambda, n+1) = \lambda e^{q(n)}\psi_1(\lambda, n),\tag{7.91}$$

$$\frac{1}{2}\psi_2(\lambda, n-1) + \frac{1}{2}\psi_2(\lambda, n+1) = \lambda e^{q(n)}\psi_2(\lambda, n),\tag{7.92}$$

then by multiplying (7.91) by $\psi_2(\lambda, n)$ and (7.92) by $\psi_1(\lambda, n)$ and subtracting the two resulting equations, we obtain the following equation which is true for all n

$$\psi_1(\lambda, n-1)\psi_2(\lambda, n) - \psi_2(\lambda, n-1)\psi_1(\lambda, n)\tag{7.93}$$
$$= \psi_1(\lambda, n)\psi_2(\lambda, n+1) - \psi_2(\lambda, n)\psi_1(\lambda, n+1),\tag{7.94}$$

or equivalently

$$\psi_1(\lambda, 0)\psi_2(\lambda, 1) - \psi_2(\lambda, 0)\psi_1(\lambda, 1)\tag{7.95}$$
$$= \psi_1(\lambda, n)\psi_2(\lambda, n+1) - \psi_2(\lambda, n)\psi_1(\lambda, n+1).\tag{7.96}$$

Next we set $\psi_1 = \psi_+$ and $\psi_2 = \psi_-$ and choose n large enough so that (7.78) is satisfied, then using (7.85) and (7.86) we find

$$\psi_-(\theta, 0)\psi_+(\theta, 1) - \psi_+(\theta, 0)\psi_-(\theta, 1) = 2i\sin\theta.\tag{7.97}$$

From Eqs. (7.89) and (7.88) we have

$$\exp\left(\frac{1}{2}q(1)\right) = A(\theta)\psi_+(\theta, 1) + B(\theta)\psi_-(\theta, 1),\tag{7.98}$$

which when combined with (7.90) can be expressed as

$$A(\theta)\psi_+(\theta,0) + B(\theta)\psi_-(\theta,0) = 0. \tag{7.99}$$

Now we can solve (7.98) and (7.99) for $A(\theta)$ and $B(\theta)$ and get

$$A(\theta) = \frac{\exp\left(-\frac{1}{2}q(1)\right)\psi_-(\theta,0)}{2i\sin\theta}, \tag{7.100}$$

and

$$B(\theta) = \frac{\exp\left(-\frac{1}{2}q(1)\right)\psi_+(\theta,0)}{2i\sin\theta}. \tag{7.101}$$

Finally we have

$$A(\theta)B(\theta) = \frac{e^{-q(1)}\psi_-(\theta,0)\psi_+(\theta,0)}{4\sin^2\theta}$$
$$= \frac{e^{-q(1)}P\left(e^{-i\theta}\right)P\left(e^{i\theta}\right)}{4\sin^2\theta}. \tag{7.102}$$

Since

$$A(\theta)B(\theta) = A(\theta)A^*(\theta) = |A(\theta)|^2, \tag{7.103}$$

we need to find $|A(\theta)|^2$ to determine $d\rho(\lambda)$.

We start our search for $\rho(\lambda)$ by considering a simple random walk, with equiprobable ±1 steps and denote the consecutive displacements by $s_0, s_1 \cdots$, then by definition the mathematical expectation we have

$$E\left\{\exp\left[-\left(\sum_{k=0}^r q(s_k)\right)\right] : s_0 > 0,\ s_1 > 0 \cdots s_{r-1} > 0 | s_r = s\right\}$$
$$= \sum_{s_1>0,\ s_2>0\cdots s_{N-1}>0} \exp\left(-\sum_{k=0}^r q(s_k)\right)$$
$$\times P(s_0|s_1)P(s_1|s_2)\cdots P(s_{r-1}|s), \tag{7.104}$$

where $P(x|y)$ is the transition probability to go from x to y, i.e.

$$P(x|y) = \frac{1}{2}\delta(x, y-1) + \frac{1}{2}\delta(x, y+1). \tag{7.105}$$

Thus we get

$$E\left\{\exp\left[-\left(\sum_{k=0}^r q(s_k)\right)\right] : s_0 > 0,\ s_1 > 0 \cdots s_{r-1} > 0 | s_r = s > 0\right\}$$
$$= \exp\left[-\frac{1}{2}(q(s_0) + q(s))\right]a_{s_0,s}^r,\quad r \geq 1, \tag{7.106}$$

where

$$a_{m,n} = \frac{1}{2}\exp\left[-\frac{1}{2}(q(m) + q(n))\right]\delta_{m,n+1}, \tag{7.107}$$

and $a_{s_0,s}^{(r)}$ is the (s_0, s) element of A^r. The matrix A is composed of the matrix elements $a_{m,n}$ given by (7.107).

Using the spectral representation, we can write (7.106) as

$$\mathrm{E}\left\{\exp\left[-\left(\sum_{k=0}^r q(s_k)\right)\right] : s_0 > 0,\ s_1 > 0 \cdots s_{r-1} > 0 | s_r = s > 0\right\}$$

$$= \exp\left[-\frac{1}{2}(q(s_0) + q(s))\right] \int \lambda^r \phi(\lambda, s_0)\phi(\lambda, s)d\rho(\lambda). \tag{7.108}$$

Now if we set $s_0 = s = 1$, we obtain

$$\int \lambda^r d\rho(\lambda) = e^{-q(1)}\mathrm{E}\left\{\exp\left(-\sum_{k=1}^{r-1} q(s_k)\right) ; s_1 > 0 \cdots s_{r-1} > 0 | s_r = 1\right\}. \tag{7.109}$$

The next step is to consider the following sum [11], [14]

$$\frac{1}{N}\sum_{s_0=1}^N \mathrm{E}\left\{\exp\left(-\sum_{k=1}^N e^{-q(s_k)}\right), s_0 > 0,\ s_1 > 0 \cdots s_{r-1} > 0,\ s_r = s_0\right\}$$

$$= \int \lambda^r \left(\frac{1}{N}\right)\sum_{s_0=1}^N e^{-q(s_k)}\phi^2(\lambda, s_0)d\bar\rho(\lambda), \tag{7.110}$$

where $\bar\rho$ is defined by (7.25). In the limit of $N \to \infty$, we have

$$\lim_{N\to\infty}\frac{1}{N}\sum_{s_0=1}^N e^{-q(s_0)}\phi^2(\lambda, s_0) = 0, \tag{7.111}$$

for every bound state and

$$\lim_{N\to\infty}\frac{1}{N}\sum_{s_0=1}^N e^{-q(s_0)}\phi^2(\lambda, s_0) = 2|A(\theta)|^2 = 2\left|A\left(\cos^{-1}\lambda\right)\right|^2, \tag{7.112}$$

for scattering states. When $N \to \infty$, $q(N) \to 0$, i.e.

$$\lim_{N\to\infty}\sum_{s_0=1}^N \mathrm{E}\left\{\exp\left(-\sum_{k=1}^r e^{-q(s_k)}\right), s_0 > 0,\ s_1 > 0 \cdots s_{r-1} > 0,\ s_r = s_0\right\}$$

$$= \int \lambda^r \frac{d\lambda}{\sqrt{1 - \lambda^2}}. \tag{7.113}$$

By combining these relations we find that for all integers r

$$2\int_{-1}^1 \lambda^r \left|A\left(\cos^{-1}\lambda\right)\right|^2 d\rho(\lambda) = \frac{1}{\pi}\int_{-1}^1 \frac{\lambda^r d\lambda}{\sqrt{1 - \lambda^2}}, \tag{7.114}$$

and thus we have

$$d\rho(\lambda) = \frac{d\lambda}{2\pi\left|A\left(\cos^{-1}\lambda\right)\right|^2 \sqrt{1 - \lambda^2}}. \tag{7.115}$$

Since we can write $|A(\theta)|^2$ as

$$|A(\theta)|^2 = e^{-q(1)} \frac{P\left(e^{-i\theta}\right) P\left(e^{i\theta}\right)}{4 \sin^2 \theta}, \tag{7.116}$$

and $P(z)$ must be a polynomial of degree $2N$ and of the form $Q\left(z^2\right)$, therefore $|A(\theta)|^2$ cannot be prescribed arbitrarily.

In the absence of bound states $P(z)$ has the following properties [11]:

(a) $P(z)$ must be analytic and nonzero in $|z| < 1$,

(b) $P(0) = \exp\left(\sum_{k=1}^{N} q(n)\right)$, and

(c) On $|z| = 1$, the S matrix is given by $S = e^{2i\delta} = \frac{P^*(z)}{P(z)} = \frac{P\left(z^{-1}\right)}{P(z)}$. The decomposition of S matrix with the properties (a)–(c) is unique, and this follows from the analytic properties of the S matrix.

Connection of $A(\theta)$ with the Phase Shift — The phase shift $\delta(\theta)$ can be introduced with the help of $P\left(e^{i\theta}\right)$ which can be written as

$$P\left(e^{i\theta}\right) = \left|P\left(e^{i\theta}\right)\right| e^{-i\delta(\theta)}, \tag{7.117}$$

then

$$P\left(e^{-i\theta}\right) = \left|P\left(e^{-i\theta}\right)\right| e^{-i\delta(-\theta)}, \tag{7.118}$$

and since $P\left(e^{i\theta}\right)$ and $P\left(e^{-i\theta}\right)$ are complex conjugates, therefore we have $\delta(-\theta) = -\delta(\theta)$. We also note that for large n

$$\phi(n, \theta) \sim \sin(n\theta + \delta(\theta)), \tag{7.119}$$

which is the discrete analogue of $\sin(kr + \delta(k))$. From the knowledge of $\delta(\theta)$, $0 \leq \theta \leq \pi$, we can find the number of bound states. If in addition we are given the positions of the bound states, then $|A(\theta)|^2$ is determined, and which we can find the continuous part of the spectral function as can be seen from (7.115).

Spectral Function in the Absence of Bound States — If the system has no bound states then the function $\Gamma_0(z)$ which we introduce by

$$\Gamma_0(z) = -\frac{1}{2\pi} \oint \frac{\ln S\left(z'\right)}{z' - z} dz' = -\frac{1}{\pi} \oint \frac{\delta\left(z'\right)}{z' - z} dz', \tag{7.120}$$

has the following analytic properties

$$\Gamma_0(z) = -\frac{\mathcal{P}}{\pi} \oint \frac{\delta\left(z'\right)}{z' - z} dz' - i\delta(z), \tag{7.121}$$

within the circle $|z| < 1$. We can write this last equation as

$$\Gamma_0(z) = -\frac{\mathcal{P}}{\pi} \int_0^\pi \frac{\sin \theta' \delta\left(\theta'\right) d\theta'}{\cos \theta - \cos \theta'} - i\delta(\theta), \tag{7.122}$$

where we have used the antisymmetry property of $\delta(\theta)$.
We have also the following properties of $\Gamma_0(z)$:

$$\Gamma_0^*(z) = \frac{\mathcal{P}}{\pi} \int_0^\pi \frac{\sin\theta' \delta\,(\theta')\,d\theta'}{\cos\theta - \cos\theta'} + i\delta(\theta), \tag{7.123}$$

and at $z = 0$ we have

$$\Gamma_0(z)|_{z=0} = -\frac{1}{\pi} \oint \frac{\delta\,(z')\,dz'}{z'} = -\frac{i}{\pi} \int_{-\pi}^\pi \delta\,(\theta')\,d\theta' = 0. \tag{7.124}$$

Thus if we set

$$P(z) = A\exp\left(\Gamma_0(z)\right), \tag{7.125}$$

with A chosen so as the normalization condition $\int d\rho(\lambda) = 1$ is satisfied, we observe
that $P(z)$ is uniquely determined and that

$$A = \exp\left[\sum_{k=1}^N q(k)\right]. \tag{7.126}$$

So we find the following equation for $|A(\theta)|^2$

$$|A(\theta)|^2 = \frac{A^2}{4\sin^2\theta}\exp\left[\frac{2\mathcal{P}}{\pi}\int_0^\pi \frac{\sin\theta'\delta\,(\theta')\,d\theta'}{\cos\theta - \cos\theta'}\right], \tag{7.127}$$

and finally the following expression for the spectral function

$$\rho(\lambda) = \begin{cases} 0 & \lambda < -1 \\[2mm] \frac{2A^2}{\pi}\int_{-1}^\lambda \sqrt{1-\mu^2}\exp\left[\frac{\mathcal{P}}{2\pi}\int_{-1}^1 \frac{\delta(\mu')d\mu'}{\mu'-\mu}\right]d\mu \\[2mm] 1 & \lambda > 1, \end{cases} \tag{7.128}$$

where $-1 < \lambda < 1$. This relation reduces to $\bar\rho(\lambda)$ given by (7.25) when $\delta(\mu) = 0$
and $A = 0$. In the presence of bound states $d\rho(\lambda)$ takes the form

$$\frac{d\rho(\lambda)}{d\lambda} = \begin{cases} \sum_{j=1}^{M/2} C_j^2\delta(\lambda + \lambda_j) & \lambda < -1 \\[2mm] \frac{2}{\pi}\sqrt{1-\lambda^2}A^2\exp\left(\frac{2\mathcal{P}}{\pi}\int_{-1}^1 \frac{\delta(\mu')d\mu'}{\mu'-\mu}\right)\times \\ \prod_{j=1}^{M/2}\left[1 - \left(\frac{\lambda+i\sqrt{1-\lambda^2}}{\lambda_j - i\sqrt{1-\lambda_j^2}}\right)^2\right]d\lambda & -1 < \lambda < 1 \\[2mm] \sum_{j=1}^{M/2} C_j^2\delta(\lambda - \lambda_j) & \lambda > 1, \end{cases} \tag{7.129}$$

and A is obtained from the normalization condition $\int d\rho(\lambda) = 1$.
Let us summarize the essential parts of this inversion method. We are given

a set of discrete eigenvalues λ_i, $(|\lambda_i| > 1)$, with the normalization constant $C_i^2 = \sum_n \phi^2(\lambda_i, n)$ and the phase shifts $\delta(\lambda)$ for $-1 < \lambda < 1$. From the phase shift we calculate $|A(\theta)|^2$ $(\cos\theta = \lambda)$, Eq. (7.116), with either (7.128) or (7.129) we can determine $d\rho(\lambda)$. We also take the reference wave function (when $v(n) \equiv 0$) to be $\bar{\phi}(\lambda, n) = \frac{\sin(n\theta)}{\sin\theta}$, and the spectral function $d\bar{\rho}(\lambda)$, Eqs. (7.24) and (7.50) respectively, and from these we construct $Q(n, m)$, Eq. (7.62) which is the input function for the kernel $K(n, m)$, Eq. (7.61). By solving the latter equation we determine $K(n, m)$ which can be used in (7.68) to find $v(\Delta n)$. We have already shown that in the limit of $\Delta \to 0$ and in the absence of bound states we have

$$\bar{\phi} \to \frac{\sin(\sqrt{2E}\, x)}{\sqrt{2E}} \tag{7.130}$$

and

$$\frac{d\rho}{dE} \to \begin{cases} 0 & E < 0 \\[2ex] \frac{2}{\pi}\sqrt{2E}\exp\left(-\frac{2}{\pi}\mathcal{P}\int_0^\infty \frac{\delta(E'dE')}{E'-E}\right) & E > 0, \end{cases} \tag{7.131}$$

and when we take this limit we recover the Gel'fand–Levitan equation (4.56).

References

[1] An early and an interesting approach for finding a general rule of construction of exactly solvable potential is due to M.F. Manning, Phys. Rev. 48, 161 (1935). In recent years many other ways of obtaining solvable and quasi-solvable potentials have been found, e.g. see R. Milson's paper in Intl. J. Theo. Phys. 37, 1735 (1998).

[2] For a full account of quasi-solvable potentials see A.G. Ushveridze, *Quasi-Exactly Solvable Models in Quantum Mechanics*, Second Edition, (Taylor and Francis, London, 200). An early example of this type of potential is given in Chapter 9 of this book.

[3] F. Calogero, *Variable Phase Approach to Potential Scattering*, (Academic Press, New York, 1967).

[4] G.C. Stokes, *On the Intensity of the Light Reflected or Transmitted through a Pile of Plates*, Mathematical and Physical Papers, 2 (University Press, Cambridge, 1883).

[5] B.N. Zakhariev and A.A. Suzko, *Direct and Inverse Problems: Potentials in Quantum Scattering*, (Springer-Verlag, Berlin 1990).

[6] M. Hron and M. Razavy, Geophys. J.R. Astr. Soc. 51, 545 (1977).

[7] M. Hron and M. Razavy, Can. J. Phys. 55, 1434 (1977).

[8] M. Hron and M. Razavy, J. Phys. A: Math. Gen. 14, 2215 (1981).

[9] K.M. Case and M. Kac, J. Math. 14, 594 (1973).

[10] For a different discussion of this discrete formulation see (B.N. Zakhariev and A.A. Suzko, *Direct and Inverse Problems: Potentials in Quantum Scattering*, (Springer-Verlag, Berlin 1990)).

[11] K.M. Case, J. Math. Phys. 14, 916 (1973).

[12] K.M. Case, J. Math. Phys. 15, 143 (1974).

[13] K. Chadan and P.C. Sabatier, *Inverse Problems in Quantum Scattering Theory*, (Second Eition), (Springer-Verlag, New York, 1989), Sec. IX.6.

[14] M. Kac, Rocky Mountain J. Math. 4, 511 (1974). Kac shows that the Gel'fand–Levitan solution of the inverse scattering problem can be included among the potential-theoretic problems which can be solved by means of Brownian motion and Wiener integrals.

Chapter 8

$\mathcal{R}$ Matrix Theory and Inverse Problems

In many problems of quantum physics the forces are appreciable only inside a given interval, e.g. in the case of the radial Schrödinger equation the potential may be very weak or be exactly zero for $r > a$. Then it is easier to deal with a set of functions forming a complete set on the finite interval $[0\ a]$, $a < \infty$. In this case, rather than working with a continuous spectrum of wave functions on semi-axis or on the entire axis. We will use the eigenfunctions of the Schrödinger operator which corresponds to the homogeneous boundary conditions at $r = 0$ and $r = a$. But first, for a simple approach to this theory, let us consider the observation of Kapur and Peierls and look at the S-wave scattering of a spinless particle described by the Schrödinger equation [1]

$$\frac{d^2u(r)}{dr^2} + \left(k^2 - v(r)\right)u(r) = 0, \quad r \le a, \tag{8.1}$$

where the potential energy $v(r) \equiv 0$ for $r > a$. That is for $r > a$, we have

$$\frac{d^2u(r)}{dr^2} + k^2u(r) = 0, \quad r > a. \tag{8.2}$$

The solution of (8.2) may be written as

$$u(r) = \frac{I}{k}\sin(kr) + Se^{ikr}, \quad r > a, \tag{8.3}$$

where I is the amplitude of the incident wave and S is that of the scattered wave. Evidently

$$\frac{du(r)}{dr} = I\cos(kr) + ikSe^{ikr}. \tag{8.4}$$

173

Therefore we can express I and S in terms of the value and the slope of the wave function at $r = a$;

$$Ie^{-ika} = \left(\frac{du(r)}{dr}\right)_{r=a} - iku(a), \tag{8.5}$$

and

$$S = \cos(ka)u(a) - \frac{1}{k}\sin(ka)\left(\frac{du(r)}{dr}\right)_{r=a}. \tag{8.6}$$

From Eq. (8.5) we can see that if there is no incident wave present, then $u(r)$ would be described by the boundary condition

$$\left(\frac{du(r)}{dr} - iku(r)\right)_{r=a} = 0. \tag{8.7}$$

This boundary condition at $r = a$ is different from (8.7) of the Kapur and Peierls formalism [2]–[6]. Now let us consider the $\mathcal{R}$ matrix formulation for elastic scattering of a spinless particle for the partial wave ℓ. Here we start with the radial wave function $u_\ell(r)$ for the ℓth partial wave, where $u_\ell(r)$ is the solution of the wave equation

$$-\frac{d^2 u_\ell(r)}{dr^2} + \left(v(r) + \frac{\ell(\ell+1)}{r^2}\right)u_\ell(r) = k^2 u_\ell(r), \quad 0 \le r \le a. \tag{8.8}$$

The complete set of states $u_\lambda^\ell(r)$ for the interior region are the solutions of [5]

$$-\frac{d^2 u_\lambda^\ell(r)}{dr^2} + \left(v(r) + \frac{\ell(\ell+1)}{r^2}\right)u_\lambda^\ell(r) = k_\lambda^2 u_\lambda^\ell(r), \quad 0 \le r \le a. \tag{8.9}$$

Note that in the present discussion we denote the ℓth partial wave by either $u^\ell(r)$ or $u_\ell(r)$. The wave function u_λ^ℓ is subject to the boundary conditions

$$u_\lambda^\ell(0) = 0, \tag{8.10}$$

and

$$\left(\frac{du_\lambda^\ell}{dr}\right)_{r=a} = 0. \tag{8.11}$$

In addition these wave functions satisfy the orthogonality condition

$$\int_0^a u_\lambda^\ell(r)u_{\lambda'}^\ell(r)dr = \delta_{\lambda\lambda'}. \tag{8.12}$$

In the region $0 \le r \le a$, we can expand $u_\ell^\ell(r)$ in terms of eigenfunctions $u_\lambda^\ell(r)$;

$$u_\ell(r) = \sum_{\lambda=1}^{\infty} c_\lambda^\ell u_\lambda^\ell(r), \quad 0 \le r \le a, \tag{8.13}$$

where

$$c_\lambda^\ell = \int_0^a u_\ell(r)u_\lambda^\ell(r)dr. \tag{8.14}$$

Using Green's theorem we have

$$-u_\lambda^\ell(a)\left(\frac{du^\ell(r)}{dr}\right)_{r=a} + (k_\lambda^2 - k^2)\int_0^a u^\ell(r)u_\lambda^\ell(r)dr = 0. \tag{8.15}$$

Now from Eqs. (8.11) and (8.14) it follows that

$$c_\lambda^\ell = \frac{1}{a}\left(\frac{u_\lambda^\ell(a)}{k_\lambda^2 - k^2}\right)\left[a\frac{du_\ell(r)}{dr}\right]_{r=a}. \tag{8.16}$$

If we substitute (8.13) in (8.14) and use (8.16) we find

$$\frac{u_\ell(r)}{\left(a\frac{du_\ell(r)}{dr}\right)_{r=a}} = \frac{1}{a}\sum_{\lambda=1}^\infty \frac{u_\lambda^\ell(r)u_\lambda^\ell(a)}{k_\lambda^2 - k^2}. \tag{8.17}$$

Let us define the $\mathcal{R}$ matrix for the ℓth partial wave $\mathcal{R}_\ell$ by

$$\mathcal{R}_\ell = \frac{1}{a}\sum_{\lambda=1}^\infty \frac{\left(u_\lambda^\ell(a)\right)^2}{k_\lambda^2 - k^2} = \frac{1}{a}\sum_\lambda \frac{(\gamma_\lambda^\ell)^2}{k_\lambda^2 - k^2}, \tag{8.18}$$

where γ_λ is the reduced width amplitude;

$$\gamma_\lambda^\ell = u_\lambda^\ell(a). \tag{8.19}$$

As we can see from (8.17) this $\mathcal{R}_\ell$ relates the amplitude of $u_\ell(r)$ to its derivative at the boundary $r = a$:

$$\mathcal{R}_\ell = \frac{u_\ell(a)}{\left(a\frac{du_\ell(r)}{dr}\right)_{r=a}}, \tag{8.20}$$

i.e. $\mathcal{R}$ is equal to the reciprocal of a times the logarithmic derivative of $u_\ell(r)$ at a.

$\mathcal{R}_\ell$ and S_ℓ **Matrices** — The scattering matrix S_ℓ can be found from the $\mathcal{R}_\ell$ matrix by equating the logarithmic derivatives of the wave functions for inside and outside the boundary at $r = a$. For the ℓth partial wave the wave function for $r > a$ can be written as

$$u_\ell(r) = r[I_\ell j_\ell(kr) + S_\ell h_\ell(kr)] \rightarrow$$
$$\rightarrow \left[\frac{I_\ell}{k}\sin\left(kr - \frac{1}{2}\ell\pi\right) + S_\ell e^{ikr - \frac{1}{2}\ell\pi}\right], \quad \text{as} \quad r \to \infty. \tag{8.21}$$

For $\ell = 0$, $u_0(r)$ reduces to (8.3) which we studied earlier. By matching the logarithmic derivatives of $u_\ell(r < a)$ and $u_\ell(r > a)$ we find

$$\mathcal{R}_\ell = \left(\frac{u_\ell(r)}{r\frac{du_\ell(r)}{dr}}\right)_{r=a} = \left\{\frac{r[j_\ell(kr) + S_\ell h_\ell(kr)]}{r\frac{d}{dr}[r(j_\ell(kr) + S_\ell h_\ell(kr))]}\right\}_{r=a}. \tag{8.22}$$

Now $\mathcal{R}_\ell$ is real and S_ℓ is expressible in terms of a phase shift $\delta_\ell(k)$ [4]

$$S_\ell = \exp(2i\delta_\ell). \tag{8.23}$$

Noting that

$$(1 - S_\ell(k)) = -2i\sin\delta_\ell e^{-i\delta_l} = -2ik f_\ell(k), \tag{8.24}$$

where $f_\ell(k)$ is the partial wave scattering amplitude, we can express the total scattering amplitude as the sum of $f_\ell(k)$ over all partial waves with the weight factor $(2\ell + 1)P_\ell(\cos\theta)$ where θ is the scattering angle;

$$f(\theta) = \frac{i}{2k} \sum_{\ell=0}^{\infty} (2\ell + 1)(1 - S_\ell)P_\ell(\cos\theta). \tag{8.25}$$

For the elastic differential cross section we find the following relation

$$\sigma(\theta) = |f(\cos\theta)|^2 = \frac{1}{4k^2} \left| \sum_{\ell=0}^{\infty} (2\ell + 1)(1 - S_\ell)P_\ell(\cos\theta) \right|^2. \tag{8.26}$$

General Form of the Boundary Condition at $r = a$ — We can consider a more general boundary condition at the boundary $r = a$ by replacing (8.11) by the homogeneous boundary condition

$$\left[\frac{r}{u_\lambda(r)} \left(\frac{du_\lambda(r)}{dr} \right) \right]_{r=a} = B, \tag{8.27}$$

where B is independent of λ. For this case (8.17) becomes

$$u_\ell(r) = \frac{1}{2a} \left[a\frac{du_\ell(r)}{dr} - Bu_\ell(r) \right]_{r=a} \sum_\lambda \frac{u_\lambda^\ell(r)u_\lambda^\ell(a)}{k_\lambda^2 - k^2}. \tag{8.28}$$

Putting $r = a$ in this expression and defining

$$\mathcal{R}_\ell(B) = \frac{1}{\mathcal{F}_\ell} - B, \tag{8.29}$$

where

$$\mathcal{F}_\ell = \left(\frac{r}{u_\ell(r)} \frac{du_\ell(r)}{dr} \right)_{r=a}, \tag{8.30}$$

is the logarithmic derivative of the radial wave function times a. We can relate $\mathcal{R}_\ell(B)$ to $\mathcal{R}_\ell(B = 0)$, where $\mathcal{R}_\ell(B = 0)$ is given by (8.20). In this way we find the following result

$$B\mathcal{R}_\ell(B) = \frac{B\mathcal{R}_\ell(0)}{1 - B\mathcal{R}_\ell(0)}. \tag{8.31}$$

Convergence of Expansion in Terms of Eigenfunctions — Let us note that either (a) the expansion (8.13) does not converge uniformly or (b) the term by term differentiation is not admissible or both. Thus the completeness of Eq. (8.13) does not guarantee the validity of (8.17). We can show this in the simple case where B tends to infinity. In this case

$$u_\ell^\lambda(0) = u_\ell^\lambda(a) = 0. \tag{8.32}$$

From Eq. (8.11) it follows that the general form of (8.28) can be written as $B \to \infty$. In this case

$$c_\ell^\lambda = \left[\frac{u_\ell^\lambda(a)}{k^2 - k_\lambda^2} \frac{du_\ell(r)}{dr} + \frac{u_\ell(a)}{k^2 - k_\lambda^2} \frac{du_\ell^\lambda(r)}{dr} \right]_{r=a}. \tag{8.33}$$

At $r = a$ this equation becomes

$$c_\ell^\lambda = \frac{u_\ell(a)}{k^2 - k_\lambda^2} \left(\frac{du_\ell^\lambda}{dr} \right)_{r=a}. \tag{8.34}$$

By substituting (8.34) in (8.13) we obtain

$$\frac{u_\ell(r)}{u_\ell(a)} = \sum_{\lambda=1}^\infty \frac{u_\ell^\lambda(r)}{k^2 - k_\lambda^2} \left(\frac{du_\ell^\lambda(r)}{dr} \right)_{r=a}. \tag{8.35}$$

By letting r to go to a, we observe that the left-hand side of (8.35) approaches one, whereas by taking the limit of term by term of the right-hand side of this equation yields a null result, and this follows from the boundary condition (8.32).

Similarly by differentiating the expansion

$$\eta(r) = a \left(\frac{d}{dr} - B \right) \sum_{\lambda=1}^\infty \frac{u_\ell^\lambda(r) u_\ell^\lambda(a)}{k_\lambda^2 - k^2}$$

$$= \frac{\left[a \frac{du_\ell(r)}{dr} - B u_\ell(r) \right]}{\left[a \frac{du_\ell(r)}{dr} - B u_\ell(r) \right]_{r=a}}, \tag{8.36}$$

and taking the limit term by term in (8.36) using Eq. (8.27) we get the null result, whereas $\eta(a) = 1$ as (8.36) shows. Thus we conclude that the series given by (8.13) is not uniformly convergent [5].

$\mathcal{R}_0$ **Matrix for a Square Well** — If we denote the depth of the square well by V_0 and its width by a, then the wave function inside the well is given by

$$u_0(r) = \sin(qr), \quad q = \left[\frac{2\mu}{\hbar^2}(E - V_0) \right]^{\frac{1}{2}}. \tag{8.37}$$

At $r = a$, $u_0(a)$ should vanish, thus we find

$$\mathcal{R}_0 = \frac{1}{qa} \tan(qa). \tag{8.38}$$

The energy eigenvalues are those energies at which $u_0(r)$ has a zero derivative at the surface $(r = a)$, i.e.

$$q_\lambda a = \pi \left(\lambda - \frac{1}{2} \right), \quad \lambda = 1, \, 2, \cdots \tag{8.39}$$

or

$$E_\lambda = V_0 + \frac{\pi^2 \hbar^2}{2\mu a^2} \left(\lambda - \frac{1}{2} \right)^2. \tag{8.40}$$

For additional information about the mathematical properties of the $\mathcal{R}$ matrix the reader is referred to references [7]–[9].

8.1   Inverse Problem for $\mathcal{R}$ Matrix Formulation of Scattering

In the last section we showed that all of the physical properties of a system when the force only acts between 0 and a are determined by the solutions given in the range $[0, \, a]$. With homogeneous boundary conditions at both boundaries 0 and a, we found an infinite set of discrete eigenfunctions which we denoted by $u_\lambda^\ell(r)$. In the following account of in the inversion, we will consider a fixed partial wave ℓ and will suppress the superscript ℓ.

From the solution of the Schrödinger operator with homogeneous boundary conditions at 0 and a (called $\mathcal{R}$-resonant states), we can determine the input for Gel'fand–Levitan equation as well as Marchenko's equation. For the $\mathcal{R}$ matrix inversion, in the case of the Gel'fand–Levitan formulation, we change the limit of integration from $[0, \, r]$ to $[r, \, 0]$, replace the normalization factor C_n^2 by γ_λ^2 and also change the boundary conditions for $u(r)$ to the following:

$$u(E, a) = \bar{u}(E, 0) = 0, \quad \left(\frac{du(E, r)}{dr} \right)_{r=a} = \left(\frac{d\bar{u}(E, r)}{dr} \right)_{r=a} = 1. \tag{8.41}$$

Thus in the Gel'fand–Levitan method we write the input as

$$g(r, r') = \sum_{\lambda=1}^{\infty} \left\{ \gamma_\lambda^2 u \left(k_\lambda^2, r \right) u \left(k_\lambda^2, r' \right) - \gamma_{0,\lambda}^2 u \left(k_{0,\lambda}^2, r \right) u \left(k_{0,\lambda}^2, r' \right) \right\}. \tag{8.42}$$

In this equation $\gamma_{0,\lambda}^2$ s and $k_{0,\lambda}^2$ s are the reduced width amplitudes and the energy eigenvalues (both measured in units of length^{-2}). The subscript 0 indicates that these quantities are derived for the reference potential $v_0(r)$. When the potential is $v(r) = v_0(r) + \Delta v(r)$, then we denote the corresponding quantities with no zero subscript. The function $u \left(k_{0,\lambda}^2, r \right)$ refers to λth eigenstate inside the potential $v_0(r)$ and $u \left(k_\lambda^2, r \right)$ refers to corresponding state when the potential is $v(r) = v_0(r) + \Delta v(r)$.

So the basic equations in this case become

$$K\left(r, r'\right) + g\left(r, r'\right) + \int_r^a K\left(r, r''\right) g\left(r'', r'\right) dr'' = 0, \tag{8.43}$$

and

$$\Delta v(r) = -\frac{1}{2}\frac{d}{dr}K(r, r). \tag{8.44}$$

Having found the equations for determining the potential, we note that in Eq. (8.42) there are no integrals and $g\left(r, r'\right)$ is degenerate, provided we restrict the summation to a finite number of terms. This can be done if in the set of $\left\{k_\lambda^2, \gamma_\lambda^2\right\}$ only a few resonance parameters differ from the set $\left\{k_{0,\lambda}^2, \gamma_{0,\lambda}^2\right\}$ given for the reference potential. If only one member of the set $\left\{k_\lambda^2, \gamma_\lambda^2\right\}$ differs from the set $\left\{k_{0,\lambda}^2, \gamma_{0,\lambda}^2\right\}$ then the terms where $k_\lambda^2 = k_{0,\lambda}^2$ and γ_λ^2 coincides with $\gamma_{0,\lambda}^2$ will cancel out in both sums and only a single amplitude of reduced width γ_ν from the set $\{\gamma_\nu\}$ will be different from $\gamma_{0,\nu}$. Thus $g\left(r, r'\right)$ has a single term

$$g\left(r, r'\right) = \left(\gamma_\nu^2 - \gamma_{0,\nu}^2\right) u\left(k_{0,\nu}^2, r\right) u\left(k_{0,\nu}^2, r'\right). \tag{8.45}$$

For this special case from the solution of Gel'fand–Levitan equation equation for a degenerate $g\left(r, r'\right)$ we find

$$K\left(r, r'\right) = -\left(\gamma_\nu^2 - \gamma_{0,\nu}^2\right) u\left(k_{0,\nu}^2, r\right) u\left(k_{0,\nu}^2, r'\right) d^{-1}(r), \tag{8.46}$$

where

$$d(r) = 1 + \left(\gamma_\nu^2 - \gamma_{0,\nu}^2\right) \int_r^a u^2\left(k_{0,\nu}^2, r\right). \tag{8.47}$$

8.2   The Finite-difference Analogue of the $\mathcal{R}$ Matrix Theory of Scattering

If the potential acts on a particle within the interval $[0, a]$, we need the wave function just in this interval, since for $r > a$ the wave function is that of a free particle which is known. The knowledge of the wave function enables us to describe the system using a countable set of eigenfunctions to parametrize the continuous dependence of energy of the logarithmic derivative of the wave function at the point a and replace it with a discrete form.

Consider the expansion of the wave function $\psi(E, n)$ in terms of $u_\lambda(r)$, [13]

$$\psi(E, n) = \sum_{\lambda=1}^N A_\lambda(E) u_\lambda(n), \quad r = n\Delta, \tag{8.48}$$

and $\Delta = \frac{a}{N}$. Since $u_\lambda(r)$ s form an orthogonal set, Eq. (8.12), we can invert (8.48) and write

$$A_\lambda(E) = \sum_{n=1}^{N} \psi(E,n)u_\lambda(n)\Delta. \tag{8.49}$$

Just as in the case of springs and masses Sec. 1.6 we can choose two sets of boundary conditions e.g.

$$u_\lambda(0) = u_\lambda(a) = 0, \tag{8.50}$$

and

$$u_\mu(a) = \left(\frac{du_\mu(r)}{dr}\right)_a = 0. \tag{8.51}$$

Then the spectra will interlace:

$$-\infty < E_{\mu 1} < E_{\lambda 1} < E_{\mu 2} < E_{\lambda 2} \cdots, \tag{8.52}$$

and that these energies have the asymptotic behaviour [15];

$$E_\lambda \to \lambda^2 - 2A + \alpha_\lambda, \quad E\mu \to \left(\mu - \frac{1}{2}\right)^2 - 2A + \beta_\mu, \tag{8.53}$$

where A is an arbitrary number and

$$\sum_{\lambda}^{\infty} \alpha_\lambda^2 < \infty \quad \text{and} \quad \sum_{\mu}^{\infty} \beta_\mu^2 < \infty. \tag{8.54}$$

Let us start with the difference equations for $\psi(n)$:

$$-\frac{1}{2\Delta^2}[\psi(n+1) - 2\psi(n) + \psi(n-1)] + v(n)\psi(n) = E\psi(n). \tag{8.55}$$

Now if we multiply (8.55) by $u_\lambda(n)$, and the equation for $u_\lambda(n)$ by $\psi(E,n)$ subtract the two resulting equations, then sum over n using (8.49), and impose the boundary condition $u_\lambda(N+1) = 0$, we find

$$A_\lambda(E) = \frac{1}{2\Delta}\left(\frac{\psi(E,N+1)u_\lambda(N)}{E_\lambda - E}\right). \tag{8.56}$$

By substituting (8.56) in (8.48) we get

$$\psi(E,n) = \frac{1}{2}\left(\sum_{\lambda}^{N} \frac{u_\lambda(n)u_\lambda(N)}{E_\lambda - E}\right)\psi(E,N+1). \tag{8.57}$$

Now if we define γ_λ by

$$\gamma_\lambda = \frac{1}{\sqrt{2a}}u_\lambda(N), \tag{8.58}$$

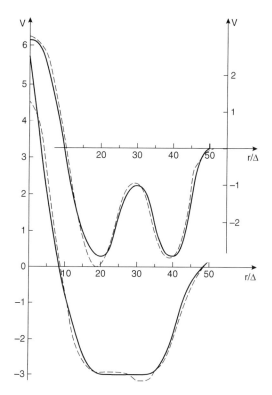

Figure 8.1: Inversion of the scattering data when they are given in terms of the energy eigenvalues and the reduced widths. Solid lines show the input potentials and dashed lines are the same potentials found from inversion [14].

then we can write (8.56) as [14]

$$\mathcal{R}(E) = \sum_{\lambda=1}^{N} \frac{\gamma_\lambda^2}{E_\lambda - E}, \tag{8.59}$$

where the eigenvalues E_ν are found by solving the homogeneous difference equation (8.55) with the boundary conditions

$$\psi_\nu(0) = 0, \quad \psi_\nu(N+1) - \psi_\nu(N-1) = B\psi_\nu(N) \equiv \frac{B\gamma_\nu}{\sqrt{2R}}, \tag{8.60}$$

and the eigenfunctions are $\psi_\nu(n) = \psi(E = E_\nu, n)$.

To calculate the reduced widths, we can solve (8.55) with two boundary conditions B and B' in (8.59), and find the corresponding sets E_ν and E_ν'. Once these eigenvalues are determined γ_ν can be calculated either from (8.61) or from

$$\gamma_\nu^2 = \frac{\prod_{\mu=1}^{N-1} (E_\nu - E_\mu')}{\prod_{\mu=1,\mu\neq\nu}^{N}(E_\nu - E_\mu)}. \tag{8.61}$$

8.3   Shell-model Hamiltonian in Tri-diagonal Form

In Sec. 1.6 we considered how the direct and inverse problems of masses connected linearly with springs can be formulated and solved in terms of tri-diagonal matrices. The same mathematical model can be used for shell model calculation in nuclear physics [16], [17]. In the latter case one selects a model space and an effective Hamiltonian operator, and then calculates the complete $N \times N$ matrix of the effective Hamiltonian in terms of the bases states in the assumed model space. This general symmetric matrix can be transformed to a tri-diagonal form by successive orthogonal transformations (rotations) of the basis vectors [17]. A different formulation of the same problem, advanced by Whitehead, assumes a Hamiltonian for the system which is tri-diagonal at the outset. Thus we start with the Hamiltonian

$$H\psi_i = E_i\psi_i, \tag{8.62}$$

where E_i s are the eigenvalues and ψ_j s are the linear combinations of the shell-model states. If we denote the matrix elements of H by h_{ij}, then from the Hermiticity of H it follows that

$$h_{ij} = h_{ji}, \tag{8.63}$$

and

$$h_{ij} = 0, \quad \text{if } |i - j| \geq 2. \tag{8.64}$$

This last condition means that H is a tri-diagonal matrix.

In the direct problem, from the $N-1$ distinct elements of h_{ij}, we can calculate N eigenvalues E_1, $E_2 \cdots E_N$ of H. Clearly knowing these eigenvalues will not be sufficient to determine the matrix elements h_{ij}. We can find the additional information for the inverse problem in the following way:

Let ψ_1, $\psi_2 \cdots \psi_N$ be the eigenvectors of the matrix h_{ij}. Now we introduce the auxiliary quantity ψ_{N+1} which enables us to relate the input parameters to the finite difference analogue of the logarithmic derivative of the wave function. Thus we introduce $\mathcal{R}(E)$ which corresponds to the finite difference form of the $\mathcal{R}$ matrix by

$$\frac{\psi_{N+1}}{\psi_N} = 1 + \frac{1}{\mathcal{R}(E)}. \tag{8.65}$$

Since in the original eigenvalue problem ψ_{N+1} does not appear, therefore the eigenvalues of H are the zeros of $1 + \mathcal{R}(E)$. Hence we conclude that $\frac{\psi_N}{\psi_{N+1}}$ has poles at E_1, $E_2 \cdots E_N$, and the residues of $\frac{\psi_N}{\psi_{N+1}}$ at these poles, Γ_n s are given by

$$\Gamma_n = \lim_{E \to E_m} (E - E_m) \left(\frac{\psi_N}{\psi_{N+1}} \right) = \frac{\mathcal{R}(E)_m}{\prod_{i \neq m}^{N}(E_m - E_i))}, \quad m = 1, 2 \cdots N. \tag{8.66}$$

These Γ_n s are the analogues of the reduced widths in the $\mathcal{R}$ matrix theory. A convenient way of writing $\frac{\psi_N}{\psi_{N+1}}$ is the following

$$\frac{\psi_N}{\psi_{N+1}} = \beta_N \frac{\prod_{n=1}^{N+1}\left(1 - \frac{E}{\varepsilon_n}\right)}{\prod_{n=1}^{N}\left(1 - \frac{E}{E_n}\right)}, \tag{8.67}$$

where β_N is a constant (independent of energy E). This form exhibits the zeros and the poles of $\frac{\psi_N}{\psi_{N+1}}$ directly. If we assume the boundary condition

$$\psi_{N+1} = 0, \tag{8.68}$$

which is just the condition for finding the eigenvalues of the $N \times N$ matrix, we can determine the N roots $E = E_1, E_2, \cdots E_N$. On the other hand $\psi_N = 0$ gives us the eigenvalues of $(N-1) \times (N-1)$ matrix $\varepsilon_1, \varepsilon_2 \cdots \varepsilon_{N-1}$. These eigenvalues interlace the E_i s as was mentioned in Sec. 1.6, and are related to Γ s by the relation

$$\Gamma_m = -E_m \beta_N \frac{\prod_{n=1}^{N-1}\left(1 - \frac{E_m}{\varepsilon_n}\right)}{\prod_{n \neq m}^{N}\left(1 - \frac{E_m}{E_n}\right)}, \quad m = 1, 2 \cdots N. \tag{8.69}$$

The set of N equations (8.69) can be solved to determine the N unknowns ε_1, $\varepsilon_2 \cdots \varepsilon_{N-1}$ in terms of Γ_m s and E_m s. Thus if the $2N$ parameters E_m and Γ_m are given we can construct $R(E)$ and $\frac{\psi_{N+1}}{\psi_N}$. The constant β_N in (8.69) is a normalization constant and multiplies the tri-diagonal Hamiltonian. Thus there are $2N - 1$ input parameters from which we can find the $(2N - 1)$ matrix elements of H uniquely.

8.4   Continued Fraction Expansion of the $\mathcal{R}$ Matrix

In a number of problems in nuclear physics and in the classical wave propagation in inhomogeneous media we find that the $\mathcal{R}$ matrix can be represented by the ratio of two polynomials in a given parameter occuring in the problem. To show the details of the mathematical technique for inversion, let us assume that the parameter is the energy of a particle, E, or the square of the wave number for waves propagating in an inhomogeneous medium [22].

Let us consider the ratio $\left(\frac{\psi_{N+1}(E)}{\psi_N(E)}\right)$ as a function of E as is given in (8.67). This ration can be written as the ratio of two polynomials of degrees $N - 1$ and N respectively. For the inverse problem, we assume that the empirical data for $\mathcal{R}(E)$ gives us $\left(\frac{\psi_{N+1}(E)}{\psi_N(E)}\right)$ in the form of a rational function of E [22],

$$1 + \frac{1}{\mathcal{R}(E)} = \left(\frac{\psi_{N+1}(E)}{\psi_N(E)}\right) = \frac{\alpha_{N\,N}E^N + \alpha_{N\,N-1}E^{N-1} + \cdots + \alpha_{N\,0}}{\alpha_{N-1\,N-1}E^{N-1} + \cdots + \alpha_{N-1\,0}}. \tag{8.70}$$

Next we expand (8.70) in terms of the continued fraction of E. The fact that such an expansion is unique enables us to find the solution to the inverse problem (see Appendix A). Thus assuming that all elements $\alpha_{i\,j}$, $i = N, N-1, j = N \cdots 0$ are known constants we write

$$\frac{\psi_N(E)}{\psi_{N+1}(E)} = \cfrac{1}{r_N E + s_N + \cfrac{1}{r_{N-1}E + s_{N-1} + \cfrac{1}{r_{N-2}E + s_{N-2}} + \cdots \cfrac{1}{r_1 E + s_1}}}. \tag{8.71}$$

We can find the constants r_j, s_j, $j = N, \cdots 1$ by the method described in detail in Appendix A (Eqs. (A.9)–(A.14)).

References

[1] P.L. Kapur and R. Peierls, Proc. Roy. Soc. A 166, 277 (1938).

[2] T.Y. Wu and T. Ohmura *Quantum Theory of Scattering*, (Prentis Hall, Englewood Cliffs, 1961) Chapter X.

[3] E.P. Wigner and L. Eisenbud, Phys. Rev. 72, 29 (1947).

[4] A.M. Lane and K.G. Thomas, Rev. Mod. Phys. 30, 257 (1958).

[5] Y.E. Kim and A.L. Zubarev, J. Phys. A 31, 6483 (1998).

[6] B.N. Zakhariev and A.A. Suzko, *Direct and Inverse Problems: Potentials in Quantum Scattering*, (Springer-Verlag, Berlin 1990).

[7] H. Le Rouzo, Am. J. Phys. 71, 273 (2003).

[8] F.J. Narcowich, J. Math, Phys. 15, 1626 (1974).

[9] K. Takanagi, Phys. Rev. A 77, 062714 (2008).

[10] P.G. Burke, *R-Matrix Theory of Atomic Collisions*, (Springer, Berlin, 2011) Chapter 4.

[11] I.S. Gradshteyn and I.M. Ryzhik, *Tables of Integrals, Series and Products*, (Academic Press, 1965) p. 36.

[12] H.S. Wall, *Analytic Theory of Continued Fractions*, (D. van Nostrand, New York, 1948).

[13] B.N. Zakhariev and A.A. Suzko, *Direct and Inverse Problems: Potentials in Quantum Scattering*, (Springer-Verlag, Berlin 1990) Sec. 1.2.6.

[14] V.N. Melnikov and B.N. Zakhariev, Rep. Math. Phys. 18, 353 (1980).

[15] V.A. Marchenko, Dokl. Akad. Nauk SSSR, 77, 557 (1951).

[16] R.R. Whitehead, Nucl. Phys. A 182, 290 (1972).

[17] J.B. McGrory, Proc. Intl. Conf. on Nucl. Phys. Munich, Vol. 2, p. 146 (1973).

[18] H. Hochstadt, Linear Algebra Appl. 8, 435 (1974).

[19] R.A. Usmani, Computers Math. Applic. 27, 59 (1994).

[20] M. Hron and M. Razavy, Can. J. Phys. 55, 1434 (1977).

[21] Gladwell, *Inverse Problems in Vibration*, (Springer, Dordrecht, 1986) Chapter 4.

[22] M. Hron and M. Razavy, Geophy. J. R. Astr. Soc. 51, 545 (1977).

Chapter 9

Solvable Models of Fokker–Planck Equation Obtained Using the Gel'fand–Levitan Method

The Gel'fand–Levitan method can be used as a method of constructing solvable cases both for nonlinear evolution equations and for linear equations such as the Fokker–Planck equation. Here we will consider only the latter type, and also limit our discussion to equations in one-dimensional space.

The Fokker–Planck partial differential equation describes the time evolution of the probability function of a particle under the action of drag forces and random forces [1]–[3]. Here we want to show that if we can solve this equation for a given potential, then with the help of the Gel'fand–Levitan transformation we can find a large number of other solvable cases of the Fokker–Planck equation. To formulate this application of the inverse scattering, we follow van Kampen's work and by using the eigenfunction expansion of the probability function, we can relate the terms of the expansion to the solution of the Schrödinger equation for confining potentials.

Let us denote the probability function for one-dimensional space by $P(x,t)$ and write the Fokker–Planck equation as [4]

$$\frac{\partial P(x,t)}{\partial t} = \frac{\partial}{\partial x}\left(\frac{dU(x)}{dx}P(x,t)\right) + \theta \frac{\partial^2 P(x,t)}{\partial x^2}, \qquad (9.1)$$

where θ is a constant. The initial condition for this equation is [4], [5]

$$P(x,t=0) = \delta(x-y), \qquad (9.2)$$

that is, we assume that the probability distribution is located sharply around the point $x = y$. There are few known exactly solvable cases of (9.1) with the initial condition (9.2). These are all related to the solvable Schrödinger equation with confining potentials. This connection was suggested and used by van Kampen to find solutions for the Fokker–Planck equation [5].

Following van Kampen [4] we observe that the time-dependence of $P(x,t)$ in (9.1) can be separated from the space-dependent part by setting

$$P(x,t) = P(x)e^{-\lambda t}, \tag{9.3}$$

and substituting this expression in (9.1) to get

$$\theta P'' + U'P' + (U'' + \lambda)\, P = 0, \tag{9.4}$$

where prime denotes derivative with respect to x. Now we introduce a function, $\phi(x)$ by

$$P(x) = \exp\left[-\frac{1}{2\theta}U(x)\right]\phi(x). \tag{9.5}$$

If we substitute $P(x)$ in (9.4) we find that $\phi(x)$ is a solution of the differential equation

$$\phi''(x) + \left[-\left(\frac{U'}{2\theta}\right)^2 + \frac{U''}{2\theta} + \frac{\lambda}{\theta}\right]\phi(x) = 0. \tag{9.6}$$

This equation can be written as a Schrödinger-like equation if we define the "potential" $v_1(x)$ to be

$$v_1(x) = \left(\frac{U'}{2\theta}\right)^2 - \frac{U''}{2\theta} + C, \tag{9.7}$$

where C is an arbitrary constant. Thus the solution of (9.4) can be related to the solution of the Schrödinger equation with the "potential" $v_1(x)$;

$$\phi''(x) + [E - v_1(x)]\phi(x) = 0. \tag{9.8}$$

Now Eq. (9.6) is a Riccati differential equation for U'. We can change this to a linear differential equation if we replace U' by

$$U' = -\frac{2\theta Z'}{Z}, \quad \text{i.e.} \quad Z = \exp\left(-\frac{U}{2\theta}\right), \tag{9.9}$$

then Z satisfies the differential equation

$$Z'' + [C - v_1(x)]Z = 0, \tag{9.10}$$

which is the same as (9.8). Clearly Z has to be a positive number, $Z > 0$. Therefore $Z(x)$ considered as a solution of the Schrödinger equation, can only be the ground state wave function, and $C = E_0$ be the lowest energy level. From this discussion it follows that we can take an arbitrary Schrödinger equation, (9.8), and use its ground state $\phi_0(x)$ to define a viscous potential $U(x)$ by (9.9), i.e.

$$U(x) = -2\theta \ln \phi_0(x). \tag{9.11}$$

Next we observe that the original equation for $P(x)$ has eigenfunctions related to the set $\phi_n(x)$, i.e. the eigenfunctions of (9.8);

$$P_n(x) = \exp\left[-\frac{1}{2\theta}U(x)\right]\phi_n(x) = \phi_0(x)\phi_n(x), \quad n = 1, 2, 3\cdots \tag{9.12}$$

Associated with these, we have the set of eigenvalues

$$\lambda_n = \theta(E_n - E_0). \tag{9.13}$$

From these relations we find the general solution of (9.1) to be

$$P(x,t) = c_0\phi_0^2(x) + \phi_0(x)\sum_{n=1}^{\infty} c_n\phi_n(x)\exp[-\theta(E_n - E_0)t]. \tag{9.14}$$

If $\phi_0(x)$ is normalized, then $c_0 = 1$ and the other c_n s can be obtained from the initial distribution function $P(x,0)$ in the usual way. Then

$$P(x,t,y) = \phi_0^2(x) + \phi_0(x)\left[\sum_{n=1}^{\infty} \frac{\phi_n(y)}{\phi_0(y)}\exp[-\theta(E_n - E_0)t]\right]. \tag{9.15}$$

If $v_1(x)$ in Eq. (9.10) is a double-well potential, then for certain values of the parameters of $v_1(x)$, $U(x)$ will also be bistable, and the symmetry (or asymmetry) of $v(x)$ will determine the symmetry (or asymmetry) of $U(x)$.

Now let us suppose that all of the eigenvalues λ_n and eigenfunctions $\phi_n(x)$ are known, then by expanding $P(x,t)$ in terms of $\phi_n(x)\phi_0(x)\exp(-\lambda_n t)$ we find that the solution of (9.3) can be expressed as an infinite series [4]

$$P(x,y,t) = \sum_{n=0}^{\infty} \frac{\phi_0(x)}{\phi_0(y)}\phi_n(x)\phi_n(y)\exp[-\theta(\lambda_n - \lambda_0)t], \tag{9.16}$$

where we have assumed that the wave functions are normalized. We also note that since $\phi_0(x)$ is the ground state, it does not have any nodes. When the potential $v_1(x)$ is bistable, then the set of eigenvalues λ_n satisfy the following important inequality

$$\lambda_1 - \lambda_0 \ll \lambda_n - \lambda_0, \quad n \geq 2. \tag{9.17}$$

This means that after a time $t \approx [\theta(\lambda_1 - \lambda_0)]^{-1}$, Eq. (9.15) reduces to

$$P(x,y,t) = \phi_0^2(x) + \phi_1(y)\phi_1(x)\exp[-\theta(\lambda_1 - \lambda_0)t]\left(\frac{\phi_0(x)}{\phi_0(y)}\right). \tag{9.18}$$

An exponential decay law of the form (9.18) is a direct result of the inequality (9.17) satisfied by the low-lying eigenvalues. This condition can be obtained directly by choosing $v_1(x)$ to be a double-well. Now let us consider the case where the Schrödinger equation for a confining potential is solvable, and at least the lowest eigenvalues λ_n and eigenfunctions $\phi_n(x)$ are known (the potential need not be a double or a multiple well). If λ_n s do not satisfy the condition (9.17), we can use the Abraham–Moses method (Sec. 4.5) to change one or a few of the lowest eigenvalues and at the same time determine the wave functions for these low-lying states. By changing one or a few of the eigenvalues we also change the potential and the wave functions according to (4.99) and (4.100).

In general we can change the the normalization of a finite number, N, of the low-lying wave functions and/or we can change a finite number, M, of the low-lying eigenvalues. The corresponding $K(x,y)$ can be obtained by solving a set of $N+M$ linear equations. For the special case of $N=1$ and $M=0$, $g(x,y)$ has a single factorizable term. This means that after a time $t \approx \theta(\lambda_1 - \lambda_0)^{-1}$, Eq. (9.15) reduces to

$$g(x,y) = (\Gamma - 1)\phi_0(x)\phi_0(y). \tag{9.19}$$

In this case the completeness relations for the normalized set of ϕ_n s and ψ_n s are

$$\sum_{n=0}^{\infty} \phi_n(x)\phi_n(y) = \delta(x-y), \tag{9.20}$$

and

$$\sum_{n=0}^{\infty} \psi_n(x)\psi_n(y) + (\Gamma - 1)\psi_0(x)\psi_0(y) = \delta(x-y), \tag{9.21}$$

and the kernel $K(x,y)$ determined from (4.98) can be found exactly (see Eqs. (4.84) and (4.85))

$$K(x,y) = \frac{(\Gamma - 1)\phi_0(x)\phi_0(y)}{1 + (\Gamma - 1)\int_{-\infty}^{x} \phi_0^2(z)dz}. \tag{9.22}$$

Now let us assume that the solution of (9.1) is known when $U(x)$ is a confining potential, but may be bistable or not, and we want to obtain a solution of the Fokker–Planck equation

$$\frac{\partial Q(x,t)}{\partial t} = \frac{\partial}{\partial x}\left(\frac{dW(x)}{dx}Q(x,t)\right) + \theta\frac{\partial^2 Q(x,t)}{\partial x^2}, \tag{9.23}$$

with the initial condition

$$Q(x,t=0) = \delta(x-y). \tag{9.24}$$

Here $W(x)$ is related to $\psi_0(x)$ by

$$W(x) = -2\theta \ln \psi_0(x) = -2\theta \ln\left[\phi_0(x) - \int_{-\infty}^{x} K(x,y)\phi_0(y)dy\right], \tag{9.25}$$

and the coefficient $W(x)$ is bistable either because $v_1(x)$ is a double-well or because v_0 is chosen to be close to λ_1, i.e. $\lambda_1 - v_0 \ll \lambda_2 - v_0$. With our choice of $W(x)$, Eq. (9.23) for $Q(x,y,t)$ becomes similar to Eq. (9.1), for $P(x,y,t)$ and therefore has a solution similar to (9.15), i.e.

$$Q(x,y,t) = \sum_{n=0}^{\infty} \frac{\psi_0(x)}{\psi_0(y)}\psi_n(x)\psi_n(y)\exp[-\theta(\lambda_n - \lambda_0)t] + (\Gamma - 1)\psi_0^2(x). \tag{9.26}$$

The last term which is also a solution of (9.12) is added to the left-hand side of (9.15) so that at $t = \infty$, the condition $\int Q(x, y, t = \infty)\, dx = (\Gamma - 1)$ can be satisfied. The new distribution function $Q(x, y, t)$ can be directly related to the old one, $P(x, y, t)$ by first replacing $\psi_n(x)$ s by $\phi_n(x)$ s using (4.100). (Note that here ϕ_n has been replaced by $\psi_n(x)$, and $\bar{\phi}_n(x)$ by $\phi_n(x)$.)

9.1 Solution of the Fokker–Planck Equation for Symmetric and Asymmetric Double-Well Potentials

We can find solvable Fokker–Planck equation either with solvable potential $v_1(x)$ and examples of these are given by van Kampen [4] and by Frisch *et al.* [6], [7]. In van Kampen's paper a symmetric bistable potential consisting of two rectangular wells separated by by a rectangular barrier is used for $v_1(x)$, whereas in the work of Frisch *et al.* a W-shaped potential is assumed. Van Kampen uses the method that we outlined in the previous section and Frisch and collaborators use the Laplace transform technique. Here we take $v_1(r)$ to be a quasi-solvable potential. The low-lying wave functions of the following symmetric double-well potential are known analytically if the parameter n is an integer [8]–[10]

$$v_1(x) = \left[\frac{1}{8}\xi^2 \cosh(4x) - (n+1)\xi \cosh(2x) - \frac{1}{8}\xi^2 \right]. \tag{9.27}$$

In this case the wave functions can be written as

$$\phi_n(x) = N_n \eta_n(x) \exp\left[-\frac{1}{4}\xi \cosh(2x) \right], \tag{9.28}$$

where $\eta_n(x)$ is a polynomial in x. For instance if we choose $n = 3$, we find $\eta_0, \cdots \eta_3$ to be given by

$$\eta_0(x) = N_0 \left\{ 3\xi \cosh x + \left[4 - \xi + 2\sqrt{4 - 2\xi + \xi^2} \right] \cosh(3x) \right\}, \tag{9.29}$$

$$\eta_1(x) = N_1 \left\{ 3\xi \sinh x + \left[4 + \xi + 2\sqrt{4 + 2\xi + \xi^2} \right] \sinh(3x) \right\}, \tag{9.30}$$

$$\eta_2(x) = N_2 \left\{ 3\xi \cosh x + \left[4 - \xi - 2\sqrt{4 - 2\xi + \xi^2} \right] \cosh(3x) \right\}, \tag{9.31}$$

and

$$\eta_3(x) = N_3 \left\{ 3\xi \sinh x + \left[4 + \xi - 2\sqrt{4 + 2\xi + \xi^2} \right] \sinh(3x) \right\}. \tag{9.32}$$

For $\xi = 2$, the normalization constants are $N_0 = 0.0653753$, $N_1 = 0.0395018$, $N_2 = 0.38529$ and $N_3 = 1.51328$ and the lowest four eigenvalues are $\lambda_0 = -11$, $\lambda_1 = -9.928$, $\lambda_2 = -3$ and $\lambda_4 = 3.9282$ respectively. The minima of the potential are at the points $x = \pm 1.03172$ and at this point min $v_1(x) = -17$.

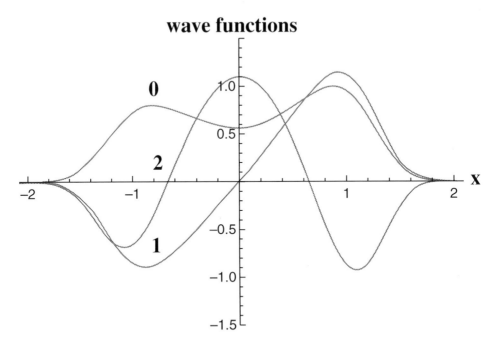

Figure 9.1: Three of the four lowest wave functions for the asymmetric double-well potential. These are found from the symmetric wave functions Eqs. (9.28)–(9.32) with the help of the
Gel'fand–Levitan method.

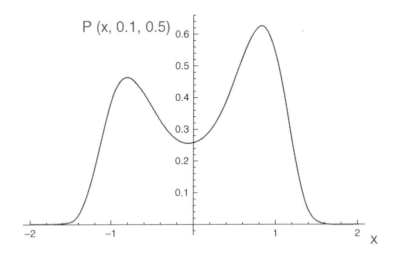

Figure 9.2: The distribution function found using van Kampen's method [4] when $U(x)$ is given
by $U(x) = -2 \ln \phi_0(x)$. After a long time this distribution becomes symmetric.

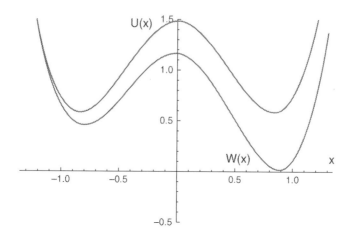

Figure 9.3: The two potentials $U(x) = -2\ln\phi_0(x)$ and $W(x) = -2\ln\psi(x)$, Eqs. (9.11) and (9.14), the first is symmetric and the second is asymmetric. These are found from the ground states of the symmetric Eq. (9.28) and the asymmetric wave functions Eq. (9.33).

Distribution Function for the Case of Symmetric Potential $U(x)$ — The distribution function $P(x, y, t)$ calculated using only the four low-lying wave functions belonging to the symmetric potential $v_1(x)$ in the Schrödinger equation (9.16) (the first three of these are shown in Fig. 9.1). This distribution is calculated for $t = 0.5$ and $y = 0.1$. For times shorter than 0.5, keeping only four terms in the expansion (9.15) is not sufficient, a result that follows from the fact that $P(x, y, t)$ can become negative at large values of $|x|$. As we can see in Fig. 9.2, $P(x, y, t)$ does not show any sign of symmetry as long as θt is small. However as $t \to \infty$, $P(x, y, t)$ tends to $\phi_0^2(x)$ and becomes symmetric. On the other hand $Q(x, y, t)$ corresponding to the asymmetric potential $W(x)$ in the Fokker–Planck equation is not symmetric for θt small, nor it becomes symmetric after a long time.

Wave Functions for an Asymmetric Double-Well — From the wave functions (9.28)–(9.32) for the symmetric double-well potential $v_1(x)$, Eq. (9.27), we can find asymmetrical wave functions using the Gel'fand–Levitan method either by changing the normalization of the ground state wave function or by changing the eigenvalues. The simplest way is to change the norm of the ground state according to Eqs. (4.84), (4.85) and Eqs. (4.100), (4.101). Thus for our bistable potential $v_1(x)$, (9.27), we get four new wave functions, $\psi_n(x)$ s, and these are the eigenfunctions of the asymmetric potential, $v(x) = v_1(x) + \Delta v(x)$.

$$\psi_n(x) = \frac{\phi_n}{1 + (\Gamma - \bar{\Gamma}) \int_{-\infty}^{x} \phi_0^2(y) dy}. \tag{9.33}$$

The lowest three wave functions $\psi_n(x)$ calculated with the parameters $\Gamma = 0.7$ and $\bar{\Gamma} = 1$ are shown in Fig. 9.1, and they clearly show the lack of symmetry or antisymmetry.

References

[1] H. Risken, *The Fokker–Planck Equation: Methods of Solutions and Applications*, (Springer, Berlin 1984).

[2] C.W. Gardiner, *Handbook of Stochastic Methods for Physics, Chemistry and Natural Sciences*, (Springer, New York, 1985) Chapter 5.

[3] L.E. Reichl, *A Modern Course in Statistical Physics*, (John Wiley & Sons, New York, 1991).

[4] N.G. van Kampen, J. Stat. Phys. 17, 77 (1977).

[5] N.G. van Kampen, *Stochastic Processes in Physics and Chemistry: Rivised and enlarged edition*, (Elsevier, Amsterdam, 1992) Chapter VIII.

[6] H.L. Frisch, V. Privman, C. Nicolis and G. Nicolis, J. Phys. A: Math. Gen. 23, L1147 (1990).

[7] H.L. Frisch, V. Privman, J. Chem. Phys. 94, 8216 (1991).

[8] M. Razavy, Am. J. Phys. 48, 285 (1980).

[9] M. Razavy, Phys. Lett. 72A, 89 (1979).

[10] M. Hron and M. Razavy, J. Stat. Phys. 38, 655 (1985).

Chapter 10

The Eikonal Approximation

In high energy elastic or inelastic scattering one needs to sum the partial wave scattering amplitudes $f_\ell(k)$ over a large number of partial waves. In many cases it is more convenient to determine the high energy scattering amplitude by replacing the sum over ℓ by an integral over the impact parameter b. When this is the case, a very useful method for calculating the scattering amplitude is by means of the eikonal approximation [1]–[6]. For the direct formulation of the problem we start with the Schrödinger equation for scattering of a spinless particle by a potential $V(r)$ and try to determine the high energy scattering amplitude as an integral over the impact parameter rather than a sum over partial waves. In this respect the eikonal approximation is similar to the semiclassical treatment that we studied in Sec. 1.5.

We start with the Schrödinger equation

$$-\frac{\hbar^2}{2m}\nabla^2\psi(\mathbf{r}) + V(r)\psi(\mathbf{r}) = E\psi(\mathbf{r}), \tag{10.1}$$

with the boundary condition

$$\psi(\mathbf{r}) \to e^{i\mathbf{k}\cdot\mathbf{r}} + f(\mathbf{k},\theta)\frac{e^{ikr}}{r}, \quad \text{as} \quad r \to \infty. \tag{10.2}$$

The formal solution of (10.1) with the boundary condition (10.2) can be written as an integral equation;

$$\psi(\mathbf{r}) = e^{i\mathbf{k}\cdot\mathbf{r}} - \frac{2m}{\hbar^2}\int V(\mathbf{r}_1)\psi(\mathbf{r}_1)G(\mathbf{r},\mathbf{r}_1)d^3r_1, \tag{10.3}$$

where $G(\mathbf{r},\mathbf{r}_1)$ which is the Green function for the problem with outgoing wave boundary condition is given by

$$G(\mathbf{r},\mathbf{r}_1) = \frac{\exp(ik|\mathbf{r}-\mathbf{r}_1|)}{4\pi|\mathbf{r}-\mathbf{r}_1|}. \tag{10.4}$$

195

Next we define the wave function $\phi(\mathbf{k}, \mathbf{r})$ by

$$\phi(\mathbf{k}, \mathbf{r}) = \psi(\mathbf{k}, \mathbf{r}) e^{-i\mathbf{k}\cdot\mathbf{r}}, \qquad (10.5)$$

and from (10.3) we deduce that $\phi(\mathbf{k}, \mathbf{r})$ satisfies the integral equation

$$\phi(\mathbf{k}, \mathbf{r}) = 1 - \frac{m}{2\pi\hbar^2} \int \frac{\exp\left\{i\left[k(\mathbf{r} - \mathbf{r}_1) - \mathbf{k}\cdot(\mathbf{r} - \mathbf{r}_1)\right]\right\}}{(\mathbf{r} - \mathbf{r}_1)} V(\mathbf{r}_1)\phi(\mathbf{k}, \mathbf{r}_1) d^3r_1$$

$$= 1 - \frac{m}{2\pi\hbar^2} \int \frac{1}{r_2} \exp\left[i(kr_2 - \mathbf{k}\cdot\mathbf{r}_2)\right] V(\mathbf{r} - \mathbf{r}_2)\phi(\mathbf{k}, \mathbf{r} - \mathbf{r}_2) d^3r_2, \quad (10.6)$$

where $\mathbf{r}_2 = \mathbf{r} - \mathbf{r}_1$. For large k values, the exponential in the integral oscillates rapidly, except where the phase is stationary, i.e. $kr_2 - \mathbf{k}\cdot\mathbf{r}_2 = 0$, or $\mathbf{r}_2$ is parallel to $\mathbf{k}$. By integrating over the angular variable we find [2]

$$\phi(\mathbf{k}, \mathbf{r}) \approx 1 - \frac{mi}{\hbar^2 k} \int_0^\infty V(\mathbf{r} - \mathbf{r}_1)\phi(\mathbf{k}, \mathbf{r} - \mathbf{r}_1)\left(1 - e^{2ikr_1}\right) dr_1$$

$$\approx 1 - \frac{ik}{2E} \int_0^\infty V(\mathbf{r} - \mathbf{r}_1)\phi(\mathbf{k}, \mathbf{r} - \mathbf{r}_1) dr_1, \qquad (10.7)$$

provided that $\mathbf{r}_1 = \frac{r_1 \mathbf{k}}{k}$.

Now we can write $\mathbf{r}$ in terms of two vectors $\hat{\mathbf{k}}$ and $\mathbf{b}$, where $\hat{\mathbf{k}}$ is the unit vector in the direction of $\mathbf{k}$, and $\mathbf{b}$ is the impact parameter which is orthogonal to $\mathbf{k}$,

$$\mathbf{r} = \mathbf{b} + \hat{\mathbf{k}}z. \qquad (10.8)$$

Replacing $\mathbf{r}$ in terms of $\mathbf{b}$ and $\hat{\mathbf{k}}$ in (10.7) we find

$$\phi(\mathbf{k}; \mathbf{b} + \hat{\mathbf{k}}z) \approx 1 - \frac{ik}{2E} \int_{-\infty}^z V\left(\mathbf{b} + \hat{\mathbf{k}}z_2\right)\phi\left(\mathbf{k}, \mathbf{b} + \hat{\mathbf{k}}z_2\right) dz_2, \qquad (10.9)$$

where $z_2 = z - z_1$. Solving this equation we find

$$\phi\left(\mathbf{b}, \mathbf{b} + \hat{\mathbf{k}}z\right) \approx \exp\left[-\frac{ik}{2E} \int_{-\infty}^z V\left(\mathbf{b} + \hat{\mathbf{k}}z_1\right) dz_1\right], \qquad (10.10)$$

and the resulting wave function in this approximation is

$$\psi\left(\mathbf{b} + \hat{\mathbf{k}}z\right) \approx \exp\left\{ik\left[z - \int_{-\infty}^z \frac{V\left(\mathbf{b} + \hat{\mathbf{k}}z_1\right)}{2E} dz_1\right]\right\}. \qquad (10.11)$$

Let us assume that the particle is incident in the direction of $\mathbf{k}_i$, if we want to consider the scattering in the direction of $\mathbf{k}_f$, we choose the path of integration along a straight line halfway between $\mathbf{k}_i$ and $\mathbf{k}_f$ parallel to the unit vector

$$\hat{\mathbf{K}} = \frac{\mathbf{k}_i + \mathbf{k}_f}{|\mathbf{k}_i + \mathbf{k}_f|}, \qquad (10.12)$$

and the vector $\mathbf{b}$ orthogonal to it

$$\psi\left(\mathbf{k}_i, \mathbf{b} + \hat{\mathbf{k}}z\right) \approx \exp\left\{i\left[\mathbf{k}_i \cdot \mathbf{b} + \mathbf{k}_i \cdot \hat{\mathbf{K}}z - \frac{k_i}{2E}\int_{-\infty}^{z} V\left(\mathbf{b} + \hat{\mathbf{K}}z_1\right)dz_1\right]\right\}.$$

(10.13)

Using this wave function we can calculate the scattering amplitude

$$\begin{aligned}
f\left(\mathbf{k}_f, \mathbf{k}_i\right) &= -\left(\frac{m}{2\pi\hbar^2}\right)\int \exp(-i\mathbf{k}_f \cdot \mathbf{r})V(\mathbf{r})\psi(\mathbf{k}_i, \mathbf{r})d^3r \\
&= -\left(\frac{m}{2\pi\hbar^2}\right)\int V(\mathbf{b} + \hat{\mathbf{K}}z)dz \\
&\quad \times \exp\left\{i\left[(\mathbf{k}_i - \mathbf{k}_f) \cdot (\mathbf{b} + \hat{\mathbf{K}}z) - \frac{k}{2E}\int_{-\infty}^{z} V(\mathbf{b} + \hat{\mathbf{K}}z_1)dz_1\right]\right\}d^2b,
\end{aligned}$$

(10.14)

where the $\mathbf{b}$ integration is two dimensional. For elastic scattering, $k_i = k_f = k$, we have

$$(\mathbf{k}_i - \mathbf{k}_f) \cdot \hat{\mathbf{K}} = 0.$$

(10.15)

Now we can carry out the integration over z in Eq. (10.14) to get

$$f(\mathbf{k}_f, \mathbf{k}_i) \approx \frac{-ik_i}{2\pi}\int e^{-i\mathbf{q}\cdot\mathbf{b}}\left\{\exp\left[\frac{-ik_i}{2E}\int_{-\infty}^{\infty} V(\mathbf{b} + \hat{\mathbf{K}}z)dz\right] - 1\right\}d^2b.$$

(10.16)

In this relation $\mathbf{q}$ is the momentum transfer which is defined by

$$\mathbf{q} = \mathbf{k}_f - \mathbf{k}_i, \qquad q = 2k_i\sin\frac{\theta}{2}.$$

(10.17)

If $V(\mathbf{r}) = V(r)$, i.e. the potential is spherically symmetric, we can integrate $f(\mathbf{k}_f, \mathbf{k}_i)$ over the angular variable and get a simpler result. Noting that

$$\frac{1}{2\pi}\int_0^{2\pi} e^{-iqb\cos\phi}d\phi = J_0(qb) = J_0\left(2kb\sin\frac{\theta}{2}\right), \quad k_i = k,$$

(10.18)

we find that

$$f(\mathbf{k}_i, \mathbf{k}_f) \approx -ik\int_0^{\infty} J_0\left(2kb\sin\frac{\theta}{2}\right)\left(e^{2i\chi(b)} - 1\right)bdb,$$

(10.19)

where

$$\begin{aligned}
\chi(b) &= -\frac{k}{2E}\int_0^{\infty} V\left(\sqrt{b^2 + z^2}\right)dz, \\
&= -\frac{k}{2E}\int_b^{\infty} \frac{rdr}{\sqrt{r^2 - b^2}}V(r).
\end{aligned}$$

(10.20)

This is the simplest way of inverting the scattering phase shift which is given as a function of the impact parameter b. By setting $E = \frac{\hbar^2 k^2}{2m}$ and by writing $V(r) = \frac{\hbar^2}{2m} v(r)$ we can write (10.20) as

$$\chi(b) = -\frac{1}{2k} \int_b^\infty \frac{rv(r)}{\sqrt{r^2 - b^2}} dr. \tag{10.21}$$

Inverse Scattering Problem Using Impact Parameter Approximation —
Here we assume that $\chi(b)$ is known for all values of b, and then we have two possible cases:
(a) That $\chi(b)$ is given for a fixed E but for high energy scattering and for all values of b.
(b) That $\chi(b, k)$ is given as a function of b as well as k, and the scattering is caused by a weak potential $V(r) \ll E$, where E is the energy of the incident particle. In the first case $\chi(b)$, and $v(r)$ are related to each other by Eq. (10.20), and we want to determine the potential function $v(r)$. As Eq. (10.21) shows we have to solve Abel's integral equation to find $v(r)$. As we have seen before the solution of (10.21) for $v(r)$ is given by

$$v(r) = \frac{2}{\pi} \int_r^\infty \frac{db}{\sqrt{b^2 - r^2}} \frac{d}{db} \chi(b). \tag{10.22}$$

For instance if we choose $\chi(b)$ to be

$$\chi(b) = -\left(\frac{\lambda}{k}\right) e^{-\mu b}, \tag{10.23}$$

where λ and μ are two constants, then from (10.22) we find the potential to be

$$v(r) \approx v_1(r) = \frac{2\mu\lambda}{\pi k} K_0(\mu r), \tag{10.24}$$

where $K_0(\mu r)$ is the Bessel function of imaginary argument and of order 0. Note that in this inversion $v(r)$ is energy dependent. When replacing the angular momentum, ℓ, by the impact parameter kb is a valid approximation we can devise other approximate ways of constructing potentials from a given a set of partial waves phase shifts $\delta_\ell(k)$ s or equivalently the set of $\tan \delta_\ell(k)$ s. Let us briefly discuss these methods.

If the impact parameter phase shifts are known as a function of both ℓ and k, and if these phases are small, i.e. for weak scattering, then the Born approximation is valid, i.e.

$$\tan \delta_\ell \approx \delta_\ell(k) \approx -k \int_0^\infty v(r) r^2 j_\ell^2(kr) dr, \quad \ell = 0, 1, \cdots \infty \tag{10.25}$$

where $v(r) = \frac{2m}{\hbar^2} V(r)$. Now let us consider the weighted sum of the partial phase shifts where the phases are found from the Born approximation;

$$I(k) = \sum_{\ell=0}^\infty w(\ell) \tan \delta_\ell(k) = -k \int_0^\infty v(r) r^2 dr \left[\sum_{\ell=0}^\infty w(\ell) j_\ell^2(kr) \right]. \tag{10.26}$$

We note that for large ℓ, $w(\ell)\tan\delta_\ell(k)$ goes to zero for short range potentials, since large ℓ implies large impact parameter, $\ell+\frac{1}{2}=kb$, and if b is larger than the range of the force, R, then the incident particle will hardly feels the force and the scattering will be very small. In fact for $w(\ell)$ growing as ℓ^3 for large ℓ, $\tan\delta_\ell(k)$ goes to zero very rapidly.

Now we interchange the order of summation and the integration in (10.26) and assume that the weight functions are such that the infinite sum

$$h(k,r) = \sum_{\ell=0}^{\infty} w(\ell)j_\ell^2(kr), \tag{10.27}$$

can be found analytically. If $h(k,r)$ has an inverse, say $h^{-1}(k,r)$ such that

$$\int_0^\infty h(k,r)h^{-1}(k,r')k^2dk = \frac{1}{r^2}\delta(r-r'), \tag{10.28}$$

then we can multiply (10.26) by $h^{-1}(k,r')k^2dk$ and integrate over k to get

$$\int_0^\infty h^{-1}(k,r')I(k)k^2dk = \int_0^\infty v(r)\frac{1}{r^2}dr \times \int_0^\infty h^{-1}(k,r')h(k,r)k^2dk$$
$$= \int_0^\infty v(r)\delta(r-r')dr = v(r) \tag{10.29}$$

where in (10.28) we have interchanged the order of the two integrations. We will study two cases of such an inversion.
(1) For the case where $w(\ell)$ is given by $(-1)^\ell(2\ell+1)$, we have

$$h(k,r) = \sum_{\ell=0}^{\infty}(-1)^\ell(2\ell+1)j_\ell^2(kr) = \frac{1}{2}j_0(2kr). \tag{10.30}$$

Using the orthogonality relation of the spherical Bessel function [7]

$$\int_0^\infty j_\ell(2kr)j_\ell(2kr')k^2dk = \frac{\pi}{16r^2}\delta(r-r'), \tag{10.31}$$

we find that

$$h^{-1}(k,r) = \frac{8r^2}{\pi}j_0(2kr). \tag{10.32}$$

Substituting $h^{-1}(k,r)$ in (10.29) we obtain $v(r)$.
(2) If we choose
$$w(\ell) = (-1)^\ell\ell(\ell+1)(2\ell+1), \tag{10.33}$$

then

$$\sum_{\ell=0}^{\infty}(-1)^\ell\ell(\ell+1)(2\ell+1)[j_\ell(kr)]^2 = -j_1(2kr). \tag{10.34}$$

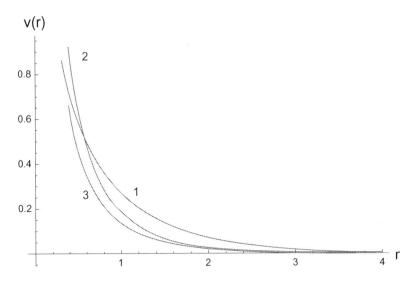

Figure 10.1: Potentials found from the inversion of the phase shift given as a function of the impact parameter phase, $\chi(b,k)$, (10.35). The labels 1, 2 and 3 refer to the potentials $v_1(r)$, $v_2(r)$ and $v_3(r)$, Eqs. (10.36), (10.37) and (10.38) respectively. For all these cases we have set $\mu = 1$, $\lambda = 1$ and for $v_1(r)$ we have chosen $k = 1$.

We can find $h_1(k,r)$ as in the first example. Of course we assume the convergence of the infinite sum (10.27) for short range potentials. Just to compare the result of these inversions we choose $\chi(b)$ to be

$$\chi(b) = -\lambda e^{-\mu b}, \tag{10.35}$$

then

$$v(r) \approx v_1(r) = \frac{2\lambda}{\pi} \int_r^\infty \frac{-\lambda \exp(\mu b)db}{\sqrt{b^2 - r^2}} = \frac{2\lambda}{\pi} K_0(\mu r). \tag{10.36}$$

For the same $\chi(b)$ from Eq. (1.20) we get

$$v(r) \approx v_2(r) = -\frac{16\lambda}{\pi r} \int_0^\infty \sin(2kr)dk \left\{ \int_0^\infty \rho \sin(\pi\rho) \left[\exp\left(\frac{-\mu\rho}{k}\right) \right] d\rho \right\}$$

$$= -\frac{16\lambda}{\pi r} \int_0^\infty \sin(2kr) \left[-\frac{2\pi k^3}{(\mu^2 + k^2\pi^2)} \right] dk$$

$$= \frac{16\lambda(\pi - \mu r)e^{-2\mu r}}{\pi^2 \mu r}. \tag{10.37}$$

When we choose $w(\ell)$ to be of the form $w(\ell) = (-1)^\ell \ell(\ell + 1)(2\ell + 1)$, then using the sum rule (10.34) we obtain

$$v(r) \approx v_3(r) = -\frac{16\lambda}{\pi r} \int_0^\infty j_1(2kr)dk$$

$$\times \left\{ \int_0^\infty \rho \sin(\pi\rho)\rho \left(\rho^2 - \frac{1}{4} \right) \sin(\pi\rho)\chi(\rho, k)d\rho \right\}. \tag{10.38}$$

In our example $\chi(\rho, k) = -\lambda \exp(-\mu b) = -\lambda \exp\left(\frac{\mu\rho}{k}\right)$, and we can calculate the last integral in (10.38) analytically:

$$J(k) = \int_0^\infty \rho\left(\rho^2 - \frac{1}{4}\right)\sin(\pi\rho)\left[-\lambda\exp\left(\frac{-\mu\rho}{k}\right)\right]d\rho$$

$$= \pi\left\{\frac{k^3\mu\pi\left[\mu^4 + 2k^2\mu^2\left(\pi^2 - 24\right) + k^4\pi^2\left(48 + \pi^2\right)\right]}{2\left(\mu^2 + k^2\pi^2\right)^4}\right\}. \tag{10.39}$$

Thus we can write

$$v(r) \approx v_3(r) = -\frac{16\lambda}{\pi r}\int_0^\infty j_1(2kr)J(k)dk. \tag{10.40}$$

In order to compare these three approximate methods of calculating $v(r)$, we have chosen $\chi(b)$ as an exponentially damped function of b, Eq. (10.35) and have obtained $v_1(r)$, $v_2(r)$ and $v_3(r)$ from Eqs. (10.36)–(10.38) and plotted them as a function of r. For a careful analysis of the error(s) made when sums are replaced by integrals see references [8], [9].

10.1 Finding the Impact Parameter Phase Shifts from the Cross Section

The empirical result found from quantum mechanical scattering experiment for elastic or inelastic phase shift is the differential cross section $\sigma(\theta)$. This measurable quantity is related to the scattering amplitude $f(\theta)$ by

$$\sigma(\theta) = |f(\theta)|^2. \tag{10.41}$$

Since $f(\theta)$ is a complex quantity, we can write it in terms of its magnitude and its phase $\nu(\theta)$;

$$f(\theta) = |f(\theta)|e^{i\nu(\theta)} = \sqrt{f(\theta)}e^{i\nu(\theta)}, \tag{10.42}$$

where $\nu(\theta)$ is a function to be determined. A well-known result in quantum scattering theory states that $f(\mathbf{k}_i, \mathbf{k}_f)$ satisfies the unitarity condition [2]

$$\frac{1}{2i}[f(\mathbf{k}_i, \mathbf{k}_f) - f^*(\mathbf{k}_i, \mathbf{k}_f)] = \frac{k}{2\pi}\int f^*(\mathbf{k}_i, \mathbf{k})f(\mathbf{k}_i, \mathbf{k}_f)d\Omega_{\mathbf{k}} \tag{10.43}$$

where the integration is over all possible directions of the vector $\mathbf{k}$. By substituting for $f(\theta)$ from (10.42) in (10.43) we obtain an integral equation for $\nu(\theta)$ [10]

$$\sqrt{\sigma(\theta)}\sin\nu(\theta) = \frac{k}{4\pi}\int\sqrt{\sigma(\theta')\sigma(\theta'')}\cos\left[\nu(\theta') - \nu(\theta'')\right]d\Omega_{k''}. \tag{10.44}$$

In this relation θ, θ' and θ'' are defined by

$$\cos\theta = \hat{\mathbf{k}}_f \cdot \hat{\mathbf{k}}_i, \quad \cos\theta' = \hat{\mathbf{k}} \cdot \hat{\mathbf{k}}_i \text{ and } \cos\theta'' = \hat{\mathbf{k}}_f \cdot \hat{\mathbf{k}}. \tag{10.45}$$

We have already found that the scattering amplitude is related to the impact parameter phase shift $\chi(b,k)$ by Eq. (10.19) or by

$$f(\theta) = -ik \int_0^\infty \left(e^{2i\chi(b)} - 1\right) J_0\left(2kb\sin\frac{\theta}{2}\right) bdb, \tag{10.46}$$

where $|\mathbf{k}_i| = k$. Thus from Eqs. (10.42) and (10.46) it follows that

$$\sqrt{\sigma(\theta)}\sin\nu(\theta) = k \int_0^\infty J_0\left(2kb\sin\frac{\theta}{2}\right)(1 - \cos(2\chi(b,k)))bdb. \tag{10.47}$$

By changing the variable from scattering angle θ to the momentum transfer variable $q = 2k\sin\frac{\theta}{2}$ we find

$$\sqrt{\sigma(q)}\sin\nu(q) = k \int_0^\infty J_0(qb)[1 - \cos 2\chi(b,k)]bdb. \tag{10.48}$$

For high energies q can be regarded as a variable which changes from zero to infinity. Therefore by multiplying (10.48) by $qJ_0(bq)$ and integrating over the range of q we find

$$2\sin^2\chi(k,b) = \frac{1}{k}\int_0^\infty \sqrt{\sigma(q)}\sin\nu(q)qJ_0(bq)dq, \tag{10.49}$$

where $\nu(q)$ can be obtained from the integral equation (10.44).

Determination of Complex Phase Shift for Inelastic Scattering — When the scattering is inelastic, the unitarity condition (10.43) has to be modified in the following way [1]:

$$\frac{1}{2i}[f(\mathbf{k}_i,\mathbf{k}_f) - f^*(\mathbf{k}_i,\mathbf{k}_f)] = \frac{k}{2\pi}\int f^*(\mathbf{k}_i,\mathbf{k})f(\mathbf{k}_i,\mathbf{k}_f)d\Omega_\mathbf{k}$$
$$+ f_{abs}(\mathbf{k}_f,\mathbf{k}_i), \tag{10.50}$$

where

$$f_{abs}(\mathbf{k}_f,\mathbf{k}_i) = -\frac{1}{2\pi}\int [\text{Im}v(r)]\,\psi^*(\mathbf{k}_f,\mathbf{r})\psi(\mathbf{k}_i,\mathbf{r})d^3r. \tag{10.51}$$

In the last equation $\psi(\mathbf{k}_i,r)$ is the exact wave function and f_{abs} is related to the absorption cross section σ_{abs} by

$$\sigma_{abs} = \frac{4\pi}{k}|f_{abs}(\mathbf{k}_i,\mathbf{k_i})|^2. \tag{10.52}$$

Equation (10.42) now only holds for the elastic scattering cross section

$$\sigma(\theta) = |f(\theta)|^2, \tag{10.53}$$

and

$$f(\theta) = \sqrt{\sigma_{sc}}\, e^{i\nu(\theta)}. \tag{10.54}$$

Therefore in this case $\nu(\theta)$ is the solution of the integral equation

$$\sqrt{\sigma_{sc}(\theta)}\sin\nu(\theta) = \frac{k}{4\pi}\int \sqrt{\sigma(\theta')\sigma(\theta'')}\cos[\nu(\theta') - \nu(\theta'')]\,d\Omega_{\mathbf{k}} + f_{abs}(k,\theta), \tag{10.55}$$

where $|k| = |k_i|$. The phase shift $\delta(k,b)$ is now a complex quantity

$$\delta(k,b) = \eta(k,b) + i\gamma(k,b), \tag{10.56}$$

where η and γ are real functions of k and b and $\gamma(k,b) > 0$ for absorptive potential. We can write the scattering amplitude, $f(\theta)$, for the complex phase shift as

$$f(\theta) = -ik\int_0^\infty \left[e^{-2\gamma(b)}e^{i\eta(b)} - 1\right] J_0\left(2kb\sin\frac{\theta}{2}\right) b\,db \tag{10.57}$$

and we can replace (10.48) by

$$\sqrt{\sigma_{sc}(\theta)}\sin\nu(\theta) = k\int_0^\infty J_0\left(2kb\sin\frac{\theta}{2}\right)\left[1 - e^{-2\gamma(b)}\cos(2\eta(b))\right] b\,db. \tag{10.58}$$

By inverting this relation we find

$$1 - e^{-2\gamma(b)}\cos(2\eta(b)) = \frac{1}{k}\int_0^\infty J_0(bq)\sqrt{\sigma_{sc}}\,[\sin\nu(q)]q\,dq, \tag{10.59}$$

where $\nu(q)$ is given by the solution of (10.44).

If the real part of the phase shift is zero we can determine $\gamma(b)$ uniquely, since in this case from (10.57) it follows that

$$\mathrm{Re}(f(\theta)) = 0, \quad \mathrm{Im}(f(\theta)) > 0, \tag{10.60}$$

and therefore

$$f(\theta) = i\sqrt{\sigma_{sc}(\theta)}, \tag{10.61}$$

and

$$\nu(\theta) = \frac{\pi}{2}. \tag{10.62}$$

In this special case (10.59) reduces to

$$1 - e^{-2\gamma(b)} = \frac{1}{k}\int_0^\infty J_0(qb)\sqrt{\sigma_{sc}(q)}q\,dq, \tag{10.63}$$

and therefore if $\sigma_{sc}(q)$ is known $\gamma(b)$ can be determined from (10.63).

As an example consider the case where

$$\sigma_{sc} = \frac{c^2 k^2}{q^2}, \tag{10.64}$$

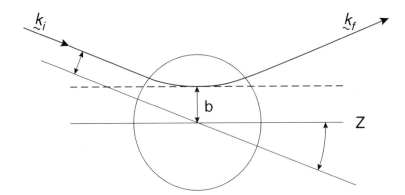

Figure 10.2: The classical trajectory used in the WKB approximation is the curve shown in the figure. The dashed line is the approximate path used in the impact parameter approximation, and b is the impact parameter.

then from (10.63) we find that

$$e^{-2\gamma(b)} = \begin{cases} 0 & b < c \\ \frac{1}{2} & b = c \\ 1 & b > c \end{cases} . \tag{10.65}$$

Therefore $\gamma(b) = \infty$ for $r < c$ and $\gamma(b) = 0$ for $b > c$ [1].

References

[1] R. Glauber, Lectures in Theoretical Physics, Vol. 1 (Interscience Publishers, New York, 1959).

[2] R.G. Newton, *Scattering of Waves and Particles*, Second Edition, (Springer-Verlag, New York, 1982) Chapter 18.

[3] For a detailed discussion of WKB and eikonal approximations and their relationship see R. Bellman and R. Vasudevan in *Mathematics and Applications*, Vol. 385, (Springer, New York, 1977).

[4] S.J. Wallace, Ann. Phys. 78, 190 (1973). This paper provides an excellent review of the eikonal approximation.

[5] M. Razavy, *Quantum Theory of Tunneling*, Second Edition, (World Scientific, Singapore, 2014) Chapter 2.

[6] M. Razavy, Can. J. Phys. 49. 1885 (1971).

[7] G. Arfken, *Mathematical Methods for Physicists*, Third Edition, (Academic Press, New York 1985) p. 653.

[8] R.P. Boas and C. Stutz, Am. J. Phys. 39, 745 (1971).

[9] W. Squire, Am. J. Phys. 41, 291 (1973).

[10] T.Y. Wu and T. Ohmura, *Quantum Theory of Scattering*, (Prentis Hall, Englewood Cliffs, 1961).

Chapter 11

Inverse Methods Applied to Study Symmetries and Conservation Laws

It is well-known that the justification for symmetry principles, like other physical laws, is based on the experimental evidence for them. In this chapter we want to examine two different quantum systems where the direct solution of the equation of motion lead to certain symmetries, and the related conservation laws, but the inverse problems, in most cases, do not lead to conservation laws. The first problem is the classically degenerate motion of a two-dimensional anharmonic oscillator with a close orbit (Sec. 1.3), and we observe that this degeneracy breaks down when the system is quantized. The second problem is about the accidental degeneracy of a hydrogen type atom, having the same spectrum as that of hydrogen, but the quantum analogue of the Runge–Lenz vector is not a constant of motion. Thus the solution of the inverse problem, for his case, generates a large number of dynamical systems, and only one is the system where the Runge–Lenz vector is a constant of motion.

In this book we will not discuss the relation between the observable symmetries and the conserved quantities, but refer the reader to review articles on this topic [1]–[3].

11.1   Classical Degeneracy and Its Quantum Counterpart

As we have discussed in Chapter 1, from the observation of a closed orbit having a degenerate Hamiltonian in action-angle variable (1.26), we can construct infinitely many Hamiltonians in phase space, Eq. (1.27). Using the semiclassical WKB approximation, we find that the energy eigenvalues of these Hamiltonians are;

$$E_{WKB}(m,n) = \hbar\omega(m+n+1), \tag{11.1}$$

where m and n are integers. Thus in this approximation E_{WKB} is degenerate, and depends on the sum of m and n. Now for the quadratic potential such as the one given in (1.27) we can solve the Schrödinger equation exactly. The total wave function for the Hamiltonian (1.27), when quantized can be written as the product

$$\Psi(x,y) = \psi_\mu(x)\psi_\nu(y), \tag{11.2}$$

where $\psi_\mu(x)$ satisfies the differential equation

$$\left[-\frac{\hbar^2}{2}\frac{d^2}{dx^2} + \Omega_x^2(x)x^2 \right] \psi_\mu(x) = \varepsilon_\mu \psi_\mu(x), \tag{11.3}$$

assuming that the particle has a unit mass.
We can write a similar relation for $\psi_\nu(y)$. The solution of (11.3) with the boundary condition $\psi_\mu(x) \to 0$ as $x \to \pm\infty$ is given by [7]

$$\psi_\mu(x) = \begin{cases} N_1 D_\mu \left(\sqrt{\frac{2\omega_1}{\hbar}}x \right) & \text{for } x \geq 0 \\[2ex] N D_\beta \left(-\sqrt{\frac{2\omega_2}{\hbar}}x \right) & \text{for } x \leq 0 \end{cases}, \tag{11.4}$$

where $D_\mu(z)$ is the parabolic cylinder function, and ω_1 and ω_2 are given by [4]–[6];

$$\omega_i = \frac{\omega}{\sqrt{1 \pm \lambda_i}}, \quad i = 1, 2. \tag{11.5}$$

The constant N_1 and N_2 are the normalization constants, and β and μ are related to each other by the relation

$$\left(\beta + \frac{1}{2} \right)(1 + \lambda_1) = \left(\mu + \frac{1}{2} \right)(1 - \lambda_1). \tag{11.6}$$

At $x = 0$ the two branches of the solution (11.4) should join each each other smoothly, and this condition gives us the eigenvalue equation

$$1 + \left(\sqrt{\frac{1 - \lambda_1}{1 + \lambda_2}} \right) \frac{\Gamma\left(\frac{1-\mu}{2} \right)\Gamma\left(-\frac{1}{2}\beta \right)}{\Gamma\left(\frac{-\mu}{2} \right)\Gamma\left(\frac{1-\beta}{2} \right)} = 0. \tag{11.7}$$

Therefore the exact quantum mechanical energy eigenvalues are given by

$$E(\mu, \nu) = \varepsilon_\mu + \varepsilon_\nu = \hbar\omega \left(\frac{\mu + \frac{1}{2}}{1 + \lambda_1} + \frac{\nu + \frac{1}{2}}{1 + \lambda_2} \right), \tag{11.8}$$

where μ and ν refer to the roots of (11.7) and the corresponding equation for the y coordinates. So unlike the approximate total energy $E(m, n)$, Eq. (11.1) which exhibits degeneracy, the exact result does not possess this property.

11.2   Inverse Problem for Angular Momentum Eigenvalues

We have argued that whereas the direct problem, in general, yields unique and unambiguous results for for the observed or measured quantities, for the inverse problem this is not true. That is, from the empirical data we cannot infer a unique rule or a set of rules that governs the physics of the problem. We have shown this for energy levels of bound systems and for scattering phase shifts.

In classical dynamics as well as in quantum mechanics we find symmetries and conservation laws which can be derived from the force laws and the equations of motion, and these can be verified experimentally. The connection between the accidental degeneracy of a system and its internal symmetry group has been carefully examined by Ui and Takeda [8]. By constructing a simple model of three-dimensional harmonic oscillator with a constant spin-orbit potential of the form $\frac{1}{2} \left(\mathbf{p}^2 + \mathbf{r}^2 \right) + \lambda \boldsymbol{\sigma} \cdot \mathbf{L}$ with $\lambda = \pm 1$, these authors show that there is no such a symmetry group in the system. We can ask what will happen to these symmetries and conservation laws when we are dealing with inverse problems. As far as the present author knows, there has not been a study of constrained inverse problems. In what follows we discuss two simple examples about the breakdown of the accidental degeneracy when we go from classical theory to quantum theory and vice versa. In the first one we start from the classical inversion of a problem discussed in Sec. 1.3. In the second one we assume the full spectrum of the hydrogen atom, including the directional (geometrical) and accidental degeneracies of the of the eigenvalues, and show that these degeneracies are possible without the conserved Runge–Lenz vector. We have already noticed that from the spectrum of the harmonic oscillator with the help of the Gel'fand–Levitan equation we can find an infinite set of potentials. In a similar way we can show that from the set of eigenvalues we can obtain the condition for the centrality of the potential in a two-body problem. But in addition we can find infinitely many non-central forces having the same set of eigenvalues.

One of the early great successes of quantum mechanics in its matrix formulation was the complete solution of the non-relativistic hydrogen atom problem by Pauli [9]. Pauli assumed that classical constants of motion remain constants of motion when quantized. Thus in addition to the conserved operators (matrices) the total Hamiltonian H and the angular momentum $\mathbf{L}$ for the motion of the electron in hydrogen atom, there is a third conserved quantity which is the Runge–Lenz vector $\mathbf{K}$ [10]. He then showed how it is possible to diagonalize the Hamiltonian and find all of its eigenvalues [9], [11]. In this way he was able to show the directional and the accidental degeneracies of the energy levels. The same results were obtained later from the solution of the Schrödinger equation. The presence of this accidental degeneracy is attributed to the constancy of the Runge–Lenz vector. Mathematically, the accidental degeneracy is due to the fact that the wave equation is separable in two coordinate systems, the spherical polar and the parabolic coordinate systems [12], [13].

For the direct problem, we start with the Schrödinger equation for central force, and by separating the angular-dependent part of the wave function from the radial part, we reach the well-known conclusion that the eigenvalues of the square angular momentum operator, $\mathbf{L}^2$, are $\hbar^2 \ell(\ell + 1)$ where ℓ is an integer.

For the inverse problem we study the eigenvalues of a force depending on the polar angle θ and has a form $\frac{f(\theta)}{r^2}$. This form allows for the separation of the variables in the Schrödinger equation which is essential for solving the inverse problem. Now we inquire wether it is possible to find the function $f(\theta)$ in such a way that the eigenvalues of the angular wave function are identical with those of a central force.

Our starting point for this derivation is the general form of the Schrödinger equation for two particles of reduced mass μ interacting by a nonlocal noncentral potential [15] $V_1(\mathbf{r}, \mathbf{r}')$,

$$-\frac{\hbar^2}{2\mu}\nabla^2 \psi(\mathbf{r}) + \int V_1(\mathbf{r}, \mathbf{r}')\, \psi(\mathbf{r}')\, d^3 r' = E\psi(\mathbf{r}), \tag{11.9}$$

where $V_1(\mathbf{r}, \mathbf{r}')$ is a real energy independent symmetric function of $\mathbf{r}$ and $\mathbf{r}'$. For the particular problem that we are studying we choose

$$V_1(\mathbf{r}, \mathbf{r}') = \frac{1}{r^2}\left[V(r)\delta(\phi - \phi') - \frac{f(\theta)}{2\pi r^2}\right]\delta(r - r')\,\delta(\cos\theta - \cos\theta'). \tag{11.10}$$

For the central force we can choose the Coulomb or the harmonic oscillator potential, and $f(\theta)$ is a function of $\cos\theta$ to be determined later. Here we choose $V_1(\mathbf{r}, \mathbf{r}')$ so that the wave equation becomes separable in spherical polar coordinates. If we write $\psi(\mathbf{r}) = \psi_m(r, \theta)e^{im\phi}$, and substitute this in (11.9) we find

$$\left[-\frac{\hbar^2}{2\mu}\nabla^2 + V(r)\right]\psi_m(r, \theta) - \frac{f(\theta)}{r^2}\psi_m(r, \theta)\delta_{m,0} = E\psi_m(r, \theta), \tag{11.11}$$

where

$$\nabla^2 = \frac{1}{r^2}\frac{\partial}{\partial r}\left(r^2\frac{\partial}{\partial r}\right) - \frac{1}{r^2\sin\theta}\frac{\partial}{\partial\theta}\left(\sin\theta\frac{\partial}{\partial\theta}\right) + \frac{1}{\sin^2\theta}\frac{\partial^2}{\partial\phi^2}. \tag{11.12}$$

Equation (11.11) shows that for $m \neq 0$, we have exactly the same set of eigenvalues as those obtained by solving the Schrödinger equation with a local potential, and therefore the same degeneracies as those associated that with $V(r)$. Now for $m = 0$, it can be seen in (11.11), that the force is noncentral. We want to know whether it is possible to choose $f(\theta)$ so that the problem retains the same set of eigenvalues for $\mathbf{L}^2$, i.e. $\hbar^2 \ell(\ell + 1)$. Since the spectrum of these eigenvalues are discrete, we can follow the work of Abraham and Moses to find $f(\theta)$. We first simplify the notation and write

$$x = \cos\theta, \quad \text{and} \quad \Delta v(x) = \frac{2\mu}{\hbar^2} f(\cos\theta), \tag{11.13}$$

then the differential equation for $x = \cos\theta$ changes from Legendre differential equation to

$$\frac{d}{dx}\left[(1 - x^2)\frac{d\Theta(x)}{dx}\right] + [\ell(\ell + 1) + \Delta v(x)]\,\Theta(x) = 0, \tag{11.14}$$

where we have chosen the separation constant to be $\ell(\ell+1)$ with ℓ being an integer. Thus we use the set of eigenvalues as in the case of $\Delta v(x) = 0$.

By changing $\Theta(x)$ to $\phi_\ell(x)$ where

$$\Theta(x) = \frac{\Phi_\ell(x)}{\sqrt{1 - x^2}}, \tag{11.15}$$

we eliminate the first derivative in (11.14) to find

$$(1 - x^2)\frac{d^2\Phi_\ell(x)}{dx^2} + \left\{\ell(\ell + 1) + \frac{1}{1 - x^2} + \Delta v(x)\right\}\Phi_\ell(x) = 0. \tag{11.16}$$

In order to find $\Delta v(x)$ we start with the Gel'fand–Levitan equation, but now it is convenient to write it in a different form

$$K(x, y) = -\Omega(x, y) - \int_{-1}^{x} K(x, z)\Omega(z, y)\frac{dz}{(1 - z^2)}, \tag{11.17}$$

where $\Omega(x, y)$ which is separable

$$\Omega(x, y) = -D\left(1 - x^2\right)\left(1 - y^2\right), \tag{11.18}$$

is, in fact, a solution of the differential equation

$$\left[(1 - x^2)\frac{d^2}{dx^2} + \frac{1}{1 - x^2}\right]\Omega(x, y) = 0. \tag{11.19}$$

Let us note that in Eq. (11.18) D is the normalization constant. If we substitute (11.18) in (11.17) we find $K(x, y)$ to be

$$K(x, y) = \frac{D\sqrt{1 - x^2}\sqrt{1 - y^2}}{1 - (x + 1)D}. \tag{11.20}$$

Having determined the kernel $K(x, y)$, we can find $\Delta v(x)$ from Eq. (4.59)

$$\Delta v(x) = -2\sqrt{1-x^2}\,\frac{d}{dx}\left(\frac{K(x,x)}{\sqrt{1-x^2}}\right), \tag{11.21}$$

with the result that $\Delta v(x)$ is given by

$$\Delta v(x) = -2\sqrt{1-x^2}\,\frac{d}{dx}\left(\frac{D\sqrt{1-x^2}}{1-D-xD}\right). \tag{11.22}$$

The angular part of the wave function, $\phi(x)$, can be related to the Legendre polynomial $P_\ell(x)$;

$$\phi_\ell(x) = \sqrt{1-x^2}\,P_\ell(x) + \int_{-1}^{x}\frac{K(x,z)}{(1-z^2)}P_\ell(z)\sqrt{1-z^2}dz. \tag{11.23}$$

We can simplify (11.23) by noting that [16]

$$\int_{\cos\theta}^{1}P_\ell(x)dx = \sin\theta P_\ell^{(-1)}(\cos\theta). \tag{11.24}$$

Then the wave function can be found analytically:

$$\Phi_\ell(x) = \sqrt{1-x^2}\Theta(x) = \sqrt{1-x^2}\left\{P_\ell(x) - \frac{D\sqrt{1-x^2}P_\ell^{(-1)}(x)}{1-D-xD}\right\}. \tag{11.25}$$

Thus we conclude that the choice

$$f(\theta) = \frac{\hbar^2}{\mu}\frac{d}{d\theta}\left(\frac{D\sin\theta}{1-D-D\cos\theta}\right), \tag{11.26}$$

in Eq. (11.10) produces the same set of angular momentum eigenvalues as those of the central force problems.

Accidental Degeneracy in the Spectrum of Hydrogen Atom — If the potential $V(r)$ in Eqs. (11.10) and (11.11) is the Coulomb potential, $V(r) = -\frac{e^2}{r}$, then the eigenvalues of the wave equation (11.11) when $F(\theta) = 0$ are obtained from the solution of radial Schrödinger equation

$$\frac{d}{dr}\left(r^2\frac{dR}{dr}\right) + \frac{2\mu r^2}{\hbar^2}\left[E + \frac{e^2}{r}\right] = \ell(\ell+1)R. \tag{11.27}$$

These eigenvalues are given by

$$E_n = -\left(\frac{\mu e^4}{2\hbar^2}\right)\frac{1}{n^2}, \quad n = \nu + \ell + m. \tag{11.28}$$

The fact that the energy is independent of ℓ is a special property of the hydrogen atom which must be attributed to the presence of the Coulomb potential. Thus for a given ℓ, all states corresponding to $\ell = 0 \to n-1$ have the same energy levels and are degenerate. Classically this degeneracy is associated with the conservation of $\mathbf{L}^2$, and in addition there is a conservation of the Runge–Lenz vector $\mathbf{A} = \mathbf{p} \wedge \mathbf{L} - mk\hat{\mathbf{r}}$. The conservation of the classical Runge–Lenz vector implies that the the semi-major axis is fixed in space, i.e. the orbit does not precess [18]–[21].

11.3 Quantum Potentials Proportional to $\hbar$

Using the method of Abraham and Moses [22], we can construct stronger potentials proportional to $\hbar$, and use these potentials to discuss the question of the relationship between the existence of classically closed orbits and the degeneracy of levels in the hydrogen atom. To this end we replace $V_1(\mathbf{r}, \mathbf{r}')$ in (11.9) by $V_2(\mathbf{r}, \mathbf{r}')$ where

$$V_2(\mathbf{r}, \mathbf{r}') = \frac{1}{r^2}\left[V(r)\delta(\phi - \phi')\,\delta(\cos\theta - \cos\theta') + \Delta V(r)Y_{L0}(\theta)Y_{L0}(\theta')\right]\delta(r - r'),$$

(11.29)

where L is a very large integer. With this potential the Schrödinger equation becomes

$$\left[-\frac{\hbar^2}{2\mu}\nabla^2 + V(r)\right]\psi_{n\ell m}(r) + \Delta V(r)\delta_{L\ell}\delta_{m0}\psi_{n\ell m}(r) = E_n\psi_{n\ell m}(r).$$

(11.30)

Here $\Delta V(r) = \frac{\hbar^2}{2\mu}v(r)$. In the present case the wave function for all ℓ and m values can be written as

$$\psi_{n\ell m}(r) = \frac{1}{r}u_{n\ell}(r)Y_{\ell m}(\theta, \phi).$$

(11.31)

Now for either $m \neq 0$ and or for $\ell \neq L$ we have the standard Schrödinger equation

$$-\frac{\hbar^2}{2\mu}\frac{d^2 u_{n\ell}(r)}{dr^2} + \left[V(r) - E_n + \frac{\hbar^2\ell(\ell+1)}{2\mu r^2}\right]u_{n\ell m}(r) = 0.$$

(11.32)

But when $m = 0$ and $\ell = L$, for the radial part of the wave function we get

$$-\frac{\hbar^2}{2\mu}\frac{d^2\chi_{nL}(r)}{dr^2} + \left[V(r) + \Delta V(r) - E_n + \frac{\hbar^2 L(L+1)}{2\mu r^2}\right]\chi_{nL}(r) = 0.$$

(11.33)

From what we have seen earlier, e.g. Eq. (4.86), $\Delta v(r)$ is related to the ground state wave function by

$$\Delta V(r) = -\frac{\hbar^2}{\mu}\frac{d^2}{dr^2}\ln\left[1 - C\int_0^r u_{L+1,L}^2(r)dr\right].$$

(11.34)

For instance for hydrogen atom where $V(r) = -\frac{k}{r}$, the ground state for $\ell = L$ has the radial quantum number $n_r =$) and therefore its principal quantum number is $n = L + 1$. In this case $u_{L+1,L}(r)$ is given by [23]

$$u_{L+1,L}^2(r) = \frac{2L}{\lambda} \left(\frac{2Lr}{\lambda}\right)^{2L+2} \frac{\exp\left(-\frac{2Lr}{\lambda}\right)}{(2L+1)!}. \tag{11.35}$$

In this relation $\lambda = \frac{L^2\hbar^2}{\mu k}$ is a constant with the dimension of length, and has a nonzero classical limit when L is large enough so that $L\hbar$ is the classical angular momentum. For small C, $\Delta V(r)$ can be approximated by

$$\Delta V(r) \approx \frac{\hbar^2 C}{\mu} \frac{d}{dr} u_{L+1,L}^2(r)$$

$$\approx \left(\frac{2C}{\mu\sqrt{\pi}}\right) L^{\frac{3}{2}}\hbar^2 \left(\frac{r}{\lambda}\right)^{2L+2} \exp\left[-2L\left(\frac{r}{\lambda}\right) + 2\right]\left(\frac{1}{\lambda r} - \frac{1}{\lambda^2}\right). \tag{11.36}$$

A more accurate estimate of the strength of the potential can be obtained by writing (11.36) as

$$\Delta V(r) \approx -\frac{2\hbar^2 C L^{\frac{3}{2}}}{\mu\sqrt{\pi}} \exp\left[-L\left(\frac{r}{\lambda} - 1\right)^2\right]\left(\frac{1}{\lambda^2} - \frac{1}{r\lambda}\right). \tag{11.37}$$

This expansion shows that for large values of the integer L, the potential is significant only in the neighbourhood of $r = \lambda\left(1 \pm \frac{1}{\sqrt{L}}\right)$, and that the strength of the potential in this range is proportional to $\frac{L\hbar^2}{\mu\lambda^2}$.

The presence of $f(\theta)$ in Eq. (11.9) shows that this equation is not separable in parabolic coordinates. Thus it seems that the separability of the Schrödinger equation in more than one coordinate system is not necessary for the presence of accidental degeneracy.

Precession of the Orbit for the Classical Limit of the Hydrogen Atom —
Let us consider the classical Hamiltonian for the hydrogen atom where the potential is the sum of two terms: Coulomb potential $-\frac{k}{r}$ plus a very small perturbation $\Delta V(r)$ given by (11.37)

$$H = \frac{1}{2\mu}\left(p_r^2 + p_\theta^2 + \frac{p_\phi^2}{r^2\sin\theta^2}\right) - \frac{k}{r} + \Delta V(r). \tag{11.38}$$

Since p_ϕ is a constant of motion, for simplicity we set it equal to zero, and then write H in terms of the action-angle variables

$$H = -\frac{2\pi^2\mu k^2}{(J_r + J_\theta)^2} + \Delta V(J_r + J_\theta), \tag{11.39}$$

with

$$\Delta V(J_r + J_\theta) = \frac{1}{T}\int_0^T \Delta V(r)dt$$

$$= \frac{2\pi\mu}{J_\theta T}\oint \Delta V(r(\theta))r^2(\theta)d\theta. \tag{11.40}$$

In this last relation T and r are the period of the motion and the equation of the orbit, which, in the absence of the perturbation $v(r)$ are given by

$$T = \frac{(J_r + J_\theta)^2}{4\pi^2 \mu k^2},$$ (11.41)

and

$$r(\theta) = \frac{J_\theta^2}{4\pi^2 \mu k} \left\{ \frac{1 + [J_r(J_r + 2J_\theta)]^{\frac{1}{2}} \cos\theta}{(J_r + J_\theta)} \right\}^{-1},$$ (11.42)

respectively.

Now the changes in the frequencies of the radial and angular variables are obtained from the relations:

$$\Delta\nu_r = \frac{\partial}{\partial J_r} \Delta V(J_r, J_\theta), \quad \text{and} \quad \Delta\nu_\theta = \frac{\partial}{\partial J_\theta} \Delta V(J_r, J_\theta).$$ (11.43)

Since $\Delta\nu_r$ is not equal to $\Delta\nu_\theta$, therefore there will be a very slow precession of the orbit. That is the paths of finite motion are no longer closed, and after each revolution the perihelion is displaced by a small angle α. The rate of precession is given by $\frac{d\alpha}{dt}$ where

$$\frac{d\alpha}{dt} = \frac{\alpha}{T} = \frac{2\pi}{T} \left(\frac{\nu + \Delta\nu_\theta}{\nu + \Delta\nu_r} - 1 \right) \approx 2\pi(\Delta\nu_\theta - \Delta\nu_r).$$ (11.44)

This rate according to (11.37) is proportional to $\frac{L\hbar^2}{\mu\lambda^2}$, which is very small. Thus with a suitable choice of $f(\theta)$, Eq. (11.26), we find that the classical orbit is not closed, and the conservation of Runge–Lenz vector is violated by a very small perturbation.

References

[1] G. Feinberg and M. Goldhaber, Proc. Natl. Acad. Sci. USA, 45, 1301 (1959).

[2] E.P. Wigner, Proc. Natl. Acad. Sci. USA, 51, 956 (1964).

[3] D.J. Gross, Phys. Today, 48, 46 (1964).

[4] G. Ghosh and R.W. Hasse, Phys. Rev. 24, 1027 (1981).

[5] M.M. Nieto, Phys. Rev. D 24, 1030, (1981).

[6] J.F. Marko and M. Razavy, Lettere Al Nuovo Cimento, 40, 533, 1984.

[7] M. Abramowitz and I.A. Stegun, *Handbook of Mathematical Functions*, (Dover Publications, New York, 1965) Chapter 19.

[8] H. Ui and G. Takeda, Prog. Theor. Phys. 72, 266 (1984).

[9] W. Pauli, Z. Phys. 36, 336 (1926). An English translation of this seminal work has been published in *Sources of Quantum Mechanics*, Edited by B.L. van der Waarden (North-Holland Amsterdam, 1967) p. 387 ff.

[10] H. Goldstein, *Classicl Mechanics*, Second Edition, (Addison Wesley, 1980) p. 102.

[11] M. Razavy, *Heisenberg's Quantum Mechanics*, (World Scientific, Singapore, 2011) p. 251.

[12] H.A. Bethe and E.E. Salpeter, *Quantum Mechanics of One- and Two- Electron Systems*, (Plenum Publishibg Corporation, New York, 1977).

[13] V. Bargmann, Z. Phys. 99, 587 (1936).

[14] J.M. Jauch and E.I. Hill, Phys. Rev. 57, 641 (1940).

[15] N.F. Mott and H.S.W. Massey, *The Theory of Atomic Collisions*, (Oxford University Press, Oxford, 1965) Chapter VIII.

[16] I.S. Gradshteyn and I.M. Ryzhik, *Tables of Integrals, Series, and the Products*, (Academic Press, New York, 1965) p. 794.

[17] M. Razavy, Phys. Letts. 88A, 215 (1982).

[18] P. Stehle and M.Y. Han, Phys. Rev. 159, 1076 (1967).

[19] H.V. McIntosh, Am. J. Phys. 27, 620 (1959).

[20] E.J. Saletan and A.H. Crombie, *Theoretical Mechanics* (Wiley, New York 1970) Chapter 7.

[21] J.M. Jauch and E.L. Hill, Phys. Rev. 57, 641 (1940).

[22] P.R. Abraham and H.E. Moses, Phys. Rev. A22. 1333 (1980).

[23] L.S. Brown, Am. J. Phys. 41, 525 (1973).

Chapter 12

Inverse Problems in Quantum Tunneling

One of the oldest and often used inverse problems of quantum theory is the calculation of the potential energy curves of molecules from the spectroscopic data. The method originally advanced by Rydberg in 1931, [1] was later developed as a powerful technique by contributions from Klein [2] and from Rees [3]. Today it forms one of the a basic tools of determination of potential functions in molecular physics [4], [5].

12.1 Nonlinear Equation for Variable Reflection Amplitude

In problems where the incident wave function is partly reflected and partly transmitted we have two boundary conditions, the reflection amplitude at one end and the transmitted amplitude at the other. To simplify the imposition of the boundary conditions we will consider a nonlinear equation for variable reflection amplitude for the one-dimensional motion of a particle. Now we derive a useful equation for the calculation of the reflection amplitude for the one-dimensional motion of a particle penetrating a barrier $V(x)$ [6], [7]. We formulate the problem for the Schrödinger equation, but the result can easily be modified for the reflection of a scalar wave. Consider the Schrödinger equation

$$\psi''(x) + \left[k^2 - v(x)\right]\psi(x) = 0, \tag{12.1}$$

where we have set $\frac{\hbar^2}{2m} = 1$ and $v(x) = \frac{2m}{\hbar^2} V(x)$. If a plane wave from the left of the barrier approaches the barrier, then the formal solution of the wave equation (12.1) can be written as

$$\psi(x) = e^{ikx} + \frac{1}{2ik} \int_{-\infty}^{\infty} e^{ik|x-x'|} v(x') \psi(x') \, dx'. \tag{12.2}$$

From the asymptotic forms of this formal solution, we define the reflection and the transmission amplitudes by

$$R(k) = \frac{1}{2ik} \int_{-\infty}^{\infty} e^{ikx'} v(x') \psi(x') \, dx', \tag{12.3}$$

and

$$T(k) = 1 + \frac{1}{2ik} \int_{-\infty}^{\infty} e^{-ikx'} v(x') \psi(x') \, dx'. \tag{12.4}$$

Now let us introduce a cut-off potential $v(y,x)$ which we define as

$$v(y,x) = v(x)\theta(x-y), \tag{12.5}$$

where

$$\theta(x) = \begin{cases} 1 & x > 0 \\ 0 & x < 0 \end{cases}, \tag{12.6}$$

and write the wave function (12.2) for this cut-off potential which is now a function of x and y as:

$$\psi(y,x) = e^{ikx} + \frac{1}{2ik} \int_{y}^{\infty} e^{ik|x-x'|} v(x') \psi(y,x') \, dx', \quad x \geq y. \tag{12.7}$$

Thus the reflection amplitude (12.3) becomes

$$R(y) = \frac{1}{2ik} \int_{y}^{\infty} e^{ikx'} v(x') \psi(y,x') \, dx'. \tag{12.8}$$

From Eqs. (12.7) and (12.8) we get

$$\psi(y,x) = e^{iky} + R(y)e^{-iky}. \tag{12.9}$$

Now we find $\frac{\partial}{\partial y}\psi(y,y)$ from (12.7);

$$\left[\frac{\partial}{\partial y}\psi(y,y)\right] B(y) = e^{ikx} + \frac{1}{2ik} \int_{y}^{\infty} e^{ik|x-x'|} \psi(x') \frac{\partial}{\partial y}\psi(y,x') B(y) dx', \quad x \geq y, \tag{12.10}$$

where

$$-\frac{1}{2ik} e^{iky} v(y)\psi(y,y) B(y) = 1. \tag{12.11}$$

Also from (12.10) we find that

$$\left[\frac{\partial}{\partial y}\psi(y,x)\right]B(y) = \psi(y,x). \tag{12.12}$$

This follows from the fact that the left-hand side of (12.12) satisfies (12.7), i.e. the integral for $\psi(y,x)$. We also note that (12.8) for $R(y)$ can be written as

$$R(y)B^{-1}(y) = \frac{1}{2ik}\int_y^\infty e^{ikx}v(x)\frac{\partial}{\partial y}\psi(y,x)\,dx, \tag{12.13}$$

where $B^{-1}(y)$ is the inverse of $B(y)$ introduced by (12.11). By differentiating $R(y)$, Eq. (12.9), with respect to y and eliminating the integral which depends on $\frac{\partial}{\partial y}\psi(y,y)$, we obtain the following differential equation for $R(y)$:

$$\frac{d}{dy}R(y) = -\frac{e^{iky}}{2ik}v(y)\left[e^{iky} + R(y)e^{-iky}\right] + R(y)B^{-1}(y). \tag{12.14}$$

Then we calculate the function $B^{-1}(y)$ from Eq. (12.11)

$$B^{-1}(y) = -\frac{1}{2ik}e^{-ky}v(y)\psi(y.y). \tag{12.15}$$

By substituting (12.15) in (12.14) and simplifying the result by expressing $\psi(y,y)$ in terms of $R(y)$, we find the nonlinear differential equation for $R(y)$;

$$\frac{d}{dy}R(y) = -\frac{v(y)}{2ik}\left[e^{iky} + R(y)e^{-iky}\right]^2. \tag{12.16}$$

12.2 Inverse One-dimensional Tunneling Problem

Now we want to formulate the one-dimensional scattering (or tunneling) problem when the barrier is given by the potential $v(x)$, and when both $|v(x)|$ and $|xv(x)|$ are integrable over the entire range of x, $-\infty < x < +\infty$. Earlier, in Sec. 5.4, we discussed the one-dimensional inverse reflection problem, and now we will consider its application to tunneling problem. We will repeat briefly the main points that we discussed in that section and see how it can be applied to tunneling.

If we assume that the wave is incident from the right of the barrier, and denote the wave function by $\psi_R(k,x)$, we have, as $x \to +\infty$

$$\begin{cases} \psi_R(k,x) = e^{-ikx} + S_{12}(k)e^{ikx} \\ \\ \psi_L(k,x) = S_{22}(k)e^{ikx} \end{cases} \tag{12.17}$$

and as $x \to -\infty$

$$\begin{cases} \psi_R(k, x) = S_{11}(k)e^{-ikx} \\ \\ \psi_L(k, x) = e^{ikx} + S_{21}(k)e^{-ikx}. \end{cases} \qquad (12.18)$$

The completeness relation for the scattering in this case is

$$\frac{1}{\pi} \int_0^\infty \left[\psi_R(k, x)\psi_R^*(k, y) + \psi_L(k, x)\psi_L^*(k, y) \right] dk$$

$$+ \sum_{n=1}^N \psi(i\gamma_n, x)\psi(i\gamma_n, y) = \delta(x - y), \qquad (12.19)$$

where the sum in (12.19) allows for the presence of bound states if they are present, and are normalized.

There are four coefficients S_{11}, S_{12}, S_{21} and S_{22} and there are four coefficients of reflection and transmission in both directions. These form a 2×2 scattering matrix,

$$S = \begin{bmatrix} S_{11} & S_{21} \\ S_{12} & S_{22} \end{bmatrix}. \qquad (12.20)$$

These equations are the same as Eqs. (5.87)–(5.91), except for the subscripts L and R replacing 1 and 2. Now the conservation of probability implies that the matrix S transforming the incoming wave to the outgoing wave be unitary, i.e.

$$S^\dagger S = \begin{bmatrix} S_{11} & S_{12} \\ S_{21} & S_{22} \end{bmatrix} \begin{bmatrix} S_{11} & S_{21} \\ S_{12} & S_{22} \end{bmatrix} = \begin{bmatrix} 1 & 0 \\ 0 & 1 \end{bmatrix} = SS^\dagger, \qquad (12.21)$$

where $\dagger$ denotes the Hermitian adjoint of the matrix S. Thus we have the following relations among the matrix elements of S;

$$|S_{12}|^2 + |S_{11}|^2 = 1, \quad |S_{21}|^2 + |S_{22}|^2 = 1, \qquad (12.22)$$

and

$$S_{12}^* S_{22} + S_{11}^* S_{21} = 0. \qquad (12.23)$$

From these relations we find the following results:

$$|S_{11}| = |S_{22}|, \quad \text{and} \quad |S_{12}| = |S_{21}|. \qquad (12.24)$$

Also from the asymptotic forms of $\psi_R(k, x)$ and $\psi_L(k, x)$, Eqs. (12.17) and (12.18) it is clear that

$$\begin{bmatrix} S_{11} & S_{12} \\ S_{21} & S_{22} \end{bmatrix} = \begin{bmatrix} T_L & R_R \\ R_R & T_R \end{bmatrix}, \qquad (12.25)$$

and (12.24) can be written as relations between the transmitted and reflected amplitudes T and R;

$$|T_L| = |T_R|, \quad |R_L| = |R_R|. \qquad (12.26)$$

For a symmetrical potential $v(x) = v(-x)$, then $\psi(k, x)$ and $\psi(k, -x)$ are both the solutions Schrödinger equation, and by replacing x and by $-x$ in (12.17) should give us the same wave function and therefore we conclude that

$$S_{11} = S_{22} \quad \text{or} \quad T_L = T_R. \tag{12.27}$$

Thus for a real symmetric potential the number of independent elements of the S matrix will be two. Now defining the Jost solution $f_1^{\pm}(k, x)$ by

$$\lim_{x \to +\infty} \left[\exp(\mp ikx) f_1^{\pm}(k, x)\right] = 1. \tag{12.28}$$

We can express both $\psi_1(k, x)$ and $\psi_2(k, x)$ in terms of $f_1^{\pm}(k, x)$, e.g. we can write the completeness relation (12.19) as

$$\frac{1}{\pi} \int_0^{\infty} \left[f_1(k, x) f_1^*(k, y) + f_1(k, x) f_1^*(k, y) S_{12}(k)\right] dk$$
$$+ \sum_{n=1}^{N} \bar{C}_{1n}^2 f_1(i\gamma_n, x) f_1(i\gamma_n, y) = \delta(x - y). \tag{12.29}$$

Here we have considered the reflection of waves incident from the right. We can write a similar equation for the waves incident from the left with the Jost solution $f_2^{\pm}(k, x)$ satisfying the boundary condition

$$\lim_{x \to -\infty} \left[\exp(\pm ikx) f_2^{\pm}(k, x)\right] = 1, \tag{12.30}$$

but now S_{12} and $\bar{C}_{1n}$ must be replaced by S_{21} and $\bar{C}_{2n}$. The normalization constants $\bar{C}_{1n}$ and $\bar{C}_{2n}$ represent the coefficients of the exponentially decaying bound state wave functions.

The integral equation for the first case (incidence from the right) has for the input $g_1(x, y)$ in the Gel'fand–Levitan equation

$$K_1(x, y) = g_1(x, y) + \int_x^{\infty} K_1(x, z) g_1(z, y) dz. \tag{12.31}$$

The input function $g(x, y)$ is related to the Jost function of the reflection coefficient, $f_1(k, x)$ and $S_{12}(k)$, the bound state energy $-\frac{\hbar^2 \gamma_n^2}{2m}$ and the normalization constant $\bar{C}_{1n}$. That is

$$g_1(x, y) = \frac{1}{2\pi} \int_{-\infty}^{\infty} \bar{f}_1(k, x) \bar{f}_1(k, y) \left[1 - S_{12}(k)\right] dk$$
$$+ \bar{C}_{1n}^2 \bar{f}_1(i\gamma_n, x) \bar{f}_1(i\gamma_n, y). \tag{12.32}$$

Just as the Gel'fand–Levitan or Marchenko formulations for the S wave considered earlier, if $g_1(x, y)$ is the sum of separable terms, then (12.32) is a degenerate integral equation and can be reduced to a set of linear equations.

If the potential has different asymptotic values at $x \to -\infty$ and $x \to +\infty$, which we denote by v_- and v_+ respectively, then there is the appearance of a second channel opening up at an energy equal to the largest of the two asymptotes v_- and v_+. If the energy of the incident particle is such that the propagation of the waves from one side is closed, then we can determine the whole S matrix from the reflection coefficient only in the case of wave coming from the side with the lower value of $\{v_-,\ v_+\}$.

For a through examination of the one-dimensional scattering in quantum mechanics the reader is referred to references [8] and [9].

The one-dimensional problem of reflection and transmission of scalar wave equation will be discussed in Sec. 13.2 of this book.

12.3   A Method for Finding the Potential from the Reflection Amplitude

The differential equation for the variable reflection amplitude for a single channel is (see Eq. (12.16))

$$\frac{dR(y,k)}{dy} = -\frac{v(y)}{2ik}\left(e^{iky} + R(y,k)e^{-iky}\right)^2,\tag{12.33}$$

where $R(y,k)$ is subject to the boundary condition

$$R(y \to \infty, k) \to 0.\tag{12.34}$$

The reflection amplitude is then obtained from the asymptotic solution of (12.33)

$$R(k) = R(y \to -\infty, k).\tag{12.35}$$

For the inverse problem we assume that (12.35) is known for all k values and we want to find $v(r)$. To this end we introduce another function $F(y,k)$ defined by

$$F(y,k) = \frac{1}{2ik}e^{-2iky}R(y,k),\tag{12.36}$$

and substitute for $R(y,k)$ in Eq. (12.33) to get a differential equation for $F(y,k)$

$$\frac{d}{dy}\left[2ikF(y,k)e^{2iky}\right] = -\frac{v(y)}{2ik}e^{2iky}\left[1 + 2iF(y,k)\right]^2.\tag{12.37}$$

Integrating (12.37) and substituting for $R(\infty, k)$, from (12.34) we find $F(y,k)$;

$$2ikF(y,k)e^{2iky} = \frac{1}{4k^2}\int_y^\infty v(y')\exp(2iky')\left[1 + 2iF(y',k)\right]^2 dy'.\tag{12.38}$$

Now replacing $F(y, k)$ by $R(y, k)$ on the left-hand side of (12.38), then taking the limit of $y \to -\infty$ and rearranging terms we obtain

$$\int_{-\infty}^{\infty} v(y)e^{2iky}dy = -2ikR(k) - 4\int_{-\infty}^{+\infty} v(y)e^{2iky}\left[iF(y, k) - F^2(y, k)\right]dy. \quad (12.39)$$

By taking the inverse Fourier transform of (12.39) we find the equation satisfied by $v(y)$,

$$v(y) = -\frac{2i}{\pi}\int_{-\infty}^{+\infty} kR(k)e^{-2iky}dk - \frac{2}{\pi}\int_{-\infty}^{+\infty} v(y')dy'$$
$$\times \int_{-\infty}^{+\infty} \left[iF(y', k) - F^2(y', k)\right]e^{2ik(y'-y)}dk. \quad (12.40)$$

Noting that (12.40) is an inhomogeneous integral equation for $v(y)$, we can solve it by iteration. Thus to the first order we have

$$v_1(y) = -\frac{2i}{\pi}\int_{-\infty}^{+\infty} kR(k)e^{-2iky}dk. \quad (12.41)$$

By substituting this expression for $v_1(y)$ in Eq. (12.38) and ignoring F on the right-hand side we get

$$e^{2iky}F_1(y, k) = \frac{1}{4k^2}\int_{y}^{\infty} v_1(y')e^{2iky'}dy'. \quad (12.42)$$

From this expression and (12.40) we find the potential $v(y)$ to the second order

$$v_2(y) = v_1(y) - \frac{2}{\pi}\int_{-\infty}^{+\infty} v_1(y')dy'$$
$$\times \int_{-\infty}^{+\infty} \left[iF_1(y', k) - F_1^2(y', k)\right]e^{2ik(y'-y)}dk. \quad (12.43)$$

While in principle we can continue this iteration to an arbitrary order in $v(y)$, in practice the instability of the numerical inversion of the Fourier transform limits the number of iterations [10].

A Simple Case of Inversion of the Reflection Amplitude — The simplest case that we want to consider is the inversion of the reflection amplitude for a δ-function potential $v(x) = s\delta(x)$ which is given by $R(k) = \frac{is}{2k+is}$. If we substitute this amplitude in (12.41) we find the approximate potential,

$$v_1(y) = \frac{s}{\pi}\int_{-\infty}^{+\infty} \frac{2k}{2k+is}e^{-2iky}dy = s\delta(y) - s^2e^{sy}\theta(-y). \quad (12.44)$$

A similar iterative technique for constructing the potential barrier from the reflection data is discussed in [9].

From the reflection and transmission amplitudes one can infer whether the barrier is of finite extent or not. The following important result is found by Portinari [11]. For a barrier of finite range the integral

$$\int_{-\infty}^{\infty} k \frac{R(k)}{T(k)} \exp(-2ikx)dk, \tag{12.45}$$

must vanish identically for all the points where $v(x) = 0$.

12.4 Finding the Shape of the Potential Barrier in One-Dimensional Tunneling

This inverse problem can be solved by using the techniques developed in quantum scattering theory [12] or by using the simpler formulation based on the semiclassical approximation. It is the latter approximate method which we will discuss now. Here the input data is the coefficient of transmission $|T(E)|^2$ which can be found from (see, for instance, [13]–[14])

$$|T(E)|^2 = \frac{1}{(1 + e^{2\sigma})}, \tag{12.46}$$

where $\sigma(E)$ is related to the phase integral $f_0(E)$, and to the mass of the particle which tunnels through the barrier, m

$$\sigma(E) = \sqrt{2m} \frac{f_0(E)}{\hbar}, \tag{12.47}$$

and where $f_0(E)$ is given by the integral

$$f_0(E) = \int_{x_1}^{x_2} [V(x) - E]^{\frac{1}{2}} dx. \tag{12.48}$$

Here $V(x)$ is the potential barrier and x_1 and x_2 are the two turning points, i.e. they are the roots of $E = V(x)$. Later we will include the next order correction, $f_2(E)$, in our determination of the potential. But for now we are interested in quantum tunneling, therefore the energy of the particle must be less than or equal to the maximum height of the potential;

$$E \le V_{max}. \tag{12.49}$$

In the following subsection we consider the minimum information which we can obtain from $f_0(E)$.

Determination of the Width of the Barrier as Seen by the Tunneling Particle — We assume that $V(x)$ has a minimum or a maximum between the turning points x_1 and x_2. If $|T(E)|^2$ is known then from (12.46) we can find $\sigma(E)$ and subsequently determine $f_0(E)$. Hence to find $V(x)$ we need to invert (12.48). For this inversion we use the identity that we have seen before in connection with solving Abel's integral equation [15]

$$\int_E^{V(x)} \frac{dE'}{\sqrt{(E'-E)\left[V(x)-E'\right]}} = \int_0^1 \frac{d\zeta}{\sqrt{\zeta}\sqrt{1-\zeta}}$$

$$= B\left(\frac{1}{2},\frac{1}{2}\right) = \left[\Gamma\left(\frac{1}{2}\right)\right]^2 = \pi, \qquad (12.50)$$

which is true for any $V(x)$. Here B (x,y) and $\Gamma(x)$ are beta and gamma functions respectively. Using this identity we find the following relation

$$x_2(E) - x_1(E) = \frac{1}{\pi}\int_{x_1}^{x_2} dx \int_E^{V(x)} \frac{dE'}{\sqrt{(E'-E)\left[V(x)-E'\right]}}. \qquad (12.51)$$

Now by changing the order of integration in (12.51) and observing that $V(x)$ has a maximum V_{max}, we find

$$x_2(E) - x_1(E) = \frac{1}{\pi}\int_E^{V_{max}} dE' \int_{x_1(E')}^{x_2(E')} \frac{dx}{\sqrt{(E'-E)\left[V(x)-E'\right]}}, \qquad (12.52)$$

or

$$x_2(E) - x_1(E) = -\frac{2}{\pi}\int_E^{V_{max}} \frac{df_0(E')}{dE'}\frac{dE\prime}{\sqrt{E'-E}}. \qquad (12.53)$$

Equation (12.53) shows that from $f_0(E)$ which is determined directly from the empirical knowledge of $|T(E)|^2$ we can find the width of the potential as a function of E.

Construction of the Potential Barrier from Tunneling Data when Additional Information Is Available — Obviously the information that we have found in (12.53) is not sufficient for the determination the shape of the barrier. The difference $x_2(E) - x_1(E)$ is the maximum information that we can obtain from $f_0(E)$. But in some problems the potential $V(x)$ may depend linearly on another parameter, say λ, i.e.

$$V(x,\lambda) = V_0(x) - \lambda\phi(x). \qquad (12.54)$$

For instance in the case of field emission in a metal (see [12]), we have a barrier like (12.54) where λ is the electric field at the surface of the metal, or in the case of α-decay [12] where $\lambda\phi(x) = \frac{l(l+1)}{x^2}$. In the latter example the function f_0 depends on E as well as λ. Knowing $f_0(E,\lambda)$ we can establish a second relation between x_1 and x_2, and by combining this second result with Eq. (12.53) we can find x_1 and x_2 separately, and thus determine $V(x,\lambda)$. For the barrier $V(x,\lambda)$ we start with the equation

$$
\begin{aligned}
\int_{x_1(E,\lambda)}^{x_2(E,\lambda)} \phi(x) dx &= \frac{1}{\pi} \int_{x_1}^{x_2} \phi(x) dx \int_{E}^{V(x)} \frac{dE'}{\sqrt{(E'-E)\left[V(x)-E'\right]}} \\
&= \frac{1}{\pi} \int_{E}^{V_{max}} dE' \int_{x_1(E',\lambda)}^{x_2(E',\lambda)} \frac{\phi(x) dx}{\sqrt{(E'-E)\left[V(x)-E'\right]}} \\
&= -\frac{2}{\pi} \int_{E}^{V_{max}} \frac{\partial f_0(E',\lambda)}{\partial \lambda} \frac{dE'}{\sqrt{E'-E}}.
\end{aligned}
\tag{12.55}
$$

Since $\phi(x)$ is known, we can calculate $V(x)$ from $f(E,\lambda)$. For instance, let us consider the motion of an electron in the potential field of ions, $V_0(x)$, and assume that there is an additional external field

$$
\lambda \phi(x) = e\mathcal{E}x,
\tag{12.56}
$$

acting on the electron, where $\mathcal{E}$ is the electric field at the surface of the metal and e is the charge of the electron. For this case from (12.55) we find

$$
x_2^2(E,\mathcal{E}) - x_1^2(E,\mathcal{E}) = -\frac{4}{\pi e} \int_{E}^{V_{max}} \left(\frac{\partial f_0(E',\mathcal{E})}{\partial \mathcal{E}}\right) \frac{dE'}{\sqrt{E'-E}}.
\tag{12.57}
$$

Combining the two equations (12.53) and (12.57) we obtain x_1 and x_2 separately. Thus by measuring the coefficient of transmission $|T(E,\mathcal{E})|^2$ we can determine $f_0(E,\mathcal{E})$ and from it the potential $V(x,\mathcal{E})$.

Higher Order Corrections for the Separation Between Turning Points — For a more accurate determination of the energy dependence of the separation between turning points we may include the first order correction (proportional to $\hbar^2$) to the WKB approximation. Let us write the transmission coefficient $|T(E)|^2$ or preferably $\sigma(E)$, as a power series in $\hbar^2$, [16]

$$
\sigma(E) = \sigma_0(E) + \hbar^2 \sigma_2(E) + \cdots.
\tag{12.58}
$$

For the sake of simplicity let us write $\sigma(E)$ in terms of $f(E)$;

$$
\sigma(E) = \frac{\sqrt{2m}}{\hbar} f(E) = \frac{\sqrt{2m}}{\hbar} \left[f_0(E) + \hbar^2 f_2(E)\right].
\tag{12.59}
$$

That is the leading correction term to the phase integral (12.59) is proportional to $\hbar^2$ (see [17], [18]);

$$
f(E) = \int_{x_1}^{x_2} \sqrt{V(x)-E}\, dx - \frac{\hbar^2}{48(2m)} \frac{d}{dE} \oint \frac{V''(x)\, dx}{\sqrt{V(x)-E}}.
\tag{12.60}
$$

Now let us study the case of symmetrical potentials, where for the inversion of the first term, $f_0(E)$, Eq. (12.53), simplifies to

$$
\bar{x}_0(E) = -\frac{1}{\pi} \int_{E}^{V_{max}} \frac{d f_0(E')}{dE'} \frac{dE'}{\sqrt{E'-E}}.
\tag{12.61}
$$

Here the subscript 0 shows that the result is correct to the zeroth power of $\hbar$. In this expansion $\bar{x}(E) = \bar{x}_0(E) + \hbar^2 \bar{x}_2(E)$, thus when we include $f_2(E)$ in the inversion, we want to determine the correction to the turning point $\bar{x}_0(E)$ to the order of $\hbar^2$. As in Eq. (12.53) we have

$$\bar{x}_2(E) = -\frac{1}{\pi} \int_E^{V_{max}} \frac{d\, f_2(E')}{dE'} \frac{dE'}{\sqrt{E' - E}}. \tag{12.62}$$

Integrating by parts give us the following expression [16]

$$\bar{x}_2(E) = -\frac{1}{\pi} \frac{d}{dE} \int_E^{V_{max}} f_2(E') \frac{dE'}{\sqrt{E' - E}}. \tag{12.63}$$

Substituting for $f_2(E')$ from (12.60) we find

$$\begin{aligned}
\bar{x}_2(E) &= -\frac{1}{24\pi} \left(\frac{1}{2m} \right) \frac{d}{dE} \left[\int_E^{V_{max}} \frac{d}{dE} \sqrt{E' - E} \right. \\
&\qquad \left. \times \left(\frac{d}{dE'} \oint \frac{V''(x)\, dx}{\sqrt{V(x) - E'}} \right) dE' \right] \\
&= -\frac{1}{24\pi} \left(\frac{1}{2m} \right) \frac{d^2}{dE^2} \left[\int_E^{V_{max}} \sqrt{E' - E} \right. \\
&\qquad \left. \times \left(\frac{d}{dE'} \oint \left(\frac{V''(\bar{x})d\bar{x}}{\sqrt{V(\bar{x}) - E'}} \right) \right) \right].
\end{aligned} \tag{12.64}$$

Again integrating by parts we obtain

$$\bar{x}_2(E) = -\frac{1}{24\pi m} \frac{d^2}{dE^2} \int_E^{V_{max}} \frac{dE'}{\sqrt{E' - E}} \int_{E'}^{V_{max}} \frac{V''(\bar{x})}{\sqrt{V(\bar{x}) - E'}} \frac{d\bar{x}}{dV} dV. \tag{12.65}$$

Next by interchanging the order of integration, we have

$$\begin{aligned}
\bar{x}_2(E) &= -\frac{1}{24\pi m} \frac{d^2}{dE^2} \int_E^{V_{max}} V''(\bar{x}) \frac{d\bar{x}}{dV} dV \\
&\qquad \times \int_E^{V} \frac{dE'}{\sqrt{(E' - E)(V(\bar{x}) - E')}} \\
&= \frac{1}{24m} \frac{d^2}{dE^2} \int_{V_{max}}^{E} V''(\bar{x}) \frac{d\bar{x}}{dV} dV.
\end{aligned} \tag{12.66}$$

Now we can write the last integral as

$$\begin{aligned}
\int_{V_{max}}^{E} \frac{d^2 V}{d\bar{x}^2} \frac{d\bar{x}}{dV} dx &= \int_{V_{max}}^{E} \frac{d}{dV} \left(\frac{1}{D_V \bar{x}(V)} \right) \frac{dV}{d\bar{x}} \frac{d\bar{x}}{dV} \\
&= \int_{V_{max}}^{E} \frac{d}{dV} \left(\frac{1}{D_V \bar{x}(V)} \right) dV = \frac{1}{D_V \bar{x}(E)} - \frac{1}{D_V \bar{x}(V_{max})},
\end{aligned} \tag{12.67}$$

where D_V denotes derivative with respect to V. In this way we find that the correction to x_0 to be

$$\bar{x}_2(E) = -\left(\frac{1}{24m}\right)\left[\frac{D_E^3\bar{x}_0 D_E\bar{x}_0 - 2\left(D_E^2\bar{x}_0\right)^2}{(D_E\bar{x}_0)^3}\right], \tag{12.68}$$

where D_E is the derivative with respect to E.

Inversion of $f(E)$ for a Quartic Barrier — Consider the simple case of tunneling through a quartic potential $V(x) = -x^4$. Setting $\hbar^2 = 2m = 1$, and using Eqs. (12.48) and (12.60) we calculate $f(E) = f_0(E) + \hbar^2 f_2(E)$:

$$f_0(E) = \int_{-|E|^{\frac{1}{4}}}^{|E|^{\frac{1}{4}}} \sqrt{|E| - x^4}\, dx = 1.748\, |E|^{\frac{3}{4}}, \tag{12.69}$$

and

$$f_2(E) = \frac{1}{2}\frac{d}{d|E|} \int_{-|E|^{\frac{1}{4}}}^{|E|^{\frac{1}{4}}} \frac{x^2}{\sqrt{|E| - x^4}}\, dx = 0.1498\, |E|^{-\frac{3}{4}}. \tag{12.70}$$

Substituting for $f_0(E)$ and $f_2(E)$ in (12.68) we obtain the zeroth order result which is

$$\bar{x}_0(E) = |E|^{\frac{1}{4}}\left(1 + \frac{1}{16|E|^{\frac{3}{2}}}\right). \tag{12.71}$$

In order to find higher order correction $\bar{x}_2$ we can substitute (12.70) in (12.68). Noting that the exact result in this case is $\bar{x}_0(E) = |E|^{\frac{1}{4}}$, from (12.71) it is clear that as we go to higher energies we get better results.

12.5 Construction of a Symmetric Double-Well Potential from the Known Energy Eigenvalues

In molecular physics there are a number of problems where the low-lying energy levels are paired. That is starting with the lowest energy levels $E_0 < E_1 < E_2 \cdots$, we have E_0 close to E_1 but E_1 is far from E_2, and E_2 is close to E_3, but E_3 is far from E_4 *etc*. If we look at the Table 12.1, we notice that in the ammonia molecule 0_S is very close to 0_A but then 0_A is far from 1_S and so on. This kind of distribution of eigenvalues for a one-dimensional problem arises when we have a symmetric or an asymmetric double-well. Again assuming that we are given all the energy levels, is it possible to find the shape of the potential? In particular we are interested in determining the details of the potential near the origin.

The approximate determination of the potential energy from the band spectroscopic data is one of the oldest inverse problems of quantum theory. The method

Table 12.1: Energy levels of NH_3 in units of cm^{-1}. The superscripts $+$ and $-$ refer to symmetric and antisymmetric states of the molecule respectively [12], [28].

	calculated	observed
E_0^+	0	0
E_0^-	0.83	0.67
E_1^+	935	932$\pm$ 0.5
E_1^-	961	964$\pm$ 0.5
E_2^+	1610	1600$\pm$ 15
E_2^-	1870	1910$\pm$ 15
E_3^+	2360	2380$\pm$ 15
E_3^-	2885.48	2840

is called RKR inversion after its originators. Today this technique forms one of the basic tools for the construction of the potential function in molecular physics. In addition to its simplicity of formulation and inversion, one can use it for the inversion of rotational as well as vibrational energy levels of molecules using either the exact solution or the semiclassical (WKB) approximation. For both of these problems the input data are all observables of the system. A later and important contribution is a method advanced by Wheeler for RKR analysis of the spectral perturbation due to tunneling in the asymmetric potentials with two minima [19]

A lucid and detailed account of the RKR can be found in [20], [21]. Our formulation follows closely the work of Pajunen and Child, [22], [23], and we refer the reader for the details given in that paper. It is worth mentioning that there are other exact and approximate methods that are available for finding the one-dimensional potential for molecular systems including Abraham–Moses method (Sec. 4.5) that we have already discussed (see references [24]–[26]).

To solve the inverse problem either using exact solution or the semiclassical (WKB) approximation we need an accurate knowledge of the low-lying energy eigenvalues, and we want to calculate these eigenvalues. For these low-lying states the energy of the particle, E, is less than the maximum value of the potential (which we assume it to be at the origin). That is, assuming that the potential has a local maximum at the origin $E < V(0)$, then there are four turning points $-b$, $-a$, a and b with $a < b$. Since the potential is symmetrical these turning points are also symmetrical.

Note that for $E > V(0)$, a would be zero and then we have two turning points, $-b$ and b, and the problem reduces to the standard WKB approximation. By writing the wave function for each region $x < -b$, $-b < x < -a$, $-a < x < a$, $a < x < b$, and $x > a$, and matching the solutions at the turning points we get an eigenvalue equation of the form [12], [22]

$$[\cos(2\alpha(E)) - \Phi(E)] = -\frac{\kappa(E)}{\sqrt{1 + \kappa^2(E)}}. \tag{12.72}$$

In this relation when $E < V_{max}$, $\alpha(E), \kappa(E)$ and $\delta(E)$ are defined by the following relations

$$\alpha(E) = \frac{1}{\hbar} \int_a^b \sqrt{2m(E - V(x))} \, dx, \tag{12.73}$$

$$\kappa(E) = \exp(\pi\delta(E)), \tag{12.74}$$

and

$$\pi\delta(E) = \frac{1}{\hbar} \int_{-a}^a \sqrt{2m(V(x) - E)} \, dx. \tag{12.75}$$

The function $\Phi(E)$, Eq. (12.72), which is a small phase correction depends only on $\delta(E)$ [22];

$$\Phi(E) = \arg \Gamma\left(\frac{1}{2} + i\delta(E)\right) - \delta(E)\ln|\delta(E)| + \delta(E). \tag{12.76}$$

At energies above the maximum of the barrier, $E > V_{max}$, and in Eq. (12.73) we set a equal to zero and also we replace (12.74) by

$$\pi\delta(E) = \frac{i}{\hbar} \int_{-ia}^{ia} \sqrt{2m(E - V(x))} \, dx, \tag{12.77}$$

where $\pm i|a|$ are the roots of

$$2m(E - V(x)) = 0. \tag{12.78}$$

We note that $\delta(E) > 0$, for $E < V_{max}$ and $\delta(E) < 0$ for $E > V_{max}$.

12.6   The Inverse Problem of Molecular Spectra

Following the work of Pajunen and Child, we formulate the inversion for double-well in three steps [22]:
(1) We try to find $\alpha(E)$ and $\delta(E)$ from the observed energy levels.
(2) From the information thus obtained we calculate $\alpha(V_{min})$ and $\delta(V_{max})$ using the relation

$$\alpha(V_{min}) = \delta(V_{max}) = 0. \tag{12.79}$$

(3) We also make use of the solution of the Abel integral equation (see Eq. (1.16)) to find $a(E)$ and $b(E)$

$$b(E) - a(E) = \frac{1}{\pi} \left(\frac{2}{m} \right)^{\frac{1}{2}} \int_{V_{min}}^{E} \left(\frac{da(E')}{dE'} \right) \frac{dE'}{(E - E')^{\frac{1}{2}}}, \tag{12.80}$$

and

$$2a(E) = \frac{1}{\pi} \left(\frac{2}{m} \right)^{\frac{1}{2}} \int_{E}^{V_{max}} \left(\frac{d\delta(E')}{dE'} \right) \frac{dE'}{(E - E')^{\frac{1}{2}}}, \tag{12.81}$$

subject to the condition that $a(E) = 0$, for $E \geq V_{max}$.

Approximate Calculation of the Phase Integrals — We have three phase integrals Eqs. (12.73)–(12.75) and we start with the calculation of $\alpha(E)$. For low-lying states the eigenvalues come in pairs, $E_s^{\pm}$ around the position of the minimum of the cosine function. These are given by

$$\alpha_s' = \alpha_s - \frac{1}{2} \Phi_s = \left(s + \frac{1}{2} \right) \pi. \tag{12.82}$$

Also we note that the energy splitting, $E_s^+ - E_s^-$, increases with increasing s, but in such a way that [22]

$$\begin{cases} \left(s + \frac{1}{4} \right) \pi < \alpha' \left(E_s^- \right) < \left(s + \frac{1}{2} \right) \pi \\ \\ \left(s + \frac{1}{2} \right) \pi < \alpha' \left(E_s^+ \right) < \left(s + \frac{3}{4} \right) \pi. \end{cases} \tag{12.83}$$

Our aim is to determine the energies E_s^0 which satisfies (12.82), and then use these points to interpolate for $\alpha' \left(E_s^{\pm} \right)$ between the energy eigenvalues, and then find $\kappa \left(E_s^{\pm} \right)$ and $\delta \left(E_s^{\pm} \right)$ from Eqs. (12.72) and (12.74). To solve this problem we use the following iterative technique:

We start with the set of energies E_s^0 first approximated by the average of the paired levels

$$E_s^0 = \frac{1}{2} \left(E_s^+ + E_s^- \right). \tag{12.84}$$

Now by linear interpolation between adjacent members of the set $\{\alpha_s', E_s^{\pm}\}$ we can estimate $\alpha' \left(E_s^{\pm} \right)$. From which we can find the corresponding values of $\delta(E)$:

$$\pi\delta \left(E_s^{\pm} \right) = \ln \left| \cot \left[2\alpha' \left(E_s^{\pm} \right) \right] \right|. \tag{12.85}$$

We note that using linear interpolation underestimates $\delta \left(E_s^- \right)$ and overestimates $\delta \left(E_s^- \right)$. Therefore

$$\delta_s \left(E_s^0 \right) = \frac{1}{2} \left(\delta_s \left(E^+ \right) + \delta_s \left(E^- \right) \right), \tag{12.86}$$

is assigned to the energy E_s^0.

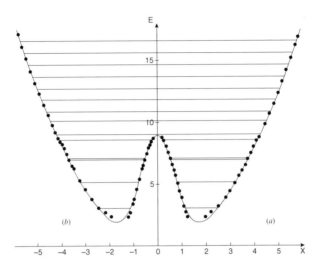

Figure 12.1: The double-well potential $V(x) = \frac{1}{2}x^2 + 9\exp(-x^2)$ which is the input as shown by the solid line. The result of inversion (a) when exact eigenvalues are used; (b) when inversion from first order eigenvalues are used. The horizontal lines represent the energy levels [22].

Table 12.2: Results for the exact and the recovered phase integrals from the first-order spectrum are tabulated for the potential $V(x) = \frac{1}{2}x^2 + 9\exp\left(-x^2\right)$ [22].

	direct calculation		results from inversion	
E	$\pi\delta(E)$	$\phi(E) - \frac{1}{2}\Phi(E)$	$\pi\delta(E)$	$\phi(E) - \frac{1}{2}\Phi(E)$
3.08459	5.57950	1.56891	5.51998	1.56879
3.08724	5.57603	1.57269	5.51712	1.57281
5.14885	3.28492	4.69368	3.29179	4.69380
5.17260	3.26172	4.73154	3.26811	4.73142
6.99053	1.61296	7.75562	1.61631	7.75594
7.10920	1.51255	7.96242	1.51510	7.96216
8.60369	0.30483	10.6779	0.32301	10.6823
8.97932	0.01577	11.3843	0.03036	11.3807
10.1449	−0.85043	13.5536	−0.85971	13.5520
10.8348	−1.34371	14.7950	−1.3320	14.7973

Next we make a correction to the estimate of $\alpha'\left(E_s^{\pm}\right)$ by linear interpolation for $\delta\left(E_s^{\pm}\right)$ from the set $\left\{\delta_s,\ E_s^0\right\}$ followed by utilizing the relation

$$\alpha'\left(E_s^{\pm}\right) = \left(s+\frac{1}{2}\right)\pi \pm \frac{1}{2}\cos^{-1}\frac{\kappa\left(E_s^{\pm}\right)}{\sqrt{1+\kappa^2\left(E_s^{\pm}\right)}}. \tag{12.87}$$

Now a better estimate for E_s^0 defined by Eq. (12.82) can be found by linear interpolation between $\alpha'\left(E_{s-1}^+\right)$ and $\alpha'\left(E_s^-\right)$ on one hand, and $\alpha'\left(E_s^+\right)$ and $\alpha'\left(E_{s+1}^-\right)$ on the other. The mean of these two values is taken to be the first correction to (12.84) except at the end points where only a single estimate is available. The same set of phase integrals (12.73)–(12.75) together with the eigenvalue equation (12.72) can be used the solve the direct problem [12].

12.7 The Inverse Problem of Tunneling for Gamow States

In the next section we will discuss the question of determination of a set of E_j s and Γ_j s from the observable survival probability $S(t)$ if this probability is measured at equal small time intervals Δt for a long time $N\Delta t$. Once these two sets are determined, then we can use an inversion method similar to the one discussed earlier to obtain the potential.

The Direct Problem — Consider a particle trapped inside an attractive potential well $v(r)$, $r_0 \leq r \leq r_1$, and there is a barrier located between $r=r_1$ and $r=r_2$ (see Fig. 12.2). This particle can escape through the barrier by the mechanism of tunneling. According to Gamow's theory, the decay of the initial state can be explained by solving the Schrödinger equation with the radiation boundary condition [12], [29], [30]. This condition for one-dimensional wave equation is

$$\lim\left(\frac{\partial\psi(x)}{\partial x} - ik\psi(x)\right) \to 0, \quad \text{as; } x \to \infty, \tag{12.88}$$

and for the n-dimensional problem it takes the form [12]

$$\lim\left[|x|^{\frac{n-1}{2}}\left(\frac{\partial\psi(x)}{\partial x} - ik\psi(x)\right)\right] \to 0, \quad \text{as } x \to \infty. \tag{12.89}$$

The wave functions satisfying the boundary condition (12.88) (or (12.89)) are called Gamow wave functions (or Gamow states) [12]. If we consider the Gamow state for the ℓ-th partial wave, we find that the eigenvalues are complex and are dependent on ℓ;

$$E = E_r - \frac{i}{2}\Gamma(E_r,\ell), \tag{12.90}$$

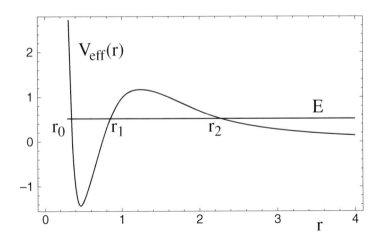

Figure 12.2: A typical potential for three-dimensional tunneling problem. For a given energy E and partial wave l, the turning points are $r_0(E)$, $r_1(E)$ and $r_2(E)$. Note that here E is real and is the same as E_r of Eq. (12.90).

where Γ is the decay width. In the WKB approximation we find that Γ is given by [12]

$$\Gamma(E_r, \ell) = \frac{\hbar}{T_0(E_r, \ell)} \exp\left[-2\sigma(E_r)\right], \qquad (12.91)$$

where

$$T_0(E_r, \ell) = 2 \int_{r_0}^{r_1} \frac{dr}{\sqrt{2m[E_r - V_{eff}(r)]}}, \qquad (12.92)$$

$$\sigma(E_r, \ell) = \frac{1}{\hbar} \int_{r_1}^{r_2} \sqrt{2m[E_r - V_{eff}(r)]}\, dr, \qquad (12.93)$$

and

$$V_{eff}(r) = V(r) + \frac{\hbar^2 \left(\ell + \frac{1}{2}\right)^2}{2mr^2}. \qquad (12.94)$$

For a metastable state of the particle, $V(r)$ must be attractive in the shorter range and become repulsive for the longer range (Fig. 12.2). The classical turning points r_0, r_1 and r_2 are all functions of E, i.e.

$$V_{eff}(r_i) = E_r, \qquad i = 0,\ 1 \ \text{and}\ 2. \qquad (12.95)$$

Now we can state the inverse problem in the following way [13]–[36]:
Let us assume that $\sigma(E, \ell)$, Eq. (12.93), is known and $L(E, \ell)$ is defined by

$$L(E, \ell) = \frac{1}{\hbar} \int_{r_0}^{r_1} \sqrt{2m(E_r - V_{eff}(r))}\, dr = \left(n + \frac{1}{2}\right)\pi + \sigma(E_r, \ell), \quad n = 0,\ 1,\ 2\cdots$$
$$(12.96)$$

and both of these are known functions of E and ℓ. We want to know if it is possible to determine $V_{eff}(r)$ or $V(r)$ from these data. In what follows we suppress the subscript r, of E_r, and thus E will denote the real part of E. We observe that $L(E, \ell)$ expresses the Bohr–Sommerfeld quantization rule for finding the energy eigenvalues of a particle bound in the well behind the barrier (see Fig. 12.2), and T_0 is the period of oscillation of this particle in the well. Let us also define $I(E, \ell)$ and $J(E, \ell)$ by the following relations:

$$I(E, \ell) = \int_{r_0(E)}^{r_1(E)} (E - V_{eff}(r)) dr, \tag{12.97}$$

and

$$J(E, \ell) = \int_{r_1(E)}^{r_2(E)} (V_{eff}(r) - E) dr. \tag{12.98}$$

By differentiating $I(E, \ell)$ and $J(E, \ell)$ with respect to E and ℓ we find the following equations:

$$\frac{\partial I(E, \ell)}{\partial E} = \int_{r_0(E)}^{r_1(E)} dr = r_1(E) - r_0(E), \tag{12.99}$$

$$\begin{aligned} \frac{\partial I(E, \ell)}{\partial \ell} &= -\int_{r_0(E)}^{r_1(E)} \frac{\partial V_{eff}(r)}{\partial \ell} dr = -\int_{r_0(E)}^{r_1(E)} \frac{\hbar^2 (2\ell + 1)}{2mr^2} dr \\ &= \hbar^2 \left(\frac{2\ell + 1}{2m} \right) \frac{r_0(E) - r_1(E)}{r_0(E) r_\ell(E)}, \end{aligned} \tag{12.100}$$

$$\frac{\partial J(E, \ell)}{\partial E} = -(r_2(E) - r_1(E)), \tag{12.101}$$

and

$$\frac{\partial J(E, \ell)}{\partial \ell} = \hbar^2 \left(\frac{2\ell + 1}{2m} \right) \frac{(r_2(E) - r_1(E))}{r_1(E) r_2(E)}. \tag{12.102}$$

When we solve these equations for the turning points $r_0(E)$, $r_1(E)$ and $r_2(E)$ we find

$$\begin{bmatrix} r_0(E) \\ r_1(E) \end{bmatrix} = \left[\frac{1}{4} \left(\frac{\partial I}{\partial E} \right)^2 - \frac{(2\ell + 1)\hbar^2}{2m} \left(\frac{\frac{\partial I}{\partial E}}{\frac{\partial I}{\partial \ell}} \right) \right]^{\frac{1}{2}} \pm \left(-\frac{1}{2} \frac{\partial I}{\partial E} \right), \tag{12.103}$$

and

$$\begin{bmatrix} r_1(E) \\ r_2(E) \end{bmatrix} = \left[\frac{1}{4} \left(\frac{\partial J}{\partial E} \right)^2 - \frac{(2\ell + 1)\hbar^2}{2m} \left(\frac{\frac{\partial J}{\partial E}}{\frac{\partial J}{\partial \ell}} \right) \right]^{\frac{1}{2}} \pm \left(\frac{1}{2} \frac{\partial J}{\partial E} \right). \tag{12.104}$$

In these equations the upper element on the left corresponds to the plus sign on the right and the lower element has the minus sign. In this way $r_1(E)$ can be calculated either from (12.103) or from (12.104). Once these turning points are known as functions of E then the shape of the potential can also be determined.

Next we want to show how $I(E, \ell)$ and $J(E, \ell)$ can be obtained from the empirical data $L(E, \ell)$ and $\sigma(E, \ell)$. For this we use the following identity:

$$E - V_{eff}(r) = \frac{2}{\pi} \int_{V_{eff}(r)}^{E} \left(\frac{E' - V_{eff}(r)}{E - E'} \right)^{\frac{1}{2}} dE'. \tag{12.105}$$

Substituting (12.105) in (12.97) for $I(E, \ell)$ we find

$$I(E, \ell) = \frac{2\hbar}{\pi \sqrt{2m}} \int_{V_{eff}^{m}}^{E} \frac{L(E', \ell) dE'}{\sqrt{E - E'}}, \tag{12.106}$$

where V_{eff}^{m} denotes the minimum of $V_{eff}(r)$ in the region of the potential well. By partial integration we can rewrite (12.106) as

$$I(E, \ell) = \frac{4\hbar}{\pi \sqrt{2m}} \int_{V_{eff}^{m}}^{E} \frac{\partial L(E', \ell)}{\partial E'} \sqrt{E - E'} \, dE'. \tag{12.107}$$

At the bottom of the well from Eq. (12.96) it follows that

$$n(l, V_{eff}^{m}) = -\frac{1}{2}, \tag{12.108}$$

and this equation can be used to find V_{eff}^{m}. Similarly using the identity

$$V_{eff}(r) - E = \frac{2}{\pi} \int_{E}^{V_{eff}(r)} \left(\frac{V_{eff}(r) - E'}{E' - E} \right)^{\frac{1}{2}} dE', \tag{12.109}$$

we find

$$J(E, l) = \frac{-4\hbar}{\pi \sqrt{2m}} \int_{E}^{V_{eff}^{M}} \frac{\partial \sigma(E', l)}{\partial E'} \sqrt{E' - E} \, dE', \tag{12.110}$$

where V_{eff}^{M} is the maximum height of the barrier which is obtained from

$$\sigma(l, V_{eff}^{M}) = 0. \tag{12.111}$$

This completes the solution of the inverse tunneling problem for the Gamow states when the WKB approximation is valid.

This method can be used to construct the nucleus-nucleus interaction from the cross section for the fusion of two nuclei at sub-barrier energies [37].

12.8   Inverse Problem of Survival Probability

For a quasi-stationary which decays in time, the survival probability $S(t)$ is defined by [12]

$$S(t) = \left| \int_{0}^{\infty} \phi^*(r, t)\phi(r, 0)dr \right|^2, \tag{12.112}$$

where $\phi(r,0)$ is the initial state of the system and $\phi(r,t)$ is the state of the system at the time t. We want to know whether from the measurement of $S(t)$ it is possible to find the discrete energy spectrum of the decaying system [38]–[42]. For this purpose we write E_j in terms of its real and imaginary parts

$$E_j = \mathrm{Re}\, E_j - \frac{i}{2}\Gamma_j, \qquad (12.113)$$

where Γ_j is positive. We also define the set of real frequencies ω_{ij} by

$$\omega_{ij} = \mathrm{Re}\, E_i - \mathrm{Re}\, E_j, \qquad \hbar = 1. \qquad (12.114)$$

Using these we write $S(t)$ as

$$S(t) = M \sum_j |L_j|^2 \exp\left(-\Gamma_j t\right)$$

$$+ M \sum_j \sum_k \exp\left[-\frac{1}{2}\left(\Gamma_j + \Gamma_k\right)t\right] \mathrm{Re}\left[L_k^* L_j \exp\left(-i\omega_{ij}t\right)\right] \qquad (12.115)$$

where M and $L_k^* L_j$ are defined by

$$M = \sum_j \sum_i \mathrm{Re}\left(L_i^* L_j\right), \qquad (12.116)$$

and

$$L_i^* L_j = C_2^*(E_i) C_2(E_j) \exp\left[-i(E_j - E_i)t\right] \int_0^\infty u_2^*(E_i, r) u_2(E_j, r) dr. \qquad (12.117)$$

The measured values of $S(t)$ are

$$S(1), S(2), \cdots S(j), \cdots S(2N), \qquad (12.118)$$

where $S(j)$ is the probability measured at time $t_j = j\Delta t$. We observe that $S(t)$ is the sum of a number of terms each with an exponential dependence on time. In order to determine Γ_j s and ω_{ij} s from the measurement of $S(t)$ we use the Prony method [42]–[43].

In this method we start by solving the set of difference equations

$$Z_N S(k) + Z_{N-1} S(k+1) + \cdots + Z_1 S(k+N-1)$$
$$+ S(k+N) = 0, \quad k = 1, 2, \cdots N \qquad (12.119)$$

where Z_j s are the unknowns. This set of N equations and N unknowns will have real solutions $Z_1 \cdots Z_N$ provided that the determinant of the coefficients in (12.119) does not vanish. From these Z_j s we get the characteristic equation

$$r^N + Z_1 r^{N-1} + \cdots + Z_N = 0. \qquad (12.120)$$

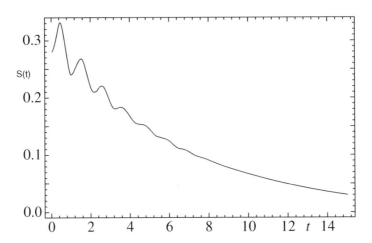

Figure 12.3: The probability $S(t)$ of finding a decaying state at time t to be in its initial state.

This equation of the Nth order will have real and complex roots.
Let $r_1, r_2, \cdots r_J$ denote the real and $r_{J+1}, r_{J+2}, \cdots r_N$ be the complex roots of (12.120), then

$$r_j = \exp\left(-\Gamma_j\right), \qquad 1 \le j \le J, \tag{12.121}$$

and

$$\ln\left[\pm r_k \exp(i\theta_k)\right] = -\frac{1}{2}\left(\Gamma_m + \Gamma_n\right) - i\omega_{mn}, \quad J+1 \le k \le N. \tag{12.122}$$

The coefficient of $\exp(-\Gamma_j)$ in Eq. (12.115) is positive, therefore r_j is also positive, however the coefficient of $\exp\left[-\frac{1}{2}\left(\Gamma_m + \Gamma_n\right)\right]$ in (12.115) can be positive or negative, hence the argument of the logarithm in (12.122) can be positive or negative. From Eqs. (12.121) and (12.122) we find

$$\Gamma_j = -\ln r_j, \qquad 1 \le j \le J, \tag{12.123}$$

and

$$\theta_k = \pm 2k\pi - \omega_{nm}, \tag{12.124}$$

or

$$\theta_k = \pm(2k+1)\pi - \omega_{nm}, \tag{12.125}$$

where $J+1 \le k \le N$. From these we find the level widths, Γ_j uniquely, and also we find the multi-valued solution for the level spacing.

For the detailed discussion of the application of Prony's method used for the determination of the complex eigenvalues and finding the survival probability the reader is referred to Ref. [12].

References

[1] R. Rydberg, Z. Phys. 73, 376 (1931).

[2] O. Klein, Z. Phys. 76, 226 (1932).

[3] A.I.G. Rees, Proc. Phys. Soc. Lond. 59, 998 (1947).

[4] W.H. Miller, J. Chem. Phys. 51, 3631 (1969).

[5] W.H. Miller, J. Chem. Phys. 54, 4174 (1971).

[6] W. van Dijk and M. Razavy, Intl. J. Quan. Chem. 16, 1249 (1979).

[7] V. Tikochinsky, Ann. Phys. (New York), 103, 185 (1977).

[8] Y. Nogami and C.K. Ross, Am. J. Phys. 64, 923 (1966).

[9] H.E. Moses, Phys. Rev. 102, 559 (1956).

[10] I.N. Sneddon, *Fourier Transform*, (McGraw-Hill, New York, 1951) Appendix B.

[11] J.C. Portinari, Ann. Phys. (New York) 45, 445 (1967).

[12] M. Razavy, *Quantum Theory of Tunneling*, Second Edition, (World Scientific, Singapore, 2014).

[13] M.W. Cole and R.H. Good Jr., Phys. Rev. A 18, 1085 (1978).

[14] K. Chadan and P.C. Sabatier, *Inverse Problems in Quantum Scattering Theory*, (Springer-Verlag, New York 1989).

[15] I.S. Gradshteyn and I.M. Ryzhik, *Table of Integrals, Series and Products*, (Academic Press, New York, 1965) p. 950.

[16] C.K. Chan and P. Lu, Phys. Rev. A 22, 1869 (1980).

[17] P. Pajunen, Mol. Phys. 40, 605 (1980).

[18] P. Pajunen, J. Mol. Phys. 88, 64 (1981).

[19] J.A. Wheeler, *Studies in Mathematical Physics: Essays in Honor of V. Bergmann*, edited by E.H. Lieb, B. Simon and A.S. Wightman (Princeton University Press, Princeton, 1976) p. 351.

[20] M.S. Child, *Molecular Collision Theory*, (Dover Publications, New York, 1996).

[21] M.S. Child, *Semiclassicl Mechanics with Molecular Applications*, (Oxford University Press, Oxford, 2014) p. 123.

[22] P. Pajunen and M.S. Child, Mol. Phys. 40, 597 (1980).

[23] P. Pajunen and M.S. Child, Molec. Phys. 40, 597 (1980).

[24] B.M. Levitan and M.G. Ghasymov, Russ. Math Survey, 19,1, (1964).

[25] J.F. Shoenfeld, W. Kwong, J.L. Rosner, C. Quigg, and H.B. Thacker, Ann. Phys. 128, 1 (1980).

[26] C. Quigg, J.L. Rosner, and H.B. Thacker, Phys. Rev. D 21, 234 (1979).

[27] L.V. Chebotarev, Am. J. Phys. 66, 1086 (1998).

[28] G. Herzberg, *Molecular Spectra and Molecular Structure*, Vol. 2. (D. van Nostrand, New York, 1966) pp. 221-227.

[29] G. Gamow, Z. Phys. 51, 204 (1928).

[30] G. Gamow, *Constitution of Atomi Nuclei and Radioactivity*, (Oxford University Press, London, 1931).

[31] See for instance: K. Chadan and P.C. Sabatier, *Inverse Problems in Quantum Scattering Theory*, Second Edition, (Springer-Verlag, New York, 1989).

[32] M.W. Cole and R.H. Good, Jr., Phys. Rev. A 18, 1085 (1978).

[33] S.C. Gandhi and E.C. Efthimiou, Am. J. Phys. 74, 638 (2006).

[34] N. Fröman and P.O. Fröman, Intl. J. Quantum Chem. 35, 751 (1989).

[35] For solving the direct problem by phase integral method, see N. Fröman and P.O. Fröman, Nucl. Phys. A147, 606 (1970) and

[36] N. Fröman and P.O. Fröman, Ann. Phys. (New York), 83, 103 (1974).

[37] A.B. Balantekin, S.E. Koonin and J.W. Negele, Phys. Rev. C 28, 1565 (1983).

[38] M. Razavy, Nuovo Cimento 111 B, 331 (1996).

[39] E.T. Whittaker and G. Robinson, *The Calculus of Observation, a Treatise on Numerical Mathematics*, (Blakie and Son, London, 1924).

[40] G.A. Korn and T.M. Korn, *Mathematical Handbook for Scientisrs and Engineers*, Second Edition, (McGraw-Hill, New York, 1968) p. 766.

[41] G.M. Pitstick, J.R. Cruz and R.J. Mulholland, Proc. IEEE, 76, 1052 (1988).

[42] M. Hron and M. Razavy, Phys. Rev. A 51, 4365 (1994).

[43] See for instance, F.B. Hilderbrand, *Introduction to Numerical Analysis*, (McGraw-Hill, New York, 1956) p. 379.

Chapter 13

Inverse Problems Related to the Classical Wave Propagation

In this chapter we want to discuss approximate methods, both analytical and numerical which can be applied to obtain certain physical properties of a medium bounded by two parallel surfaces. In particular we want to construct the velocity and the density profiles of a layered medium from the reflection coefficient at two different frequencies [1], [2].

13.1 Determination of the Wave Velocity in an Inhomogeneous Medium from the Reflection Coefficient

A scalar wave, $p(x,t)$, propagating along the z-axis in a medium where the velocity is assumed to be a function of the coordinate z. The wave amplitude satisfies the equation of wave motion:

$$\frac{\partial^2 p(x,t)}{\partial z^2} = \frac{1}{c^2(z)}\frac{\partial^2 p(x,t)}{\partial t^2}. \tag{13.1}$$

For simplicity, we assume that as $|z|$ becomes very large $c(z)$ approaches a well-defined value which we denote by c, i.e.

$$c = \lim c(z) \quad \text{as} \quad z \to \pm\infty. \tag{13.2}$$

If we substitute

$$p(z, t) = p(z, k)e^{ikct}, \tag{13.3}$$

in (13.1) we find an ordinary differential equation for $p(x, t)$;

$$p''(z, k) + k^2[1 - V(z)]p(z, k) = 0, \tag{13.4}$$

where prime denotes differentiation with respect to z and the function $V(z)$ is related to $c(z)$ by

$$1 - V(z) = \frac{c^2}{c^2(z)} \geq 0. \tag{13.5}$$

The differential equation (13.4) looks like the Schrödinger equation but differs from it in one important aspect, viz, here the "potential" is $k^2V(z)$, which depends on the wave number quadratically. This dependence affects the normalization condition and also the spectral function of the differential equation (13.4).

Solution of the Direct Problem — To solve the direct problem, we first try to find an integral representation of the reflection coefficient [4]–[7]. To this end, we start with the Green's function for the differential equation (13.4);

$$G_0'' (z, z') + k^2 G_0 (z, z') = -\delta (z - z'). \tag{13.6}$$

The solution of this equation with the outgoing boundary condition yields the one-dimensional Green's function

$$G_0 (z, z', k) = \frac{i}{2|k|} \exp (i|k| |z - z'|). \tag{13.7}$$

Now we can write the solution of (13.4) in terms of this Green's function for $k > 0$

$$p(z, k) = \frac{1}{\sqrt{2\pi}} e^{ikz} - \frac{i}{2}|k| \int_{-\infty}^{+\infty} \exp (i|k| |z - z'|) V (z') p (z') dz'. \tag{13.8}$$

This solution has the asymptotic forms at $z \to \pm\infty$:

$$p(z, k) = \frac{1}{\sqrt{2\pi}} \left[e^{ikz} + R(k)e^{-ikz} \right], \quad z \to -\infty, \tag{13.9}$$

and

$$p(z, k) = \frac{1}{\sqrt{2\pi}} T(k)e^{ikz}, \quad z \to +\infty. \tag{13.10}$$

In these relations $R(k)$ and $T(k)$ are the amplitudes of reflection and transmission respectively. Now by comparing (13.9) and the asymptotic form of (13.8), we find that for $k > 0$, $R(k)$ is related to $p(z, k)$ by

$$R(k) = -i\sqrt{\frac{\pi}{2}}|k| \int_{-\infty}^{+\infty} e^{ikz} V(z)p(z, k)dz. \tag{13.11}$$

For negative values of k, $R(k)$ is determined by analytic continuation. Thus from Eqs. (13.8) and (13.11) we can verify that

$$R(-k) = R^*(k), \quad k > 0. \tag{13.12}$$

From Eqs. (13.5), (13.8) and (13.9) we can find $R(k)$ for any inhomogeneous medium if $c(z)$ is known. To solve the inverse problem, we modify the method of Jost and Kohn so that it becomes applicable to the scalar wave equation [3]–[9]. In the formulation of the inverse problem, Jost and Kohn utilized the integral equation for the transition matrix of quantum theory. Following their lead, we derive a similar integral equation for (13.4). For the construction of the Green function we need the orthogonality condition for the solutions of (13.4). From the equation for $p(z, k)$, we find this condition to be

$$\int_{-\infty}^{+\infty} p^*(z, k)[1 - V(z)] p(z, k') \, dz = \delta(k - k'). \tag{13.13}$$

Here the orthogonality condition is different from the quantal form due to the presence of the weight function $[1 - V(z)]$. The completeness relation for the solutions of (13.4) has also the same weight factor, i.e.

$$\int_{-\infty}^{+\infty} p^*(z', k)[1 - V(z')] p(z, k) \, dk = \delta(z - z'). \tag{13.14}$$

When $V(z) = 0$, i.e. for a homogeneous medium, let us denote the solution of (13.4) by $\phi(z, k)$, which in this case is the normalized plane wave

$$\phi(z, t) = \frac{1}{\sqrt{2\pi}} e^{ikz}. \tag{13.15}$$

Using this solution we can write the Green function Eq. (13.6) as

$$G_0(z, z', k) = \int_{-\infty}^{+\infty} \frac{\phi(z, q) \phi^*(z', q)}{q^2 - k^2 - i\epsilon} \, dq. \tag{13.16}$$

We can also express the solution of the wave equation, i.e. (13.8) in terms of $G_0(z, z', k)$;

$$p(z, k) = \phi(z, k) - k^2 \int_{-\infty}^{+\infty} G_0(z, z', k) V(z') p(z', k) \, dz'. \tag{13.17}$$

By multiplying this equation by $V(z)\phi^*(z, j)$ and integrating over z we get;

$$\mathcal{T}(j, k) = \mathcal{V}(j, k) - k^2 \int_{-\infty}^{+\infty} \frac{dq}{q^2 - k^2 - i\epsilon} \mathcal{V}(j, q) \mathcal{T}(q, k) dq, \tag{13.18}$$

where

$$\mathcal{V}(j, k) = \int_{-\infty}^{+\infty} \phi^*(z, j) V(z) \phi(z, k) dz, \tag{13.19}$$

and

$$T(j,k) = \int_{-\infty}^{+\infty} \phi^*(z,j)V(z)p(z,k)dz. \tag{13.20}$$

A different integral equation for $T(j,k)$ can be found in the following way:
    We start with the full Green function $G(z,z',k)$ defined as the solution of the inhomogeneous differential equation

$$G''(z,z',k) + k^2 (1 - V(z'))G(z,z',k) = -\delta(z-z'), \tag{13.21}$$

with the boundary condition

$$G(z,z',k) \to e^{i|k||z|}, \quad z \to \infty. \tag{13.22}$$

This Green's function can be expanded in terms of the eigenfunctions of the differential equation (13.4). Thus from the orthogonality and the completeness of the set of $p(z,k)$ s, Eqs. (13.13) and (13.14) we find

$$G(z,z',k) = \int_{-\infty}^{+\infty} \frac{dq}{q^2 - k^2 - i\epsilon} p(z,q)p^*(z',q). \tag{13.23}$$

Now we will solve the differential equation

$$\phi''(z,k) + k^2[1 - V(z)]\phi(z,k) = -V(z)\phi(z.k), \tag{13.24}$$

using the Green function (13.23);

$$\phi(z,k) = p(z,k) + k^2 \int_{-\infty}^{\infty} \int_{-\infty}^{+\infty} \frac{dq\, dx'}{q^2 - k^2 - i\epsilon}\phi(x',k)\,p^*(x',q)\,V(x')\,p(x,q). \tag{13.25}$$

Next we multiply this equation by $\phi^*(x,j)V(x)$ and integrate over x and we find that

$$\mathcal{V}(j,k) = T(j,k) + k^2 \int_{-\infty}^{+\infty} \frac{dq}{q^2 - k^2 - i\epsilon}T^*(k,q)T(j,q). \tag{13.26}$$

From the definition of, $R(k)$, Eq. (13.11) it follows that the reflection coefficient is related to $T(j,k)$ matrix by the following relation:

$$R(k) = -i\pi T(-k,k). \tag{13.27}$$

If we denote $\mathcal{V}(-k,k)$ by $W(k)$ and use Eq. (13.19) we find

$$W(k) = \mathcal{V}(-k,k) = \frac{1}{2\pi} \int_{-\infty}^{+\infty} V(z)e^{2ikz}dx. \tag{13.28}$$

Now we combine Eqs. (13.26)–(13.28) to obtain

$$W(k) = \frac{i}{\pi k}R(k) + k^2 \int_{-\infty}^{+\infty} T(-k,q)\frac{dq}{q^2 - k^2 - i\epsilon}T^*(k,q), \quad k > 0, \tag{13.29}$$

and by analytic continuation we can find $W(k)$ for negative values of k.

$$W(k) = W^*(-k) = \frac{i}{\pi k}R^*(-k) + k^2 \int_{-\infty}^{+\infty} \mathcal{T}^*(k,q)\frac{dq}{q^2 - k^2 + i\epsilon}\mathcal{T}(-k,q), \quad k < 0.$$

(13.30)

We multiply (13.29) by $\theta(k)$ and (13.30) by $\theta(-k)$, where $\theta(k)$ is the step function and add the resulting expressions to get

$$W(k) = \frac{i}{\pi k}R(k) + k^2 \int_{-\infty}^{+\infty} \mathcal{T}^*(k,q)\mathcal{T}(-k,q)dq$$

$$\times \left[\theta(k)\frac{1}{q^2 - k^2 + i\epsilon} + \theta(-k)\frac{1}{q^2 - k^2 + i\epsilon}\right].$$

(13.31)

Equations (13.18), (13.26) and (13.31) together with the relation

$$\mathcal{V}(k,q) = W\left(\frac{q-k}{2}\right),$$

(13.32)

form the basic relations of the inverse problem.

Iterative Method For Solving the Inverse Problem — This method of finding the solution to our basic equation has been advanced by Moses [9]. Here we replace $R(k)$ in Eq. (13.31) by $\varepsilon R(k)$, where ε is a parameter and we write

$$W(k) = \sum_{n=1}^{\infty} \varepsilon^n W_{(n)}(k),$$

(13.33)

$$\mathcal{T}(k,j) = \sum_{n=1}^{\infty} \varepsilon^n \mathcal{T}_{(n)}(k,j),$$

(13.34)

and

$$\mathcal{V}(k,j) = \sum_{n=1}^{\infty} \varepsilon^n \mathcal{V}_{(n)}(k,j).$$

(13.35)

By substituting these in Eqs. (13.18), (13.31) and (13.32) and equating the coefficients of the same power of ε, for the lowest terms we find

$$W_{(1)}(k) = \frac{i}{\pi k}R(k),$$

(13.36)

$$\mathcal{V}_{(1)}(k,j) = \frac{4}{2\pi(j-k)}R\left(\frac{j-k}{2}\right) = \mathcal{T}_{(1)}(k,j),$$

(13.37)

$$W_{(2)}(k) = \frac{4k^2}{\pi^2} \int_{-\infty}^{+\infty} R\left(\frac{q+k}{2}\right)R^*\left(\frac{q-k}{2}\right)\frac{1}{q^2 - k^2}$$

$$\times \left[\theta(k)\frac{1}{q^2 - k^2 - i\epsilon} + \theta(-k)\frac{1}{q^2 - k^2 + i\epsilon}\right]dq.$$

(13.38)

We observe that the function $V(z)$ is related to the sum of $W_{(n)}(k)$ s;

$$V(z) = 2 \sum_{n=1}^{\infty} \varepsilon^n W_{(n)}(k) e^{-2ikz} dk, \tag{13.39}$$

a result that follows from Eq. (13.29). In principle we can use Eqs. (13.33)–(13.35) and (13.39) to calculate $V(z)$ to the desired accuracy by successive iterations. However in practice because of the numerical computation of the multiple integrals involved in the calculation, the results become unstable and inaccurate.

13.2   Solvable Examples

Let us consider a few examples where we can determine one or two of the leading terms of the infinite series which give us the unknown $V(z)$, and thus $c(z)$, Eq. (13.5).
We first consider the simple but nonrealistic example where $V(z)$ is given by

$$V(z) = 2\lambda\delta(z), \quad \lambda > 0. \tag{13.40}$$

For this $V(z)$, $p(z, k)$ can be readily found from Eq. (13.8);

$$p(z, k) = \frac{1}{\sqrt{2\pi}} \left(e^{ikz} - \frac{i\lambda|k|}{1 + i\lambda|k|} e^{i|k||z|} \right). \tag{13.41}$$

From the asymptotic form of $p(z, k)$ we find $R(k)$ to be

$$R(k) = \frac{-i\lambda k}{1 + i\lambda k}, \quad k > 0. \tag{13.42}$$

Now let us assume that $R(k)$ is given and we want to find $V(z)$. Using the iteration scheme outlined in the last section, we get the following results

$$W_{(1)}(k) = \frac{i}{\pi k} R(k), \tag{13.43}$$

$$T_{(1)}(k, j) = \frac{\lambda}{\pi} \left[1 + \frac{i}{2}\lambda(k - j) \right]^{-1}, \tag{13.44}$$

and

$$W_{(2)}(k) = \frac{\lambda k}{\pi(k - \frac{i}{\lambda})}. \tag{13.45}$$

With the help of these relations we can calculate $V(z)$ up to the order of ε^2:

$$V_{(1)}(z) = 4 \exp\left(\frac{2z}{\lambda}\right) \theta(-z), \tag{13.46}$$

and

$$V_{(2)}(z) = 2\lambda\delta(z) - 4 \exp\left(\frac{2z}{\lambda}\right) \theta(-z). \tag{13.47}$$

Thus in this case the second order term is more important than the first order term, and that the resulting $V_{(1)}(z)$ does not look like the input, but the sum $V_{(1)}(z) + V_{(2)}(z)$ in this case gives the correct potential.

Reflection from a Layered Medium — Consider a flat surface of thickness a in which the wave velocity is $\frac{c}{\nu}$, i.e.

$$c(z) = \begin{cases} c, & z > a, \quad z < 0 \\ \frac{c}{\nu}, & 0 \le z \le a, \end{cases} \tag{13.48}$$

where ν is the analogue of the index of refraction in optics. Here we assume that $\nu > 1$, but for mechanical waves it can have values less than one. For the direct problem, i.e. determining the reflection coefficient, we solve (13.4) for three regions $z < 0$, $0 \le z \le a$ and $z > a$, and join the solutions in the neighboring regions smoothly to each other, noting that for $z > a$ only the outgoing wave is present. The reflection coefficient that we find in this way is

$$R(k) = \frac{(\nu^2 - 1)(e^{2iak\nu} - 1)}{(\nu - 1)^2 e^{2iak\nu} - (1 + \nu)^2}. \tag{13.49}$$

We can expand $R(k)$ as a power series in $e^{2iak\nu}$;

$$R(k) = \left(\frac{\nu - 1}{\nu + 1}\right) - 4\nu \sum_{j=1}^{\infty} \left[\frac{(\nu - 1)^{2j+1}}{(n + 1)^{2j+1}}\right] \left(e^{2iak\nu}\right)^j. \tag{13.50}$$

To the first order in the expansion of $V(z)$, we have according to Eqs. (13.36) and (13.39);

$$V_{(1)}(z) = 2 \int_{-\infty}^{+\infty} W_{(1)}(k) e^{-2ikx} dk = \frac{2i}{\pi} \int_{-\infty}^{+\infty} \frac{R(k)}{k} dk. \tag{13.51}$$

By substituting $R(k)$ from (13.50) in (13.51) and noting that

$$\text{sgn}(x) = \int_{-\infty}^{+\infty} \left(\frac{i}{\pi k}\right) e^{-ikx} dk = \begin{cases} +1, & x > 0 \\ -1, & x < 0, \end{cases} \tag{13.52}$$

we find $V_1(z)$:

$$\begin{aligned} V_{(1)}(z) &= \frac{2i}{\pi} \int_{-\infty}^{\infty} \frac{R(k)}{k} e^{-2ikz} dk \\ &= \sum_{n=0}^{\infty} \theta(x) \left(\frac{\nu - 1}{\nu + 1}\right)^{2n+1} \{\theta[(z - n\nu a)] - \theta[z - (n + 1)\nu a]\}. \end{aligned} \tag{13.53}$$

Thus in this case as in the first problem, the second and higher order terms are important, but because of multiple Fourier transforms, their calculation is difficult.

13.3 Extension of the Inverse Method to Reflection from a Layered Medium where the Asymptotic Values of $c(t)$ at $t \to \pm\infty$ are Different

In the previous section we considered the inversion of the reflection data from a layered medium of finite thickness (a slab). Now we want to consider the same problem, viz, a scalar wave and normal incidence, but when the asymptotic values of the wave velocity are not the same, $c(-\infty) \neq c(+\infty)$, but both are known. Here we try to determine the variable velocity of the wave $c(x)$ from the frequency (or wave number) dependence of the reflected wave. Since both $c(-\infty) = c$ and $c(-\infty) = c'$ are assumed to be known we introduce an auxiliary function $V_0(z)$ with the following properties:
(a) that the differential equation

$$\chi''(z,k) + k^2[1 - V_0(z)]\chi(z,k) = 0, \tag{13.54}$$

can be solved exactly and
(b) that the limit of $V_0(z)$ as $z \to \infty$ satisfies the relation

$$V_0(z) \to 1 - \frac{c^2}{c'^2}, \quad z \to \infty. \tag{13.55}$$

Again we start with Eq. (13.1), and substitute $p(z,t) = p(z,k)e^{ikct}$ $(c = c(-\infty))$ to find Eq. (13.4). For the present problem we prefer to write this equation as

$$p''(z,k) + k^2[1 - V_1(z) - V_0(z)]p(z,k) = 0, \tag{13.56}$$

where $V_1(z)$ is defined in terms of $c(z)$;

$$V_1(z) = 1 - V_0(z) - \left(\frac{c^2}{c^2(z)}\right). \tag{13.57}$$

Now as $z \to \pm\infty$, $V_1(z)$ goes to zero, i.e. it behaves as $V(z)$ in Eq. (13.5). Since $V_0(z)$ is known, we want to relate the reflection coefficient from the combined effects of $V_0(z) + V_1(z)$ to the coefficient of reflection found from $V_1(z)$ only. To this end we replace the plane wave $\phi(z,k)$, Eq. (13.5) by $\chi(zk)$. This replacement gives us a modified Green's function which is the solution of the differential equation

$$G_1''(z,\tilde{z}) + k^2[1 - V_1(z)]G_1(z,\tilde{z}) = -\delta(z - \tilde{z}). \tag{13.58}$$

The solution of this equation with the "outgoing" and "ingoing" boundary conditions which we denote by $G_1^-(z,\tilde{z})$ and $G_1^+(z,\tilde{z})$ respectively, can be written in terms of $\chi^-(z)$ and $\chi^+(z)$, the eigenfunctions of the differential equation (13.54)

$$G_1^\pm(z,\tilde{z}) = \int_{-\infty}^{+\infty} \frac{dj}{j^2 - k^2 \pm i\varepsilon}\chi^-(z,j)\chi^+(z,j). \tag{13.59}$$

In this expression $\chi^-(z,j)$ and $\chi^+(z,j)$ are two independent solutions of the differential equation (13.54) and

$$\left(\chi^-\right)^* = \chi^+. \tag{13.60}$$

Using this Green's function we can write the solution of the differential equation

$$\phi''^*(z,q) + q^2(1 - V_0(z))\phi^*(z,q) = -q^2 V_0(z)\phi^*(z,q), \tag{13.61}$$

in terms of the ingoing solution $p*(z,k)$ of Eq. (13.56);

$$\phi^*(z,q) = \chi^+(z,q) + q^2 \int_{-\infty}^{+\infty} \frac{dj}{j^2 - q^2 + i\varepsilon} \chi^+(z,j)\langle\phi_q|V_0(z)|\chi_j^+\rangle, \tag{13.62}$$

where we have used the notation

$$\langle\phi_q|V_0(z)|\chi_j^+\rangle = \int_{-\infty}^{+\infty} \phi^*(z,q)V_0(z)\chi^+(z,j)dz. \tag{13.63}$$

Now we use the Green's function (13.59) to write the formal solution of Eq. (13.56) in terms of the outgoing wave $\chi^-(z,k)$;

$$p(z,k) = \chi^-(z,k) - k^2 \int_{-\infty}^{+\infty} \frac{dj}{j^2 - k^2 + i\varepsilon} \chi^+(z,j)\langle\chi^+(z,j)|V_0(z)|p(z,j)\rangle. \tag{13.64}$$

The $\mathcal{T}$-matrix for the combined effects of $V_0(z)$ and $V_1(z)$ is given by

$$\mathcal{T}(q,j) = \langle\phi(z,q)|V_0(z)|p(z,k)\rangle + \langle\phi(z,q)|V_1(z)|p(z,k)\rangle. \tag{13.65}$$

Now by substituting $p(z,k)$ from (13.64) in the first term of (13.65), and $\phi^*(z,p)$ from (13.62) in the second term of (13.65), we obtain the following equation for $\mathcal{T}(q,j)$

$$\begin{aligned}
\mathcal{T}(q,k) &= \langle\phi(z,q)|V_0(z)|\chi^-(z,k)\rangle + \langle\chi^+(z,q)|V_1(z)|p(z,k)\rangle \\
&\quad + \mathcal{P}\int_{-\infty}^{+\infty} \frac{(q^2 - k^2)\,dj}{(j^2 - q^2)(j^2 - k^2)}\langle\phi(z,q)|V_0(z)|\chi^+(z,j)\rangle \\
&\quad \times \langle\chi^+(z,j)|V_1(z)|p(z,k)\rangle.
\end{aligned} \tag{13.66}$$

By replacing q by $(-k)$ in (13.65) we get a simple relation for $\mathcal{T}(-k,k)$;

$$\mathcal{T}(-k,k) = \langle\phi(z,-k)|V_0(z)|\chi^-(z,k)\rangle + \langle\chi^+(z,-k)|V_1(z)|p(z,k)\rangle. \tag{13.67}$$

Since the reflection coefficient $R(k)$ is related to $\mathcal{T}(-k,k)$, therefore we can write (13.67) as

$$\begin{aligned}
\langle\chi^+(z,-k)|V_1(z)|p(z,k)\rangle &= \frac{i}{\pi k}R(k) - \langle\phi(z,-k)|V_0(z)|\chi^-(z,k)\rangle \\
&= \frac{i}{\pi k}[R(k) - r(k)], \tag{13.68}
\end{aligned}$$

where

$$\frac{i}{\pi k} r(k) = \langle \phi(z, -k) | V_0(z) | \chi^+(z, k) \rangle, \tag{13.69}$$

can be determined from the solution of Eq. (13.54).

Since we are assuming that $V_0(z)$ does not vanish, a direct calculation of the right-hand side of (13.70) without proper limiting procedure will lead to an infinite result. Therefore here it is simpler to solve the differential equation (13.54) for $\chi^-(z, k)$ and compare the solution with

$$\frac{1}{\sqrt{2\pi}} \left[e^{ikr} + r(k)e^{-ikr} \right]. \tag{13.70}$$

Having $\mathcal{T}(-k, k)$, we want to write two integral equations to replace the basic equations (13.18) and (13.26). For finding an equation similar to (13.18), we multiply (13.64) by $\chi^{+*}(z, q)V_1(z)$ and integrate over z and we get

$$\mathcal{T}_1(q, k) = \mathcal{V}_1(q, k) - \int_{-\infty}^{+\infty} \frac{j^2 dj}{j^2 - k^2 + i\varepsilon} \mathcal{V}_1(q, j)\mathcal{T}_1(j, k), \tag{13.71}$$

where

$$\mathcal{T}_1(q, k) = \int_{-\infty}^{+\infty} \chi^{+*}(z, q)V_1(z)p(z, k)dz, \tag{13.72}$$

and

$$\mathcal{V}_1(q, k) = \int_{-\infty}^{+\infty} \chi^{+*}(z, q)V_1(z)\chi^+(z, k)dz. \tag{13.73}$$

For the second integral equation we start with the Green function

$$G''(z - z', k) + k^2[1 - V_0(z) - V_1(z)]G = -\delta(z - z'), \tag{13.74}$$

and we expand this $G(z - z', k)$ in terms of the eigenfunctions of (13.4);

$$G(z, z', k) = \int_{-\infty}^{+\infty} \frac{dj}{j^2 - k^2 - i\varepsilon} p(z, j)p^*(z', j). \tag{13.75}$$

With the help of this Green's function we can write an integral equation for the solution of Eq. (13.54),

$$\chi^\pm(z, k) = p(z, k) + k^2 \int_{-\infty}^{+\infty} \frac{dj}{j^2 - k^2 \pm i\varepsilon} p(z, j)$$
$$\times \langle p(z, j)|V_1(z)|\chi^\pm(z, k)\rangle. \tag{13.76}$$

Now by multiplying (13.76) by $\chi^-(z, q)V_1(z)$ (which is the same as $\chi^{+*}(z, q)V_1(z)$) and integrating over z, we obtain the desired result, viz,

$$\mathcal{V}_1(q, k) = \mathcal{T}_1(q, k) + k^2 \int_{-\infty}^{+\infty} \frac{dj}{j^2 - k^2 - i\varepsilon} \mathcal{T}_1(q, j)\mathcal{T}_1^*(k, j). \tag{13.77}$$

The two integral equations (13.71) and (13.77) replace Eqs. (13.18) and (13.26) as the basic relations that are needed for solving the inverse problem. The third relation which is the analogue of (13.28) follow from (13.73):

$$\mathcal{W}_1(k) = V_1(-k, k) = \int_{-\infty}^{+\infty} \chi^+(z, -k)V_1(z)\chi^-(z, k)dz. \tag{13.78}$$

Introducing the ϵ expansion as we did in the previous section, for the first order term we have

$$\mathcal{W}_1^{(1)}(k) = \int_{-\infty}^{+\infty} \chi^+(z, -k)V_1^{(1)}(z)\chi^-(z, k)dz \tag{13.79}$$

$$= \frac{1}{\pi k}[R(k) - r(k)].$$

This last relation can be regarded as an integral equation for $V_1^{(1)}(z)$ with the kernel $\chi^+(z, -k)\chi^-(z, k)$. Once $V_1^{(1)}(z)$ is determined from (13.78), then the function

$$\mathcal{V}_1^{(1)}(q, k) = \int_{-\infty}^{+\infty} \chi^-(z, q)\chi^+(z, k)V_1^{(1)}(z)dz, \tag{13.80}$$

can be found, and then Eq. (13.78) enables us to calculate $V^{(1)}(q, k)$ to the second order in ϵ:

$$\mathcal{V}_2^{(1)}(q, k) = k^2 \int_{-\infty}^{+\infty} \frac{dj}{j^2 - k^2 + i\epsilon} \mathcal{T}_1^{(1)}(q, j)\mathcal{T}_1^{(1)*}(k, j). \tag{13.81}$$

In this way, at least formally, we can determine $\mathcal{V}^{(1)}(q, k)$:

$$\mathcal{V}^{(1)}(q, k) = \mathcal{V}_1^{(1)}(q, k) + \mathcal{V}_2^{(1)}(q, k) + \cdots. \tag{13.82}$$

By substituting $\mathcal{W}_1(k) = V_1(k, -k)$ in (13.78) we obtain $V_1(z)$, and hence the velocity profile of the wave in the inhomogeneous medium.

An Example of the Auxiliary Function $V_0(z)$ for Epstein's Model — Let us consider the wave propagation in an inhomogeneous medium where the wave velocity is of Epstein form [4], [5]

$$c(z) = \frac{c}{\sqrt{1 + \lambda(1 + \tanh \alpha z)}}, \tag{13.83}$$

and the equation of wave motion is given by

$$p''(z, k) + k^2\{1 + \lambda[1 + \tanh(\alpha z)]\}p(z, k) = 0. \tag{13.84}$$

This equation is exactly solvable in terms of hypergeometric function, and the reflection coefficient can be found from the asymptotic form of the analytic solution (see Eq. (13.9)). Thus we find the reflection coefficient to be expressible in terms of Γ function [5]

$$R(k) = \frac{\Gamma\left(\frac{ik}{\alpha}\right)\Gamma\left(-\frac{ik}{2\alpha}(1+\nu)\right)\Gamma\left(-\frac{ik}{2\alpha}(1+\nu)+1\right)}{\Gamma\left(\frac{-ik}{\alpha}\right)\Gamma\left(\frac{ik}{2\alpha}(1-\nu)\right)\Gamma\left(\frac{ik}{2\alpha}(1-\nu)+1\right)}, \tag{13.85}$$

where $\nu = \sqrt{1 + 2\lambda}$.

From the expression for velocity, (13.83), we find that

$$c(-\infty) = c, \quad \text{and} \quad c(+\infty) = \frac{c}{\nu}. \tag{13.86}$$

Now for this variable velocity of the wave we choose $V_0(z)$ to be

$$V_0(z) = \left(1 - \nu^2\right)\theta(z), \tag{13.87}$$

where $\theta(z)$ is a step function, and with this choice of $V_0(z)$ we find that the solution to the equation of $\chi(z, k)$, is

$$\chi''(z, k) + k^2\left[1 - \left(1 - \nu^2\right)\theta(z)\right]\chi(z, k) = 0. \tag{13.88}$$

The solution consists of two parts

$$\chi^-(z, k) = \begin{cases} \frac{1}{\sqrt{2\pi}}\left(e^{ikz} + \left(\frac{1-\nu}{1+\nu}\right)e^{-ikz}\right), & z < 0 \\[2mm] \frac{1}{\sqrt{2\pi}}\left(\frac{2}{1+\nu}\right)e^{ik\nu z}, & z > 0 \end{cases} \tag{13.89}$$

In this relation $\frac{1-\nu}{1+\nu}$ and $\frac{2\nu}{1+\nu}$ are the reflection and transmission amplitudes respectively. The conservation of energy implies that the sum of the power reflection coefficient

$$R_p = \left(\frac{1-\nu}{1+\nu}\right)^2, \tag{13.90}$$

and the power transmission coefficient

$$T_p = \left(\frac{c}{c'}\right)\left(\frac{2\nu}{1+\nu}\right)^2 = 4\nu(1+\nu)^2, \tag{13.91}$$

should add up to one as they do. The complete outgoing wave for all values of k (positive or negative) is given by

$$\chi^-(z, k) = \frac{1}{\sqrt{2\pi}}\left[\left(e^{ikz} + \frac{1-\nu}{1+\nu}e^{-ikz}\right)\theta(-z) + \frac{2}{1+\nu}e^{ik\nu z}\theta(z)\right]\theta(k)$$

$$+ \frac{1}{\sqrt{2\pi}}\left[\left(e^{ikz} + \frac{1-\nu}{1+\nu}e^{-ikz}\right)\theta(z) + \frac{2}{1+\nu}e^{ik\nu z}\theta(-z)\right]\theta(-k). \tag{13.92}$$

The incoming wave $\chi^+(z,k)$ is the complex conjugate of $\chi^-(z,k)$.

With the choice of $V_0(z)$, Eq. (13.87) and $c(z)$ defined by (13.83), $V_1(z)$ takes the form

$$V_1(z) = \begin{cases} \frac{-2\lambda e^{-\alpha z}}{e^{\alpha z}+e^{-\alpha z}}, & z < 0 \\[4mm] \frac{2\lambda z e^{\alpha z}}{e^{\alpha z}+e^{-\alpha z}}, & z > 0 \end{cases}, \tag{13.93}$$

and $V_1(z)$ is discontinuous at $z = 0$,

$$V_1\left(z = 0^+\right) - V_1\left(z = 0^-\right) = 2\lambda, \tag{13.94}$$

since it is the difference between $V(z)$ which is continuous and V_0 which is discontinuous.

A Method of Inversion to Find the Velocity Profile — Here we assume that $\mathcal{T}(-k,k)$ is given by

$$\mathcal{T}(-k,k) = \langle \chi^+(z,-k)|V_1(z)|p(z,k)\rangle = \frac{i}{\pi k}[R(k) - r(k)], \tag{13.95}$$

where for the velocity profile $c(z)$ of Eq. (13.83) $R(k)$ is a combination of Γ functions Eq. (13.85) and $r(k)$ is

$$r(k) = \frac{1-\nu}{1+\nu}. \tag{13.96}$$

For our choice of $V_0(z)$, $\chi^-(z,k)$ is already determined, Eq. (13.89), therefore the kernel in (13.80) is known

$$\chi^+(z,-k)\chi^-(z,k) = \frac{1}{\pi}\left[\frac{1}{1+\nu}\exp[ik(1+\nu)z] + \frac{1-\nu}{1+\nu}\exp[ik(\nu-1)z]\right]. \tag{13.97}$$

The relation (13.80) is an integral equation with the kernel given by (13.97), but we observe that $V_1(z)$ cannot be found by a simple inverse Fourier transform.

To invert (13.80) we note that

$$\frac{1}{2}(1+\nu)^2 \int_{-\infty}^{+\infty} \chi^+(z,-k)\chi^-(z,k)\exp\left[-ik(1+\nu)z'\right]dk$$

$$= \delta\left(z - z'\right) + \delta\left(z - \frac{\nu-1}{\nu+1}z'\right). \tag{13.98}$$

At this stage we can use the ϵ expansion of the previous section, Eqs. (13.33), (13.34) and replace $W_1^{(1)}$ by the difference between the reflection coefficients $R(k)$ and $r(k)$;

$$V_1(z) + V_1\left(\frac{\nu+1}{\nu-1}z\right) = \frac{1}{2}(1+\nu)^2 \int_{-\infty}^{+\infty} W_1(k)e^{-ik(1+\nu)z}dk. \tag{13.99}$$

Therefore, in the first order of expansion $V_1^{(1)}(z)$ satisfies the functional equation

$$V_1^{(1)}(z) + V_1^{(1)}\left(\frac{\nu+1}{\nu-1}z\right) = \frac{1}{2}(1+\nu)^2 \int_{-\infty}^{+\infty} \frac{i}{\pi k}[R(k) - r(k)]e^{-ik(1+\nu)z}$$

$$= I(z). \tag{13.100}$$

By iterating this equation we find

$$V_1^{(1)}(z) = \sum_{n=0}^{N-1} I\left[\left(\frac{\nu+1}{\nu-1}\right)^n z\right] + (-1)^{N-1}V_1^{(1)}\left[\left(\frac{\nu+1}{\nu-1}\right)^{N-1}z\right]. \tag{13.101}$$

As N becomes large, the last term in (13.101) can be ignored for all values of z except for $z = 0$. At this point from (13.100) we have $V_1^{(1)}(0) = \frac{1}{2}I(0)$. Thus, in principle, we can continue the ϵ expansion and calculate $V_1^n(z)$ and $V_1(z) = \sum_{n=1}^{\infty} V_1^{(n)}(z)$, but in practice we can calculate one or two terms of the expansion [4].

13.4   Direct and Inverse Problems of Wave Propagation Using Travel Time Coordinate

An inverse method of considerable interest in seismic wave propagation which we can solve without any reference to the inverse scattering problem of Gel'fand–Levitan is due to Goupillaud [10], [11]. Now in this section we want to formulate both the direct and the inverse problems of wave propagation in a heterogeneous medium when the assumed variable is the travel time coordinate. We first discuss the direct problem, and how we can write the equation determining the phase of the wave as a nonlinear first order equation. The advantage of this approach is that the inverse is rather easy to solve by iteration.

For a scalar wave propagating in an inhomogeneous medium in the z direction we can write the equation for $p(z, k)$ as (see Eq. (13.4))

$$\frac{d^2 p(z,k)}{dz^2} + k^2[1 - V(z)]p(z,k) = 0. \tag{13.102}$$

Let us assume that $V(z)$ changes smoothly, and let us transform (13.102) to the travel time coordinate. For this transformation we introduce the travel time coordinate $\xi(z)$ which is defined by [12], [13]

$$\xi(z) = \int_0^z [1 - V(z')]^{\frac{1}{2}} dz' = \int_0^z \frac{c}{c(z')} dz', \tag{13.103}$$

where as before we define $V(z)$ by

$$V(z) = \begin{cases} 1 - \frac{c^2}{c^2(z)} & \text{if} \quad 0 \le z \le a \\ \\ 0 & \text{if} \quad z > a \end{cases}. \tag{13.104}$$

Liouville Transformation — For the sake of simplicity we assume that the medium is bounded by a perfectly reflecting surface at $z = 0$, therefore the boundary condition in this case is

$$p(z = 0, k) = 0. \tag{13.105}$$

We also introduce a new wave amplitude $\phi(z)$ which is related to $p(z, k)$ by

$$\phi(\xi, k) = [1 - V(\xi(z))]^{\frac{1}{4}} p(z(\xi), k). \tag{13.106}$$

By substituting (13.103) and (13.106) in (13.102) we find the following differential equation

$$\phi''(\xi, k) + k^2\phi(\xi, k) + \left\{\frac{1}{4}\left(\frac{V''}{1 - V(\xi)}\right) + \frac{3}{16}\left[\frac{V'^2}{(1 - V(\xi))^2}\right]\right\}\phi(\xi, k) = 0, \tag{13.107}$$

where primes denote derivatives with respect to ξ.

Now we divide the inhomogeneous medium $0 \le z \le a$ into a group of N equal travel time layers where we keep the quantity λ defined by

$$\lambda = (1 - V_m)^{\frac{1}{2}}\Delta_m, \tag{13.108}$$

constant, i.e. independent of m. With this division, the range a of $V(z)$ is the sum of Δ_m s

$$a = \sum_{m=0}^{N}\Delta_m. \tag{13.109}$$

Now we replace the differential equation (13.107) by a difference equation in ϕ;

$$\phi_{j+1} + \phi_j - 2\cos(k\lambda)\phi_j - W_j\phi_j = 0, \tag{13.110}$$

where W_j is given by

$$Wj = -\left\{\left(\frac{1}{4}\right)\frac{(V_{j+1} + V_{j-1} - 2V_j)}{1 - V_j} + \left(\frac{3}{16}\right)\frac{(V_{j+1} - V_j)^2}{(1 - V_j)^2}\right\}, \tag{13.111}$$

and ϕ_j is subject to the boundary condition

$$\phi_0 = 0. \tag{13.112}$$

This boundary condition follows from (13.105).

Solving the Direct Problem with Difference Equation — Let us consider the asymptotic solution of the difference equation

$$p_{j+1} + p_{j-1} - 2\cos(k\Delta)\phi_j = \Delta^2 k^2 V_j\phi_j, \quad j = 0, 1, \cdots N, \tag{13.113}$$

which is a discrete version of the differential equation (13.102). We prefer to work with (13.113) rather than other forms of finite difference equation approximating

(13.102), because as we explained earlier, this form is simpler for solving the inverse problem. The solution of this difference equation is subject to the boundary condition

$$p_0 = 0. \tag{13.114}$$

Now for a homogeneous medium, i.e. when $V_j = 0$ for all j s, this solution for $j = N$ is given by

$$p_N^H = B\sin(kN\Delta), \quad N\Delta = a, \tag{13.115}$$

where the superscript H indicates that the solution is for a homogeneous medium. On the other hand the asymptotic solution of (13.110) for $j = N$ is

$$\phi_N = B\sin(kN\lambda + \delta(k)), \tag{13.116}$$

a result which follows directly from (13.110) for $j \geq N$, by setting $W_j = 0$. The phase shift $\delta(k)$ is caused by the presence of W_j in (13.110), and can be calculated by solving (13.110) subject to the boundary condition (13.112). By comparing the arguments of the two sine functions (13.115) and (13.116), we find the total phase shift η_B to be

$$\eta_R = kN(\lambda - \Delta) + \delta(k)$$

$$= k\sum_{m=0}^{N}\left[(1 - V_m)^{\frac{1}{2}} - 1\right]\Delta_m + \delta(k). \tag{13.117}$$

The first term in the right-hand side of (13.117) which is proportional to k is the WKB phase shift

$$\eta_{WKB} = k\sum_{m=0}^{N}\left[(1 - V_m)^{\frac{1}{2}} - 1\right]\Delta_m, \tag{13.118}$$

which for large k values gives the major contribution to η_B.

To calculate $\delta(k)$ we first derive a nonlinear difference equation for $\tan\delta(k)$ which is the discrete analogue of the Calogero equation of quantum scattering theory [14].

To obtain a nonlinear but first order equation for ϕ_j we write (13.110) as a linear set of equations for ϕ_j

$$\phi_j = \sin(jk\lambda) + \sum_{m=0}^{j}\frac{\sin[(j - m)k\lambda]}{\sin(k\lambda)}W_m\phi_m. \tag{13.119}$$

We can verify that (13.119) is a formal solution of the difference equation (13.110) by direct substitution in the latter equation. Also we note that ϕ_j satisfies the boundary condition (13.112). Equation (13.113) can also be written as

$$\phi_j = \sin(jk\lambda)C_j + \cos(jk\lambda)S_j \tag{13.120}$$

where

$$C_j = 1 + \sum_{m=0}^{j}\frac{\cos(km\lambda)}{\sin(k\lambda)}W_m\phi_m, \tag{13.121}$$

Table 13.1: Exact and approximate phase shifts in radians for the variable wave velocity $c(z) = c\left[1 + \alpha^2 \exp(-\beta z)\right]^{-\frac{1}{2}}$. For the parameters $\alpha = 1$ and $\beta = 0.5\ L^{-1}$, the phase shift, $\eta_B(k) = \eta_{WKB} + \delta(k)$ Eq. (13.117), is given as a function of the number of layers N. Here L is an arbitrary unit of length [17].

$k(L^{-1})$	$\eta(k)$	$\eta_B(N = 20)$	$\eta_B(N = 50)$	$\eta_B(N = 200)$
		$a = 13.36\ L$	$a = 13.29\ L$	$a = 13.25\ L$
0.01	0.16×10^{-4}	0.53×10^{-3}	-0.17×10^{-3}	0.60×10^{-3}
0.1	0.14×10^{-1}	0.14×10^{-1}	$0.86 \times 10^{-2} S$	0.52×10^{-2}
0.5	0.399	0.328	0.355	0.369
1.0	0.879	0.749	0.818	0.852
5.0	4.515	3.976	4.261	4.445
10.0	9.037	7.804	8.510	8.900

and

$$S_j = -\sum_{m=0}^{j} \frac{\sin(km\lambda)}{\sin(k\lambda)} W_m \phi_m. \tag{13.122}$$

Noting that

$$W_{N+1} = W_{N+2} = \cdots = 0, \tag{13.123}$$

we have

$$C_N = C_{N+1}, \qquad S_N = S_{N+1}, \tag{13.124}$$

and therefore the asymptotic form of ϕ_j is given by

$$\phi_N = C_N \left[\sin(kN\lambda) + \frac{S_N}{C_N} \cos(kN\lambda)\right]. \tag{13.125}$$

By comparing this relation with (13.116) we find the important result that

$$\frac{S_N}{C_N} = \tan \delta(k). \tag{13.126}$$

Now returning to Eqs. (13.121) and (13.122) we have the following difference equations. For C_j and S_j:

$$C_{j+1} - C_j = \frac{\cos(kj\lambda)}{\sin(k\lambda)} W_j[\sin(kj\lambda)C_j + \cos(kj\lambda)S_j], \tag{13.127}$$

and

$$S_{j+1} - S_j = -\frac{\sin(kj\lambda)}{\sin(k\lambda)} W_j[\sin(kj\lambda)C_j + \cos(kj\lambda)S_j]. \tag{13.128}$$

By multiplying (13.127) by S_j and (13.128) by C_j and subtracting the resulting two equations, we obtain a new equation which is

$$\frac{S_j C_{j+1} - S_{j+1} C_J}{C_j^2} = \frac{W_j}{\sin(k\lambda)}\left[\sin(kj\lambda) + \frac{S_j}{C_j}\cos(kj\lambda)\right]^2. \tag{13.129}$$

Let us define T_j by

$$T_j = \frac{S_j}{C_j}, \tag{13.130}$$

then (13.129) can be written as a difference equation for T_j [17];

$$T_{j+1} - T_j = -W_j\left\{\frac{[\sin(kj\lambda) + T_j\cos(kj\lambda)]^2}{[\sin(k\lambda) + W_j[\sin(kj\lambda) + T_j\cos(kj\lambda)]\cos(kj\lambda)]}\right\}. \tag{13.131}$$

From Eqs. (13.125) and the boundary condition (13.112) we find that (13.131) is subject to the boundary condition

$$T_0 = 0. \tag{13.132}$$

Also according to Eq. (13.126) we have

$$T_{N+1} = T_N = \tan\delta(k). \tag{13.133}$$

Thus the nonlinear difference equation (13.131) can be solved numerically to yield the phase shift $\delta(k)$.

If W_j s are small for all k values, then we can keep the leading terms in W_j in (13.131) and find a simple expression for T_{N+1};

$$T_{N+1} = T_N = -\sum_{j=1}^{N}\frac{\sin^2(kj\lambda)}{\sin(k\lambda)}W_j = \tan\delta^B(k). \tag{13.134}$$

This is the discrete version of the Born approxintion for the phase shift.

The Accuracy of the Travel Time Approximation — To test the accuracy of the phase shift calculated using the travel time coordinate, let us consider a model where in the inhomogeneous medium the speed of the wave is given by

$$c(z) = \frac{c}{[1 + \alpha^2\exp(-\beta z)]^{\frac{1}{2}}}, \tag{13.135}$$

where α and β are constants, α is dimensionless and β has the dimension L^{-1}, when z is measured in units of L. For simplicity we also assume that this medium is bounded by a perfectly reflecting surface at $z = 0$, and for $z \geq a$, the velocity approaches c. By changing z to ζ where

Table 13.2: Comparison of the approximate values of the phases $\delta(k)$, and $\delta^B(k)$ calculated from Eqs. (13.131) and (13.134) for an inhomogeneous medium where the wave velocity is dependent on the depth z according to the relation $c(z) = c\left[1 + \alpha^2 e^{-\beta t}\right]^{-\frac{1}{2}}$. The parameters used in this calculation are $\alpha = 1$, $\beta = 0.5L^{-1}$, and $N = 20$ [17].

$k(L^{-1})$	$\delta(k)$	$\delta^B(N = 20)$
0.01	-0.731×10^{-2}	-0.817×10^{-2}
0.1	-0.646×10^{-1}	-0.703×10^{-1}
1.0	-0.351×10^{-1}	-0.350×10^{-1}
10.0	-0.322×10^{-1}	-0.321×10^{-1}
100.0	-0.228×10^{-1}	-0.228×10^{-1}

$$\zeta = 2\beta\alpha \exp\left(-\frac{1}{2}\beta z\right), \tag{13.136}$$

we can write the Jost type solution of (13.102) Eq. (4.8) in terms of the Bessel function of complex order and argument

$$f(z,k) = \exp\left[\frac{ik}{\beta}\left(\ln\left(\frac{\alpha^2 k^2}{\beta^2}\right)\right)\right] J_{-\frac{2ik}{\beta}}\left(\frac{2\alpha k}{\beta}\right)\exp\left[-\frac{1}{2}\beta z\right]. \tag{13.137}$$

We can find the physical solution which satisfies the condition $p(z = 0, k) = 0$ from the Jost solution and the Jost function according to the relation:

$$p(z,k) = \frac{i}{2k}[f(k)f(-k,z) - f(-k)f(k,z)], \quad f(k) \equiv f(k, z = 0). \tag{13.138}$$

Also the phase shift can be obtained from the asymptotic form of $f(z,k)$;

$$\eta(k) = \arg\left[\Gamma\left(1 - \frac{2ik}{\beta}\right)\left(\frac{k\alpha}{\beta}\right)^{\frac{2ik}{\beta}} J_{-\frac{2ik}{\beta}}\left(\frac{2k\alpha}{\beta}\right)\right]. \tag{13.139}$$

Unlike the phase shift in potential scattering $\eta(k) + n\pi$ (n is an integer) increases linearly as a function of k except for discontinuities at regular intervals. In Table 13.1 numerical results obtained for the exact phase shift with the added $n\pi$ are tabulated

Table 13.3: The direct calculation of V_j and W_j and comparison with their inverses. These results are obtained for the parameters $\alpha = 1$, $\beta = 0.5\ L^{-1}$ and $N = 10$.

z(L)	V_{dir}	W_{dir}	W_{inv}	V_{inv}
1.129	-0.5686	-0.636×10^{-1}	-0.590×10^{-1}	-0.6145
2.366	-0.3063	0.122×10^{-1}	0.122×10^{-1}	-0.3145
3.680	-0.1588	0.788×10^{-2}	0.788×10^{-2}	-0.1569
5.041	-0.804×10^{-1}	0.454×10^{-2}	0.454×10^{-2}	-0.767×10^{-1}
6.427	-0.402×10^{-1}	0.244×10^{-2}	0.244×10^{-2}	-0.361×10^{-1}
7.828	-0.200×10^{-1}	0.125×10^{-2}	0.125×10^{-2}	-0.264×10^{-1}
9.235	-0.988×10^{-2}	0.632×10^{-3}	0.632×10^{-3}	-0.688×10^{-2}
10.646	-0.488×10^{-2}	0.315×10^{-3}	0.315×10^{-3}	-0.249×10^{-2}
12.058	-0.241×10^{-2}	0.156×10^{-3}	0.156×10^{-3}	-0.618×10^{-3}
13.472	-0.119×10^{-2}	0.77210^{-4}	0.772×10^{-4}	0.0

for a range of the wave number k, measured in the unit of inverse length L^{-1}. In the same table the approximate phase shifts η_B obtained from using travel time coordinate formulation, Eq. (13.117), are given for different k and N values. The result of this approximation depends on the number of layers, N into which we have divided the range of the inhomogeneous medium. This range should be chosen large enough so that $c(z = a) \approx c$. For the velocity profile given by (13.135) with the parameters $\alpha = 1$ and $\beta = 0.5$, a will be around $13.25\ L$. As we expect by increasing the number of constant travel time layers where $c(z)$ is changing, we increase the accuracy of the calculated phase shift. At the same time we observe that for smaller wave numbers the errors are larger, and in fact, for larger N at these low wave numbers, the sign of the phase shift may not be predicted correctly (see Table 13.1). For these cases the phase shifts are extremely small, and their numerical computation is not reliable. What is remarkable is that for larger k values the exact phase shift $\eta(k)$ and $\eta^B(k) \approx \eta_{WKB}(k)$ are close to each other, as long as k remains less than 2π, and the number of layers is large, e.g. 200.

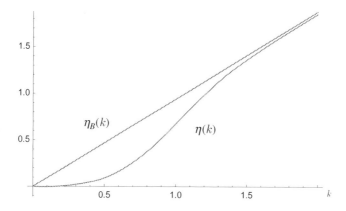

Figure 13.1: Exact and approximate phase shifts η and η_B calculated for the velocity profile $c(z)$, Eq. (13.135), with $\alpha = 3$ and $\beta = 3$. The number of travel time layers is chosen to be 200.

The Direct Problem Using Travel Time Coordinate — Assuming that we know the velocity profile, $c(z)$, in the inhomogeneous medium, using (13.103) we calculate $\xi(z)$. Then by dividing the inhomogeneous medium into N equal travel time layers Δ_1, Δ_2, $\cdots \Delta_n$ and their corresponding V_j according to (13.108), we calculate W_j from (13.111). Now we can solve the nonlinear difference equation (13.131) and find T_1, T_2, $\cdots T_N$. Finally from Eq. (13.133) we obtain the phase shift $\delta(k)$.

A very useful approximation can be found by noting that since $\delta(k)$ is small for all k values, we may keep the leading term in (13.131). This corresponds to a discrete form of the Born approximation for calculating the phase shift.

Inverse Problem Based on the Travel Time Formulation — The starting point in this inversion is the discrete version of the Born approximation (13.134). This relation can be readily inverted to yield W_j;

$$W_j = \frac{8\lambda}{\pi} \int_0^{\frac{2\pi}{\lambda}} T_N^B \sin(k\lambda) \cos(2k\lambda j) dk. \tag{13.140}$$

Having found W_j s we can calculate V_j s in the following way: we write Eq. (13.111) as

$$V_{j-1} = 2V_j - Vj + 1 - 4(1 - V_j)W_j - \frac{3}{4} \frac{(V_{j+1} - V_j)^2}{(1 - V_j)^2}. \tag{13.141}$$

This equation can be solved by iteration provided that V_{N+1} and V_N are known quantities. Since for $j > N$, $V_j = 0$, therefore assuming the continuity of the wave velocity at the boundary, we have the boundary conditions

$$V_N = V_{N+1} = 0. \tag{13.142}$$

These initial conditions can also be used when V_N is very small to obtain an approximate solution to (13.141) and calculate V_{N-1}, $V_{N-2} \cdots V_1$ by iteration.

Results of Numerical Calculation by Travel Time Method — As we discussed earlier, we will consider an inhomogeneous medium where the wave velocity $c(z)$ is given by Eq. (13.135), and it depends on two arbitrary parameters α and β. We have used two sets of these parameters: For the first we choose $\beta = 3\ L^{-1}$ and $\alpha = 1$ and for the second we choose $\alpha = 1$ and $\beta = 0.5\ L^{-1}$. In the case of the first set the results of calculation of the exact phase shifts $\eta(k)(\pm 2n\pi)$ are given in Table 13.1 where the same phase shifts (denoted by $\eta_A(k)$) found from the travel time approximation appear in columns three to five of the table. In each column the number of layers N is given and also the range of the inhomogeneity of the medium a is tabulated. This range changes slightly, so that if Δ_j is the width of each layer then $\sum_{j=1}^{N} \Delta_j = a$. The phases in these columns show that as we increase the number of layers, the agreement between the exact and approximate results improves.

Now let us consider Table 13.2 where the phases $\delta(k)$ calculated from (13.131) and (13.133) and δ^B obtained from (13.133) are given as functions of k. Here the second set of parameters $\alpha = 1$ and $\beta = 0.5$ and $N = 20$ have been used in the computation. These phases are close to each other and show the validity of the discrete Born approximation. Finally in Table 13.3, results from the direct calculations of V_j and W_j, and the subsequent inversion of $T_N^B(k)$ which leads to V_{inv} are presented. These results are encouraging considering that here we have divided the inhomogeneous medium into only 10 parts.

13.5 $\mathcal{R}$ Matrix and the Inverse Problems of Wave Propagation

We have already seen that the amplitude of a wave propagating in an inhomogeneous medium is given by Eq. (13.102) where $V(z)$ is related to the wave velocity by Eq. (13.104). The direct problem gives us the $\mathcal{R}$ matrix at the boundary $z = a$;

$$\left(\frac{1}{p(k,z)} \frac{dp(k,z)}{dz} \right)_{z=a} = k\mathcal{R}(k), \qquad (13.143)$$

where on the right-hand side we have written $k\mathcal{R}(k)$ rather than $\frac{1}{a}\mathcal{R}(k)$ to simplify the resulting equations. For the boundary condition of the differential equation (13.102) we have assumed that the wave is totally reflected at $z = 0$, i.e.

$$p(k, z = 0) = 0. \qquad (13.144)$$

To solve the direct problem we integrate (13.102) from the origin to the point a where we can determine the $\mathcal{R}$ matrix from (8.20). For solving the inverse problem we can use the inversion of the $\mathcal{R}$ matrix by approximating the differential equation for the wave propagation by its corresponding difference equation. Noting that the solution of (13.102) is given by

$$p(z,k) = A\sin[kz + \eta(k)], \qquad (13.145)$$

then from (13.144) we find $\mathcal{R}(k)$ to be

$$\mathcal{R}(k) = \cot[ka + \eta(k)]. \tag{13.146}$$

In Eqs. (13.145) and (13.146), $\eta(k)$ characterizes the effect of the medium and is one-half of the phase spectrum of the medium response. This phase $\eta(k)$ is related to $\mathcal{R}(k)$ by

$$\cot \eta(k) = \frac{\mathcal{R}(k)\cos(ka) + \sin(ka)}{\cos(ka) - \mathcal{R}(k)\sin(ka)}. \tag{13.147}$$

Thus if we know the $\mathcal{R}$ matrix we can calculate $\eta(k)$ from (13.147), and if $c(z) \equiv c$ for $z > a$, the $\eta(k)$ will be exact.

The Finite Difference Approximation to the $\mathcal{R}$ Matrix — Suppose that we divide the range of the inhomogeneity a into N equal parts each of length Δ, $\Delta = \frac{a}{N}$ and use finite differences to replace $\frac{d^2 p(k,z)}{dz^2}$ in (13.102) we find the following set of equations [15], [16]:

$$p_2(k) - 2p_1(k) + k^2\Delta^2(1 - V_1)p_1(k) = 0 \tag{13.148}$$

$$p_{j+1}(k) - 2p_j(k) + p_{j-1}(k) + k^2\Delta^2(1 - V_j)p_j(k) = 0, \quad j = 1, 2 \cdots N - 1, \tag{13.149}$$

$$[k\Delta\mathcal{R}_{DE}(k) - 1]p_N(k) + \frac{k^2\Delta^2}{2}(1 - V_N)p_N(k) = 0, \tag{13.150}$$

where $\mathcal{R}_{DE}$ is the $\mathcal{R}$ matrix at the end point $z_N = N\Delta$ calculated by solving the difference equations (13.148)–(13.150), and

$$V_j = V(z_j) = V(\Delta j), \quad j = 1, 2 \cdots N. \tag{13.151}$$

From Eqs. (13.149) we find the recurrence relation

$$\frac{p_{j+1}(k)}{p_j(k)} = 2 - \Delta^2 k^2 - \frac{1}{\left(\frac{p_j(k)}{p_{j-1}(k)}\right)} \tag{13.152}$$

and with the help of this recurrence relation we can get $k\Delta\mathcal{R}_{DE}$ as a finite continued fraction

$$k\Delta\mathcal{R}_{DE}(k) = 1 - \frac{k^2\Delta^2}{2}(1 - V_N)$$

$$+ \cfrac{1}{-2 + k^2\Delta^2(1 - V_{N-1}) + \cdots + \cfrac{1}{(-1)^j[2 - k^2\Delta^2(1 - V_{N-j})]\cdots}}. \tag{13.153}$$

This expression can be used to calculate $\mathcal{R}_{DE}(k)$ when $V(z)$ is known.

The Inverse Problem — We have seen that the phase spectrum is given by Eq. (13.147) in the limit of $\Delta \to 0$, $(N \to \infty)$, and $\mathcal{R}(k)$ is also calculated in this limit. Now let us first examine some of the properties of this exact $\mathcal{R}$ matrix. To this end we write the $\mathcal{R}(k)$ matrix as the ratio of two infinite products [18]

$$\mathcal{R}(k) = \cot(ka + \eta(k)) = \frac{\prod_{i=1}^{\infty}\left(1 - \frac{k^2}{k_i^2}\right)}{(ka + \eta(k))\prod_{j=1}^{\infty}\left(1 - \frac{k^2}{q_j^2}\right)}, \tag{13.154}$$

where

$$k_i^2 = \frac{\left(i - \frac{1}{2}\right)^2}{a^2}, \quad \text{and} \quad q_j^2 = \frac{j^2\pi^2}{a^2}, \quad i,j = 1,2,\,3\cdots\infty. \tag{13.155}$$

These k_i s and q_j s are the roots of $\cos(ka + \eta(k))$ and $\sin(ka + \eta(a))$ respectively. Now let us find the discrete version of (13.154) which we can relate to $k\Delta\mathcal{R}(k)$.

This can be achieved if we note that by solving (13.148)–(13.150) for $V(n) = 0$ for all (finite) n, we find

$$S(k,\Delta) = kN\Delta \prod_{i=1}^{N-1}\left\{1 - \frac{1}{4}k^2\Delta^2\sin^2\left[\frac{\pi\Delta i}{2a}\right]\right\}, \tag{13.156}$$

and

$$C(k,\Delta) = \prod_{j=0}^{N-1}\left\{1 - \frac{1}{4}k^2\Delta^2\sin^2\left[\frac{\pi\Delta}{2a}\left(j + \frac{1}{2}\right)\right]\right\}. \tag{13.157}$$

Table 13.4: The phase of the reflection coefficient calculated for the variable wave velocity $c(z) = c[1 + \lambda^2\exp(-\beta z)]^{-\frac{1}{2}}$. The parameters used are $\lambda = 1$, $\beta = 0.5$ in units of $(length)^{-1}$, and the length of the interval is assumed to be $L = 20$. Here for different values of the wave number k the exact results $\eta(k)$, are compared with those found from the continued fraction solution for different Δ s [20].

$k(\text{length}^{-1})$	$\eta(k)$	$\eta_{DE}(N = 20)$	$\eta_{DE}(N = 10)$
0.01	0.160×10^{-4}	-0.159×10^{-4}	0.159×10^{-4}
0.10	0.141×10^{-1}	141×10^{-1}	0.141×10^{-1}
0.05	0.399	0.413	0.617
1.0	0.879	0.604	0.453×10^{-4}
5	-1.768	0.181×10^{-4}	0.907×10^{-5}

By replacing $\sin(ka + \eta(k))$ by $S(k, \Delta)$ and $\cos(ka + \eta(k))$ by $C(k, \Delta)$ in (13.154) we obtain $\mathcal{R}_{DE}(k)$ which is the discrete analogue of (13.154) [20];

$$\mathcal{R}_{DE}(k) = \frac{C(k, \Delta) \cot_{DE} \eta_{DE}(k) - S(k, \Delta)}{(ka + \eta(k)) [S(k, \Delta) \cot_{DE} \eta_{DE}(k) + C(k, \Delta)]}. \tag{13.158}$$

As Eq. (13.154) shows the exact $R(k)$ is the ratio of two power series in k. However once we introduce a step size Δ, then $\mathcal{R}$ becomes the ratio of two polynomials. The resulting rational approximation can be found from the solution of the differ-ence equations (13.148)–(13.150). In Table 13.4 the exact result from the phase of the reflected wave $\eta(k)$ when $c(z)$ given by (13.135) is compared the solution for η_{DE} found from (13.153). For small wave number k and large N, the results are comparable. But for large k and small N this approximation breaks down.

13.6 Inverse Problem for Acoustic Waves: Determination of the Wave Velocity and Density Profiles

For an acoustic plane wave of frequency ω which is incident upon a plane stratified half-space $z > 0$, the acoustic pressure $p(x, z)$ satisfies the following differential equation [19]

$$\left(\frac{\partial^2}{\partial x^2} + \frac{\partial^2}{\partial z^2} + \frac{\omega^2}{c^2(z)} - L(z)\frac{\partial}{\partial z} \right) p(x, z) = 0, \tag{13.159}$$

where $c(z)$ is the wave velocity and $L(z) = \frac{d}{dz} \ln \rho(z)$, with $\rho(z)$ being the density of the medium. The half-space $z \leq 0$ is assumed to be homogeneous with $\rho(z) = \rho(0)$ and $c(z) = c(0)$. In this half-space, the acoustic pressure can be expressed in terms of the incident and reflected waves [19],

$$p(x, z) = e^{ikx \sin \theta} \left[e^{ikz \cos \theta} + R(k, \theta)e^{-ikz \cos \theta} \right]. \tag{13.160}$$

In this relation $k = \frac{\omega}{c(0)}$ is the wave number associated with the incident and reflected waves, θ is the angle of incidence, and $R(k, \theta)$ is the reflection coefficient.

13.7 Inversion of Travel Time Data in the Geometrical Acoustic Limit

Now let us consider the case where the incident wave is of the form of a wave packet and the amplitude of this packet is peaked around the wave number $k = k_0$. For his problem we can find the well-known travel time inversion in the geometrical acoustic limit analytically. Here the wave packet is obtained from the superposition of plane waves in the homogeneous half-space $z < 0$ where the velocity is c_0,

$$p(x, z, t) = \int_{-\infty}^{\infty} A(k) \left\{ e^{ikx \sin\theta - c_0 t} \left[e^{ikz \cos\theta} + R(k, \theta) e^{-ikz \cos\theta} \right] \right\} dk, \quad (13.161)$$

where

$$A(k) = |A(k)| e^{i\alpha(k)}, \quad (13.162)$$

is the complex amplitude of the wave. The center of the in- and the out-going wave packets can be found from

$$\left| \frac{d}{dk} \alpha(k) + k(z \cos\theta - c_0 t + x \sin\theta) \right|_{k=k_0} = 0, \quad (13.163)$$

and

$$\left\{ \frac{d}{dk} \left[-i \ln R(k, \theta) + \alpha(k) + k(-z \cos\theta - c_0 t + x \sin\theta) \right] \right\}_{k=k_0} = 0, \quad (13.164)$$

respectively. Thus the time-delay caused by transmission through the medium is related to $R(k, \theta)$ by

$$\tau(k_0, \theta) = -\frac{i}{c_0} \left[\frac{d}{dk} \ln R(k.\theta) \right]_{k=k_0}. \quad (13.165)$$

Now let us calculate this time delay in geometrical acoustic limit. In this limit the pressure in the inhomogeneous medium can be written as [2]

$$p(x, z, k) \approx \left[a^2(z) - \beta^2 \right]^{-\frac{1}{4}} e^{i\omega\beta x} \left\{ (1 - i) \exp\left(i\omega \int_z^\zeta \left[a^2(z) - \beta^2 \right]^{\frac{1}{2}} dz \right) \right\}$$

$$+ \left[a^2(z) - \beta^2 \right]^{-\frac{1}{4}} e^{i\omega\beta x} \left\{ (1 + i) \exp\left(-i\omega \int_z^\zeta \left[a^2(z) - \beta^2 \right]^{\frac{1}{2}} dz \right) \right\},$$

$$z < \zeta, \quad (13.166)$$

and

$$p(x, z, k) \approx \sqrt{2} \left[\beta^2 - a^2(z) \right]^{-\frac{1}{4}} e^{i\omega\beta x} \exp\left(-\omega \int_\zeta^z \left[\beta^2 - a^2(z) \right]^{\frac{1}{2}} dz \right),$$

$$z > \zeta, \quad (13.167)$$

where

$$\beta = \frac{\sin\theta}{c_0}, \quad \text{and} \quad \alpha(z) = \frac{1}{c(z)}, \quad (13.168)$$

and ζ is the turning point

$$a(\zeta) = \pm\beta. \quad (13.169)$$

In order to find the coefficient of reflection $R(k, \theta)$, we match (13.166) to the solution in the homogeneous medium at $z = 0$ and we find

$$R(k, \theta) = \exp\left\{ 2ic_0 k \int_0^\zeta \left[a^2(z) - \beta^2\right]^{\frac{1}{2}} dz - \frac{i\pi}{2} \right\}. \qquad (13.170)$$

Now if we substitute (13.170) in (13.165) we obtain the desired result

$$\tau(k_0, \theta) = 2 \int_0^\zeta \left[a^2(z) - \beta^2(\theta)\right]^{\frac{1}{2}} dz. \qquad (13.171)$$

This relation can be regarded as an integral equation for $a(z) = \frac{1}{c(z)}$, which is the slowness function, and thus the wave velocity can be found if the time delay is known (or measured) as a function of θ. We observe that in this geometrical acoustic limit we can find the velocity profile but not the variation of the density $\rho(z)$.

13.8   Riccati Equation for Solving the Direct Problem for Variable Velocity and Density

The partial differential equation for the acoustic pressure, Eq. (13.159) is separable in Cartesian coordinate system, and thus we can reduce it to an ordinary differential equation. To this end we define a new function $\phi(z)$ by the relation

$$\phi(z) = p(x, z) \exp(-ikx \sin \theta), \qquad (13.172)$$

and we substitute (13.172) in (13.159) and find the following equation for $\phi(z)$

$$\left\{ \frac{d^2}{dz^2} + k^2 \left[\eta^2(z) - \sin^2 \theta\right] - L(z)\frac{d}{dz} \right\} \phi(z) = 0, \qquad (13.173)$$

where

$$\eta(z) = \frac{c_0}{c(z)}. \qquad (13.174)$$

Let us assume that the medium becomes homogeneous for $z \geq d$, where d is an arbitrary finite depth, and let us denote the wave velocity and the density for $z \geq d$ by c_d and ρ_d respectively. Then for $z \geq d$, the transmitted wave, $\phi(z)$, can be written as

$$\phi(z) = T(\theta, \omega) \exp\left\{ -\gamma k \left|\sin^2 \theta - \eta^2(d)\right|^{\frac{1}{2}} z \right\}, \qquad (13.175)$$

where $T(\theta, \omega)$ is the transmission amplitude and γ is defined by

$$\gamma = \begin{cases} 1 & \text{if } \sin^2 \theta - \eta^2(d) \geq 0 \\ -i & \text{if } \sin^2 \theta - \eta^2(d) < 0 \end{cases}. \qquad (13.176)$$

To simplify the numerical integration of the direct problem, we introduce the variable acoustic impedance

$$\Omega(z) = \frac{\phi(z)}{\frac{d\phi(z)}{dz}}, \tag{13.177}$$

to replace $\phi(z)$. From the definition of $\Omega(z)$, (13.177), and Eq. (13.175) for $\phi(z)$ we find that $\Omega(z)$ is the solution of the differential equation

$$\frac{d\Omega(z)}{dz} = k^2 \left[\eta^2(z) - \sin^2\theta\right]\Omega^2(z) - L(z)\Omega(z) + 1, \tag{13.178}$$

and thus reduce the second order linear differential equation to a first order Riccati equation. The value of $\Omega(z)$ at $z = 0$ is obtained from Eqs. (13.160) and (13.172):

$$\Omega(0) = \frac{\rho(0)c_0[1 + R(\theta,\omega)]}{i\omega\rho\,(0^+)\,[1 - R(\theta,\omega)]\cos\theta}, \tag{13.179}$$

where $\rho(0)$ and $\rho(0^+)$ are the densities of the homogeneous and the inhomogeneous media at $z = 0$ respectively. We have assumed that the medium is homogeneous for $z \geq d$, and that all of the physical properties are continuous at $z = d$, therefore from Eqs. (13.175) and (13.177) it follows that

$$\Omega(d) = -\frac{1}{\gamma k}\left|\sin^2\theta - \eta^2(d)\right|^{\frac{1}{2}}. \tag{13.180}$$

By using (13.180) as the boundary condition, we can integrate (13.178) to find $\Omega(0)$, and subsequently the reflection coefficient from (13.179).

13.9   Finite Difference Equation for Acoustic Pressure in an Inhomogeneous Medium: Direct and Inverse Problems

As we discussed in Chapter 7 there are many ways of approximating a differential equation by a difference equation. Let us consider the following finite difference equation first and then discuss the reasons for our choice.

We replace $\phi(z)$ in Eq. (13.173) by $\psi(z)$ which is defined by

$$\psi(z) = \frac{e^{iqz}\phi(z)}{T(\theta,\omega)}, \tag{13.181}$$

where

$$q = \gamma k \left|\sin^2\theta - \eta^2(d)\right|^{\frac{1}{2}}. \tag{13.182}$$

By substituting for $\phi(z)$ in (13.173) we obtain the following equation for $\psi(z)$:

$$\frac{d^2\psi(z)}{dz^2} - [2q + L(z)]\frac{d}{dz}\psi(z) + \left[k^2V(z) + qL(z)\right]\psi(z) = 0, \tag{13.183}$$

where

$$V(z) = \eta^2(z) - \eta^2(d). \tag{13.184}$$

Now we change $\phi(z)$ to $\psi(z)$, and the boundary conditions at $z = d$ takes a very simple form of

$$\psi(z = d) = 1, \quad \text{and} \quad \left(\frac{d\psi(z)}{dz}\right)_{z=d} = 0. \tag{13.185}$$

Finally we replace the differential equation (13.183) by the difference equation

$$(1 - q\Delta)\left(1 - \frac{1}{2}\Delta L_n\right)\psi_{n+1} + (1 + q\Delta)\left(1 + \frac{1}{2}L_n\Delta\right)\psi_{n-1}$$
$$- 2\psi_n + k^2\Delta^2 V_n\psi_n = 0, \tag{13.186}$$

where $\Delta = \frac{d}{N}$, $\psi_n = \psi(n\Delta)$, $L_n = L(n\Delta)$, $V_n = V(n\Delta)$, and N is the number of divisions used. The boundary conditions (13.185) in this discrete form are also very simple

$$\psi_N = 1, \quad \text{and} \quad \psi_{N+1} - \psi_N = 0. \tag{13.187}$$

We know that there are many forms of difference equations approximating the second order differential equation (13.186), many of them like the Runge–Katte and Numerov's methods [3], [20] are more accurate than our choice, but these are not suitable for the inversion of the reflection data [19], [20]. An important property of (13.186) is that it gives exact solution of the wave equation for a homogeneous medium, i.e. when $V(z) = L(z) = 0$, the solutions of (13.184) and (13.186) coincide for $z = n\Delta$. In addition a property which is essential for the method of inversion, based on the continued fraction expansion, is the fact that the reflection coefficient has a rational representation. Let us study this last property in some detail. To simplify the resulting expressions we define the following functions:

$$y = 1 - \frac{2}{1+q\Delta} = 1 - \frac{2}{\left[1 + \gamma k\Delta\sqrt{|\sin^2\theta - \eta^2(d)|}\right]}, \tag{13.188}$$

$$F_n = e_n\psi_n, \tag{13.189}$$

$$e_n = \frac{W_n}{1 + \frac{1}{2}L_n\Delta}e_{n-1}, \tag{13.190}$$

and

$$W_n = 1 - \frac{1}{2}k^2 V_n\Delta^2 V_n. \tag{13.191}$$

With the help of these functions we can write (13.186) in the simple form of a linear difference equation

$$F_{n-1} = (1-y)F_n + U_n y F_{n+1}, \tag{13.192}$$

where the coefficient U_n denotes the following expression

$$U_n = \frac{\left(1 - \frac{1}{2}L_n\Delta\right)\left(1 + \frac{1}{2}L_{n+1}\Delta\right)}{W_n W_{n+1}}. \tag{13.193}$$

Using Eq. (13.192), we can write $\left(\frac{F_0}{F_1}\right)$ which depends on θ and ω in the following way:

$$G(\theta, \omega) = \frac{F_0}{F_1} = 1 - y + \frac{U_1 y}{1 - y +} \ \frac{U_2 y}{1 - y +} \ \cdots \ \frac{U_{N-1} y}{1 - y + U_{N-1} y}, \qquad (13.194)$$

where we have used the boundary conditions (13.187) to obtain (13.194). The function $G(\theta, \omega)$ contains all the information about the reflection coefficient and is related to $\Omega(0)$, defined by Eq. (13.179).

$$q + \frac{1}{\Omega(0)} \approx \frac{\psi_{+1} - \psi_{-1}}{2 \Delta \psi_0} = -\frac{W_0(1 - y)}{2\Delta + \Delta^2 L_0}$$
$$+ \frac{1}{2\Delta} \left(\frac{1 + \frac{1}{2} L_1 \Delta}{W_1} - \frac{W_0 U_0 y}{1 + \frac{1}{2} L_0 \Delta} \right) \frac{1}{G(\theta, \omega)}. \qquad (13.195)$$

This relation shows that if both the velocity and the logarithmic derivative of the density are known for the inhomogeneous medium at $z = 0$ and at $z = \Delta$, then $G(\theta, \omega)$ can be found from $\Omega(0)$. Thus it is convenient to take the empirical values of $G(\theta, \omega)$ as the input data, since according to (13.195) $G(\theta, \omega)$ has a very simple continued fraction expansion. But as we can see from (13.194), $G(\theta, \omega)$ is expressible as a ratio of two polynomials in y:

$$G(\theta, \omega) = \frac{A_{N-1} y^{N-1} + A_{N-2} y^{N-2} + \cdots + A_1 y + A_0}{B_{N-2} y^{N-2} + \cdots + B_1 y + 1} = \frac{P_{N-1}(y)}{Q_{N-2}(y)}. \qquad (13.196)$$

Assuming for a moment that a representation for $G(\theta, \omega)$ is known for a given frequency, then from a continued fraction expansion of (13.196), the coefficients U_n in (13.194) can be found uniquely. Once the values of U_n have been determined, we can find W_n s according to (13.193), and V_n s using Eq. (13.191) provided $L(z)$ (or L_n) is assumed to be known.

13.10 Determination of the Wave Velocity and the Density of the Medium

If both V_n s and L_n s are unknown, then the rational approximation to $G(\theta, \omega)$ should be given at two different frequencies, say ω_1 and ω_2. In this case the continued fraction expansion yields two sets of unique profile coefficients $U_n(\omega_1)$ and $U_n(\omega_2)$. Then from Eq. (13.193) it follows that

$$V_{n+1} = \frac{Q_n(\omega_2) - Q_n(\omega_1)}{\frac{1}{2} \{\Delta^2 [k_2^2 Q_n(\omega_2) - k_1^2 Q_n(\omega_1)]\}}, \qquad (13.197)$$

where

$$Q_n(\omega_j) = U_n(\omega_j)W_n(\omega_j), \quad j = 1, 2. \tag{13.198}$$

From these relations we can calculate V_n s successively and determine the velocity profile from Eq. (13.184). Once V_n s have been found then L_n can be obtained one after the other using Eq. (13.193) and remembering that W_n s are given by (13.191). Thus from the knowledge of $G(\theta, \omega)$ for two different frequencies, both $c(z)$ and $L(z)$ can be calculated at the points $z = n\Delta$, $n = 2, 3, \cdots N - 1$.

13.11 Rational Representation of the Input Data

The next problem that we have to consider is how to get a representation for $G(\theta, \omega)$ which is compatible with the expansion (13.194). Such a representation which is a special form of Padé approximant [22] can be found in a number of ways. For instance, suppose that $G(\theta, \omega)$ is given for N different values of y, then a simple way to find G is by expressing it as the continued fraction

$$G = 1 - y + a_0 y_0 + \frac{(y - y_0)a_1}{1 - y + a_1 y_1 +} \frac{(y - y_1)a_2}{1 - y + a_2 y_2 +} \cdots$$
$$\cdots \frac{(y - y_{N-3})a_{N-2}}{1 - y + a_{N-2} y_{N-2} + (y - y_{N-2})a_{N-1}}. \tag{13.199}$$

In this relation the coefficients a_n are simply found from the values of G at the points y_n, at $n = 0, 1, 2 \cdots N - 1$. This approach gives us the same result as the one found from Eq. (13.192) provided that $U_n F_{n+1}$ can be approximated by $U_{n-1} F_n$, an approximation which is valid when N is large. Regarding this representation of G, we must note that there are more accurate ways of obtaining G, but they are more complicated and here we have chosen the simplest [22], [23].

13.12 Direct and Inverse Methods Based on Continued Fraction Expansion Applied to Two Simple Models

In order to illustrate the accuracy and also the limitations of the method described in this chapter, we consider two models each with a profile for the wave velocity and another profile for the density of the inhomogeneous medium. But first to test the accuracy in the direct method, for a number of profiles for $c(z)$ and $\rho(z)$ we integrate Eq. (13.178) to obtain the impedance $\Omega(z)$ exactly. In these calculations we assume that $c_d > c(0)$, and we also choose the angle of incidence so that only damped waves are present in the region $z \geq d$. However these assumptions are not in any way essential to the inversion algorithm.

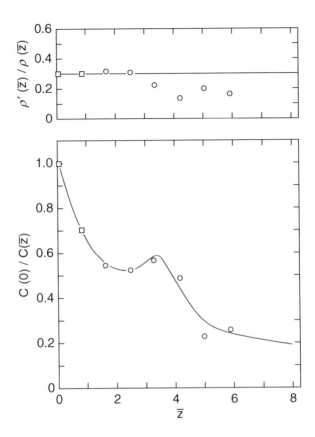

Figure 13.2: Input profiles for velocity and density are shown as a function of $\bar{z} = (10z/d)$, where d is the depth of the inhomogeneous medium. Using these profiles we have calculated the reflection coefficient for two different frequencies $\omega_1 = (13c(0)/d)$ and $\omega_2 = 1.2\omega_1$. Here the input profiles are $\frac{c(0)}{c(\bar{z})}$ and $\frac{d}{d\bar{z}}\rho(\bar{z})$. These are given by Eqs. (13.200), (13.201), and are plotted as solid curves. Using the method of continued fraction expansion, we have found both of these profiles from the reflection coefficient at two frequencies ω_1 and ω_2. For $N = 12$, the calculated values are shown by circles.

Then by approximating Eq. (13.183) and replacing it by (13.186) we again calculate $\Omega(0)$. The results obtained in this way agree with the exact results to within one part in 10^4, if N is taken to be 500, and if $0 < \omega \leq \frac{20c(0)}{d}$. Thus we conclude that for solving the direct problem our discrete method is indeed accurate and stable. Since in the present formulation the same set of equations have been used to solve both direct and inverse problems, there is the expectation that similar accuracies can be obtained for the inverse problem. However in practice, because of the accumulation of the round off errors [23] only small values of N, of the order of 10 can be used for continued fraction expansion, and this severely limits the accuracy of the found profiles [19].

In the first example let us assume the following form for $\frac{c(0)}{c(z)}$ and for $\rho(z)$:

$$\frac{c(0)}{c(\bar{z})} = \begin{cases} \left\{ 1 + 0.5\bar{z} - \exp\left[-(\bar{z} - 3.5)^2\right] \right\}^{-1} & \text{for } \bar{z} \le 10 \\ 1 & \text{for } \bar{z} > 10 \end{cases}. \tag{13.200}$$

$$\frac{d}{d\bar{z}} \ln \rho(\bar{z}) = \begin{cases} 0.3 & \text{for } \bar{z} \le 10 \\ 0 & \text{for } \bar{z} > 10 \end{cases}. \tag{13.201}$$

As Fig. 13.2 shows for small values of $\bar{z}$ the result of inversion follows the input data closely, but then the results deviates from the exact profile.

Now let us consider the case where $\frac{d}{d\bar{z}} \ln \rho(\bar{z})$ is more complicated. In this case the difference between the input profile and the points found from inversion becomes more pronounced as can be seen in Fig. 13.3.

$$\frac{c(0)}{c(\bar{z})} = \begin{cases} 0.2 + 0.8\exp\left[-\frac{\bar{z}}{1.2}\right] + 0.2\exp\left[-\frac{(\bar{z}-4)^2}{0.64}\right] & \text{for } \bar{z} \le 10 \\ 1 & \text{for } \bar{z} > 10 \end{cases}. \tag{13.202}$$

$$\frac{d}{d\bar{z}} \ln \rho(\bar{z}) = \begin{cases} \dfrac{-2\left[\frac{(\bar{z}-2.6)}{(1.5)^2}\right]}{\left\{1 + 0.125\exp\left[\frac{-(\bar{z}-2.6)^2}{2.25}\right]\right\}} & \text{for } \bar{z} \le 10 \\ 0 & \text{for } \bar{z} > 10 \end{cases}. \tag{13.203}$$

Let us briefly discuss some of the main features of this approach to the inverse problem:

(1) We can find the wave velocity and the density profiles in terms of the depth of the layer z, not as a a function of travel time.

(2) The boundary conditions at $z = d$ can be changed, e.g. we can solve this problem when at this point there is a perfect reflector.

(3) With a minor modification we can apply this method to invert the reflection data for an absorbing medium. Such a problem arises in determining the optical potential in nuclear collisions, a subject that will be studied later in this book.

(4) In the method that we discussed, we found two polynomials whose ratio gives us the reflection coefficient. But we know of other methods, such as the least-square fit to a set of measured data points to determine A_n s and B_n s in Eq. (13.196). It may be possible to get a better rational representation of $R(\theta, \omega)$, and thus a more accurate profiles of $c(z)$ and $\rho(z)$.

(5) This method, in its present form, can be applied to other problems where it can work better (see the next chapter).

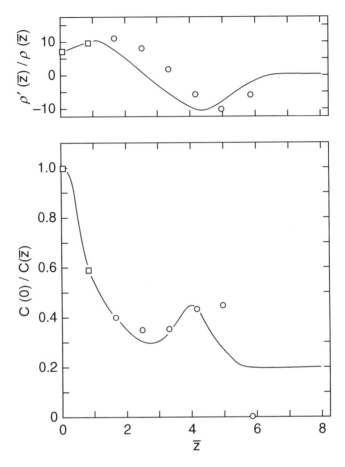

Figure 13.3: By choosing the profiles to be those given by Eqs. (13.202) and (13.203) which in this graph are shown by solid lines, we have calculated the reflection coefficient at two different frequencies $\omega_1 = \frac{15c(0)}{d}$ and $\omega_2 = 1.2\omega_1$. The result of inversion with $N = 12$ are shown by circles.

13.13   Inverse Problem of Wave Propagation Using Schwinger's Approximation

In Sec. 3.1 we studied an iterative procedure for obtaining the velocity of a scalar wave propagating in an inhomogeneous medium and we pointed out that because of the use of multiple Fourier transforms the numerical calculation becomes unstable. A different approach considered by Hooshyar *et al.* remedies this difficulty by first transforming the equation of wave motion in a medium where the velocity is a function of the coordinate z, $c(z) = \frac{c}{n(z)}$, to a Schrödinger like equation [25], [26]. This is achieved by a Liouville transformation (or travel time coordinate), where $z \to \xi(z)$ according to (13.103) and $p(z, k) \to \phi(\xi, k)$ according to (13.106) [12].

The resulting equation is given by (13.107) for $\phi(\xi, k)$ and the potential in the latter equation depends on the derivatives of $V(\xi)$ or $n(\xi)$, i.e.

$$\frac{d^2\phi(\xi, k)}{d\xi^2} + \left[k^2 - w(\xi) \right] \phi(\xi, k) = 0. \tag{13.204}$$

In this relation

$$\xi = \int_0^z n(z') \, dz', \tag{13.205}$$

and

$$w(\xi) = -n^{-\frac{3}{2}}(\xi) \frac{d^2}{d\xi^2} n^{-\frac{1}{2}}(\xi). \tag{13.206}$$

The physical index of refraction $n(z) > 0$ and $n(z) \to 1$ as $z \to \infty$, therefore the new variable ξ will have the desired form $\xi \in [0, \infty)$. That is just like the Schrödinger equation (13.204) is defined on the half-line $\xi \geq 0$. In other words it means that the wave is completed reflected at the $\xi = 0$. As we noticed in Sec. 4.1, the sufficient condition for existence of a physically acceptable solution of (13.204) is the integrability conditions

$$\int_0^\infty |w(\xi)| d\xi < \infty, \quad \text{and} \quad \int_0^\infty \xi |w(\xi)| d\xi < \infty. \tag{13.207}$$

If we want to express these results in terms of the wave speed $c(z)$ and also be able to relate the asymptotic behaviour of $\phi(\xi, k)$ to the asymptotic from of $p(z, k)$, we note that from (13.103) it follows that

$$\xi(z) = \alpha + z \quad \text{as} \quad z \to \infty, \quad \text{where} \quad \alpha = \int_0^\infty [n(y) - 1] dy. \tag{13.208}$$

Since $d\xi(z) = n(z) dz$, the integrability conditions (13.207) can be expressed in terms of $n(z)$ or $c(z)$;

$$\int_0^\infty |w(\xi)| d\xi = \int_0^\infty \left| n^{-\frac{1}{2}}(z) \frac{d^2}{dz^2} n^{-\frac{1}{2}}(z) \right| dz = \frac{1}{c} \int_0^\infty c^{\frac{1}{2}}(z) \left| \frac{d^2}{dz^2} c^{\frac{1}{2}}(z) \right| dz, \tag{13.209}$$

and

$$\int_0^\infty \xi |w(\xi)| d\xi = \int_0^\infty \xi(z) \left| n^{-\frac{1}{2}}(z) \frac{d^2}{dz^2} n^{-\frac{1}{2}}(z) \right| dz = \frac{1}{c} \int_0^\infty \xi(z) c^{\frac{1}{2}}(z) \left| \frac{d^2}{dz^2} c^{\frac{1}{2}}(z) \right| dz$$

$$\leq \frac{|\alpha|}{c} \int_0^\infty c^{\frac{1}{2}}(z) \left| \frac{d^2}{dz^2} c^{\frac{1}{2}}(z) \right| dz + \frac{1}{c} \int_0^\infty z c^{\frac{1}{2}}(z) \left| \frac{d^2}{dz^2} c^{\frac{1}{2}}(z) \right| dz. \tag{13.210}$$

From these relations and Eq. (13.208) it follows that sufficient conditions for the existence of a solution to Eq. (13.204) can be stated in the following way:

$$|\alpha| < \infty, \quad c(z) < \infty, \quad \frac{d^2}{dz^2} \sqrt{c(z)} \quad \text{exists for} \quad z \in [0, \infty), \tag{13.211}$$

$$\int_0^\infty \left| \frac{d^2}{dz^2} \sqrt{c(z)} \right| dz < \infty \quad \text{and} \quad \int_0^\infty z \left| \frac{d^2}{dz^2} \sqrt{c(z)} \right| dz < \infty. \tag{13.212}$$

When these conditions are satisfied, then $\phi(\xi, k)$ can be found from the solution of the Fredholm integral equation

$$\phi(\xi, k) = \phi_0(\xi, k) + \int_0^\infty G(\xi, \eta, k) w(\eta) \phi(\eta, k) d\eta, \tag{13.213}$$

where $\phi_0(k, \xi) = \sin(k\xi)$ and $G(\xi, \eta, k)$ is the Green function

$$G(\xi, \eta, k) = \frac{2}{\pi} \int_0^\infty \frac{\sin(q\xi)\sin(q\eta)}{k^2 - q^2} dq \tag{13.214}$$

$$= -\frac{1}{k} \begin{cases} \sin(k\xi)\cos(k\eta) & \text{if } \xi \leq \eta, \\ \cos(k\xi)\sin(k\eta) & \text{if } \eta \leq \xi. \end{cases} \tag{13.215}$$

The boundary condition on $\phi(k, \xi)$ can be found from Eqs. (13.213) and (13.215);

$$\lim \phi(\xi, k) \to 0, \quad \text{as} \quad \xi \to 0 \tag{13.216}$$

and this is consistent with the boundary condition for the regular solution of Eq. (13.204). From (13.213) and (13.215) we find the asymptotic form of $\phi(\xi, k)$;

$$\phi(\xi, k) \to \sin(k\xi) - \frac{D(k)}{k} \cos(k\xi) \quad \text{as} \quad \xi \to \infty \tag{13.217}$$

where $D(k)$ is directly related to the velocity of the wave by the relation

$$D(k) = \langle \phi_0 | w | \phi \rangle \equiv \int_0^\infty \phi_0(\xi, k) w(\xi) \phi(\xi, k) d\xi. \tag{13.218}$$

To establish a relation between $D(k)$ and the wave velocity we start with the Born series, and using the operator notation we write ϕ as a series [26]

$$\phi = \phi_0 + Gw\phi_0 + (Gw)^2 \phi_0 + (Gw)^3 \phi_0 + \cdots . \tag{13.219}$$

By substituting (13.219) in (13.217) we obtain

$$D(k) = \sum_{n=0}^\infty M_n(k), \quad \text{with} \quad M_n(k) = \langle \phi_0 | w(Gw)^n | \phi_0 \rangle . \tag{13.220}$$

Now if the Born approximation given by (13.219) is convergent, then a finite sum of $M_n(k)$ s would approximate $D(k)$ very well [28]. However in the direct problem it is difficult to calculate higher order Born approximation, and it is even harder to solve the inverse problem, partly because of the sensitivity of the outcome to the input data.

The Schwinger Variational Principle — To overcome the difficulty of calculating higher terms in the Born series (13.219) Hooshyar *et al.* advanced the idea of using Schwinger's variational problem which had been highly successful in solving the direct problem of calculating the partial phase shift from a potential to obtain the wave velocity $c(z)$ [29]. The aim here is to find a better estimate of the function $D(k)$. Thus we start with a quadratic functional $S[\phi]$, [30], [31]:

$$S[\phi] = 2\langle\phi_0|w|\phi\rangle - \langle\phi|w - wGw|\phi\rangle. \tag{13.221}$$

For the functional $S[\phi]$ to be extremum, ϕ has to satisfy the integral equation (13.221). However other trial functions for ϕ can also be used in (13.221) to find approximate solutions. In particular if we replace ϕ by $\zeta_0\phi_0$, where ζ_0 is a constant, then $S[\phi]$ becomes a quadratic form in ζ_0. By setting $\frac{d}{d\zeta_0}S[\zeta_0\phi_0]$ equal to zero we obtain

$$\zeta_0 = \frac{M_0(k)}{M_0(k) - M_1(k)}. \tag{13.222}$$

The trial plain wave used in Schwinger's functional yields accurate results even at low energies (or small k) for the phase shift [32]. Using plane wave as the trial function, we replace $\phi(\xi, k)$ by $\phi_0(\xi, k)$, and we find the Schwinger approximation to $D(k)$;

$$D_S(k) = \frac{M_0^2(k)}{M_0(k) - M_1(k)}. \tag{13.223}$$

Since the second Born approximation as given by (13.220) is

$$D_{2B}(k) = M_0(k) + M_1(k). \tag{13.224}$$

Therefore if $e_S = D(k) - D_S(k)$ and $e_2 = D(k) - D_2(k)$ denote the differences between the Schwinger and the exact and the second Born and the exact $D(k)$, we find that

$$e_S = e_{2B} - \frac{M_1^2}{M_0(k) - M_1(k)}. \tag{13.225}$$

Thus if the Born series converges rapidly so that $M_1(k)$ is small compared to $M_0(k)$, then the difference between $e_S(k)$ and $e_{2B}(k)$ will be small.

As k becomes large, then the asymptotic forms of ϕ, $D_S(k)$ and $e_S(k)$ have the following asymptotic values [26]

$$\phi(\xi, k) \to \sin(k\xi) + \mathcal{O}\left(\frac{1}{k}\right), \quad M_1(k) \to \mathcal{O}\left(\frac{1}{k}\right), \tag{13.226}$$

$$D(k) \to M_0(k) + \mathcal{O}\left(\frac{1}{k}\right), \quad D_S(k) \to M_0(k) + \mathcal{O}\left(\frac{1}{k}\right), \tag{13.227}$$

$$e_S = e_B + \mathcal{O}\left(\frac{1}{k}\right), \quad e_S = e_{2B} + \mathcal{O}\left(\frac{1}{k^2}\right), \tag{13.228}$$

where

$$e_B = D(k) - D_B(k) = D(k) - M_0(k). \tag{13.229}$$

Thus as k tends to infinity all of the above approximations tend to the first Born approximation, M_0, which is defined by as the first term in the Born series (13.220) or

$$D_B(k) = M_0(k) = \langle \phi_0 | w | \phi_0 \rangle. \tag{13.230}$$

By taking $\phi_0(\xi, k)$ to be a plane wave $\phi_0(\xi, k) = \sin(k\xi)$, we find

$$M_0(k) = \frac{1}{2} \left(\tilde{w}(0) - \tilde{w}(2k) \right), \tag{13.231}$$

where

$$\tilde{w}(0) = \int_0^\infty w(\xi) d\xi, \quad \text{and} \quad \tilde{w}(2k) = \int_0^\infty w(\xi) \cos(2k\xi) d\xi. \tag{13.232}$$

From the Riemann–Lebesgue theorem [33], [34] it follows that

$$\lim D(k) = \lim D_S(k) = \lim M_0(k) = \frac{1}{2} \tilde{w}(0), \quad \text{as} \quad k \to \infty. \tag{13.233}$$

Equation (13.232) shows that if we know the function $M_0(k)$, then by taking the inverse cosine Fourier transform, we can determine $w(\xi)$. Therefore let us try to relate $D(k)$ to $D_0(k)$. First we note that $p(z, k) = \frac{1}{\sqrt{n(\xi)}} \phi(\xi, k)$, and $\xi = \alpha + z$ as z tends to infinity. Since the asymptotic form of the wave amplitude $p(z, k)$ is a linear combination of $\sin(kz)$ and $\cos(kz)$, and the limit of $n(z)$ as $z \to \infty$ is one, therefore

$$p(z, k) \to \phi(z + \alpha) \to A(k) \sin(kz) + B(k) \cos(kz), \quad \text{as} \quad z \to \infty. \tag{13.234}$$

Thus from Eqs. (13.233) and (13.217) we get

$$D(k) = -k \frac{D_0(k) - \tan(k\alpha)}{1 + D_0(k) \tan(k\alpha)}, \tag{13.235}$$

where $D_0(k) = \frac{B(k)}{A(k)}$. This function, $D_0(k)$, is related directly to the phase of the wave and therefore measurable.

Next let us consider the solution of the differential equation (13.204) when the solution is specified from the two boundary conditions given at the origin;

$$\tilde{\phi}(\xi = 0, k) = 0, \quad \text{and} \quad \left(\frac{d\tilde{\phi}(\xi, k)}{d\xi} \right)_{\xi=0} = 1. \tag{13.236}$$

This solution which we denote by $\tilde{\phi}(\xi, k)$ satisfies the integral equation

$$\tilde{\phi}(\xi, k) = \frac{\sin(k\xi)}{k} + \int_0^\xi \frac{\sin k(\xi - \eta)}{k} w(\eta) \tilde{\phi}(\eta, k) d\eta. \tag{13.237}$$

For large values of k, $\tilde{\phi}(\xi, k)$ tends to

$$\tilde{\phi}(\xi, k) = \frac{\sin(k\xi)}{k} + \mathcal{O}\left(\frac{1}{k^2} \right). \tag{13.238}$$

When $\xi \to \infty$, from Eq. (13.237) we find that

$$\tilde{\phi}(\xi, k) = \tilde{A}(k) \sin(k\xi) + \tilde{B}(k) \cos(k\xi), \quad \text{as} \quad \xi \to \infty, \tag{13.239}$$

where $\tilde{A}(k)$ and $\tilde{B}(k)$ are given by

$$\tilde{A}(k) = \frac{1}{k} + \int_0^\infty \frac{\cos(k\eta)}{k} w(\eta)\tilde{\phi}(\eta, k)d\eta, \tag{13.240}$$

and

$$\tilde{B}(k) = -\int_0^\infty \frac{\sin(k\eta)}{k} w(\eta)\tilde{\phi}(\eta, k)d\eta. \tag{13.241}$$

By comparing (13.239) with (13.234) we obtain $\phi(\xi, k)$ in terms of $\tilde{\phi}(\xi, k)$ and also $D(k)$ in terms of $\tilde{B}(k)$

$$\phi(\xi, k) = \frac{\tilde{\phi}(\xi, k)}{\tilde{A}(k)}, \quad \text{and} \quad D(k) = -k\frac{\tilde{B}(k)}{\tilde{A}(k)}. \tag{13.242}$$

In addition we get

$$A(k) = \tilde{A}\cos(k\alpha) - \tilde{B}(k)\sin(k\alpha), \tag{13.243}$$

and

$$B(k) = \tilde{A}\sin(k\alpha) + \tilde{B}(k)\cos(k\alpha). \tag{13.244}$$

For large k values we can find the asymptotic values of $\tilde{A}(k)$ and $\tilde{B}(k)$ by noting that from Eqs. (13.238), (13.240) and (13.241) we have the following relations:

$$\tilde{A} \to \frac{1}{k} + \mathcal{O}\left(\frac{1}{k^2}\right), \quad \text{and} \quad \tilde{B}(k) \to \mathcal{O}\left(\frac{1}{k^2}\right), \tag{13.245}$$

as $k \to \infty$. Now by substituting (13.245) in (13.243) and (13.244) we obtain

$$\lim k \to \infty \begin{cases} A(k) \to k^{-1}\cos(k\alpha) + \mathcal{O}\left(k^{-2}\right) \\ \\ B(k) \to \tilde{A}k^{-1}\sin(k\alpha) + \mathcal{O}\left(k^{-2}\right) \end{cases}. \tag{13.246}$$

Thus

$$D_0(k) = \frac{B(k)}{A(k)} = \tan(k\alpha) + \mathcal{O}\left(\frac{1}{k}\right), \quad \text{as} \quad k \to \infty. \tag{13.247}$$

To find an estimate for α, let us consider the relation

$$\cos^2(k\alpha)\frac{d}{dk}[D_0(k) - \tan(k\alpha)] = \frac{1}{A^2(k)}\left[2\alpha\tilde{A}(k)\tilde{B}(k)\sin(k\alpha)\cos(k\alpha)\right]$$

$$+ \frac{1}{A^2(k)}\left[\tilde{B}'(k)\tilde{A}(k) - \tilde{B}(k)\tilde{A}'(k) + \alpha\tilde{B}^2(k)\right]\cos^2(k\alpha). \tag{13.248}$$

In this relation prime indicates derivative with respect to k. Taking the limit of (13.249) as k tends to infinity leads to

$$\alpha \to \lim \cos^2\left[\tan^{-1}(D_0(k))\right]\frac{d}{dt}[D_0(k)], \quad \text{as} \quad k \to \infty. \tag{13.249}$$

With the help of this relation together with the earlier equations we can establish relations between the function $D(k)$ and the empirical data given by $D(k)$, and thus solve the inverse problem.

Inverse Problem Using Schwinger's Approximation — To find the potential $w(\xi)$, we start with Eqs. (13.231) and (13.232) and write them as

$$\tilde{w}(2k) = \int_0^\infty w(\xi) \cos(2k\xi)d\xi = \tilde{w}(0) - 2M_0(k). \tag{13.250}$$

By taking the inverse Fourier cosine transform of (13.250) we find

$$w(\xi) = \frac{4}{\pi} \int_0^\infty [\tilde{w}(0) - 2M_0(k)] \cos(2kx)dk. \tag{13.251}$$

Assuming that $M_0(k)$, the first Born term can approximate the exact data $D(k)$, we can write

$$M_0(k) = D_B(k) \approx M_0^{(0)} = D(k). \tag{13.252}$$

With this assumption we find a rough estimate of the potential $w_B(\xi)$

$$w_B(\xi) = \frac{4}{\pi} \int_0^\infty \left[\tilde{w}(0) - 2M_0^{(0)}(k) \right] \cos(2k\xi)dk. \tag{13.253}$$

From what was discussed earlier we expect a better approximation to $w(\xi)$ can be found if we assume that the data are well approximated by the Schwinger function $D_S(k)$;

$$D_S(k) \approx D(k), \tag{13.254}$$

and use Eq. (13.223) to find a better approximation to $M_0^{(0)}(k)$ as compared to Born's $M_0^{(0)}(k)$. But to calculate $D_S(k)$ we need to know $M_1(k)$ which is the second term in the Born series (13.219) and (13.220). Noting that $M_1(k) = \langle \phi_0 | wGw | \phi_0 \rangle$ and that $G(k, \xi, \eta)$ is given by (13.215), we find

$$M_1(k) = \frac{2}{\pi} \mathcal{P} \int_0^\infty \frac{1}{k^2 - q^2} \left[M_0 \left(\frac{k+q}{2} \right) - M_0 \left(\frac{k-q}{2} \right) \right]^2 dq. \tag{13.255}$$

That is for S-wave scattering by local potentials, the off-shell matrix elements of the S matrix in the Born approximation can be found from the knowledge of the on-shell elements only [35].

Let us consider the solution of Eq. (13.225) which is equivalent to the problem of finding the zero of the functional $\tilde{F}$;

$$\tilde{F}[M_0(k)] = D_0(k) \left[M_0(k) - M_1(k) - M_0^2(k) \right] = 0. \tag{13.256}$$

Rewriting this problem as a fixed point problem, the generator of its iterative solution takes the form

$$M_0^{(n+1)}(k) = M_0^{(n)}(k) + \left(\frac{D(k)}{|D(k)|} \right)$$
$$\times \left\{ D(k) \left[M_0^{(n)}(k) - M_1^{(n)}(k) \right] - \left(M_0^{(0)}(k) \right)^2 \right\}. \tag{13.257}$$

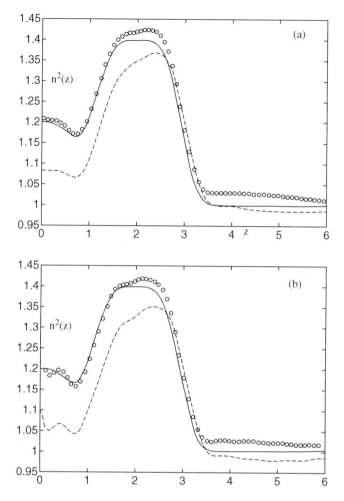

Figure 13.4: The plot of the input $n^2(z)$ as a function of z and the result of inversion. Here the square of the refractive index $n^2(z) = \left[1 + 0.2\exp\left(-z^3\right) + 0.4\exp-(z-2)^4\right]$ is shown by the solid line. In (a) the dashed line is the result of inversion obtained from Born approximation. The o s are found from the inversion technique discussed in this work. In (b) is the recovered data with 5 percent random error added to the exact result.

Once we have found $M_0^{(n+1)}(k)$, we can calculate $M_1^{(n+1)}(k)$ from (13.255). To obtain $M_1^{(n+1)}(k)$, we again use (13.255) but we replace $M_0^{(0)}(k)$ and $M_1^{(0)}(k)$ by $M_0^{(n+1)}(k)$ and $M_1^{(n+1)}(k)$. We continue this iteration until the resulting $M_0^{(n+1)}(k)$ dos not vary appreciably from $M_0^{(n)}(k)$, i.e.

$$\int_0^\infty \left| M_0^{(n+1)}(k) - M_0^{(n)}(k) \right| < \epsilon, \tag{13.258}$$

where ϵ is a small positive number. Now having determined $M_0^{(n+1)}(k)$ by successive iterations, we find $w(\xi)$ from

$$w_S(\xi) = \frac{4}{\pi} \int_0^\infty \left[\tilde{w}_0(0) - 2M_0^{(n)}(k) \right] \cos(2k\xi) dk. \tag{13.259}$$

The final step is to relate ξ to z. Since from Eq. (13.206) $\sqrt{n(\xi)}$ satisfies the differential equation [26]

$$\frac{d^2}{d\xi^2} n^{\frac{1}{2}}(\xi) = w(\xi) n^{\frac{1}{2}}(\xi), \tag{13.260}$$

with the boundary conditions

$$\sqrt{n(\xi)} = 1, \quad \text{and} \quad \left[\frac{d}{d\xi} \sqrt{n(\xi)} \right]_\xi = 0, \quad \text{as} \quad x \to \infty, \tag{13.261}$$

thus the solution yields the refractive index $n(\xi)$ or $n(z)$. The wave velocity is determined from

$$c(z) = \frac{c}{n(\xi(z))}. \tag{13.262}$$

An Example of Inversion Using Schwinger's Variational Principle — We can use the formulation discussed in this section to determine the velocity profile from the asymptotic form of the wave amplitude when the propagation takes place in an inhomogeneous medium, the inhomogeneity being limited to a finite range. In their work Hooshyar *et al.* have provided analytical as well as numerical examples for this method of inversion. The reader is referred to the original paper for these examples. Here we will consider only one of the cases solved for a profile with smooth variation of $n^2(z)$, where $n(z)$ is the index of refraction. Let us assume a model where

$$n^2(z) = \left\{ 1 + 0.2 \exp\left(-z^3\right) + 0.4 \exp\left[-(z-2)^4\right] \right\}. \tag{13.263}$$

By integrating Eq. (13.204) or the Volterra equation (13.237) we solve the direct problem and compute $D_0(k)$ and then solve the inverse problem, i.e. we start with $D_0(k)$ and find the variable refractive index. Let Z be a point on the z axis far enough from the origin so that for $z > Z$, $n(z) = 1$, and let $\tilde{D}(Z,k)$ denote the logarithmic derivative of the wave amplitude at Z; i.e.

$$\tilde{D}(Z,k) = \frac{1}{\psi(Z,k)} \left(\frac{d\psi(z,k)}{dz} \right)_{z=Z}, \tag{13.264}$$

then $D_0(k)$ can be found from the relation

$$D_0(k) = \frac{k \cos(kZ) - \tilde{D}(Z,k) \sin(kZ)}{k \sin(kZ) + \tilde{D}(Z.k) \cos(kZ)}. \tag{13.265}$$

The logarithmic derivative of $\psi(z,k)$ is found by solving Eq. (13.107) numerically for different k values. Once $D_0(k)$ is calculated, from its asymptotic value for large

k we find α Eq. (13.249). Once α is known then from (13.235) we find $D(k)$ for different k values, and from (13.233) for large k we estimate $\tilde{w}_0$ which is defined by (13.232). The final step is to determine $w(\xi)$ from Eq. (13.251). The wave velocity as a function of z, i.e. $c(z)$ can be obtained by numerically solving (13.260) with the boundary condition (13.261) and $\xi = z + \alpha$. Details of calculation for several profiles can be found in Ref. [26]. In Fig. 13.4(a) for the input we have used $n(z)$ as given by (13.263), $n^2(z)$ is shown by the solid line, and the result of inversion obtained by the Born approximation is shown by the dashed line. If we use Schwinger's approximation we find the points shown by circles ∘ s. The result for $n^2(z)$ shows a noticeable improvement particularly for small z values. A rough idea about the stability of the method can be obtained by adding 5 per cent random error to the exact result, and then we get the profile for $n^2(z)$ shown in Fig. 13.4(b). We observe that the result does not change much and this is true for different profiles [26].

References

[1] When the incident wave is normal to the surface separating the two media, there are a number itersting methods avialable for solving the inverse problem. For instance see the review by R.G. Newton in Geophys. J.R. astron. Soc. 65, 1901 (1981).

[2] L.M. Brekhouskikh, *Waves in Layered Media*, (Academic, New York, 1980), Chapter 3.

[3] H.E. Moses, Phys. Rev. 102, 559 (1956).

[4] M. Razavy, J. Acoust. Soc. Am. 58, 956 (1975).

[5] P. Epstein, Proc. Natl. Acad. Sci. 18, 627 (1930).

[6] R. Clayton and R. Stolt, Geophysics 46, 1559 (1981).

[7] R.G. Keys and A.B. Weglein, J. Math. Phys. 24, 1444 (1983).

[8] R. Jost and W. Kohn, Phys. Rev. 87, 977 (1952).

[9] J.A. Ware and K. Aki, J. Acoust. Soc. Am. 45, 911 (1969).

[10] P. Goupillaud, Geophys. 26, 754 (1961).

[11] J. Claerbout, Geophys. 33, 264 (1968).

[12] See for instance D. Zwillinger's, *Handbook of Differential Equations*, (Academic Press, New York, 1989) Sec. 22.

[13] This transformation which, in mathematical litrature is known as Liouville transformation is for one-dimensional spatial problems. However attempts have been made to extend it to two-dimensional problems, see R. Bellman's paper in Bollottino dell Unione Mathematical Italiano, 13, 535 (1958).

[14] F. Calogero, *Variable Phase Approach to Potential Scattering*, (Academic Press, New York, 1967).

[15] M. Hron and M. Razavy, Geophys. J. R. astr. Soc. 51, 545 (1977).

[16] M. Hron and M. Razavy, Can. J. Phys. 55, 1434 (1977).

[17] M. Hron and M. Razavy, Can. J. Phys. 57, 1843 (1979).

[18] M. Abramowitz and I.A. Stegun, *Handbook of Mathematical Functions*, (Dover Publications, New York, 1965).

[19] M.A. Hooshyar and M. Razavy, J. Acoust. Soc. Am. 73, 19 (1983).

[20] For methods of numerically solving ordinary differential equations by Runge–Kutta method a good source is *The Numerical Solution of the Differential-Algebraic Systems by Runge-Kutta Methods*, (Springer-Verlag, Berlin, 1989) by E. Hairer, C. Lubich and M. Roche.

[21] The numerical solution of the Schrödinger equation by Numerov's technique is discussed e.g. in B.R. Johnson, J. Chem. Phys. 67, 4086 (1997).

[22] G.A. Baker, Jr., *Essentials of Padé Approximants*, (Academic Press, New York, 1975) Chapter 8.

[23] F.S. Acton, *Numerical Mehods That Work*, (Harper and Row New York, 1970) Chapter 11.

[24] M. Hron and M. Razavy, Can. J. Phys. 57, 1843 (1979).

[25] M.A. Hooshyar, T.H. Lam and M. Razavy, Can. J. Phys. 70, 282 (1992).

[26] M.A. Hooshyar, T.H. Lam and M. Razavy, J. Acous. Soc. Am. 107, 404 (2000).

[27] D. Zwillinger, *Handbook of Differential Equations*, (Academic Press, New York) p. 88.

[28] R.T. Prosser, J. Math. Phys. 23, 2127 (1982).

[29] R.K. Nesbet, *Variational Methods in Electron-Atom Scattering Theory*, (Plenum Press, New York, 1980) Sec. 2.7.

[30] B.A. Lippmann and J. Schwinger, Phys. Rev. 79, 469 (1950).

[31] C.J. Joachain, *Quantum Collision Theory*, (North-Holland, Amsterdam, 1975) Sec. 11.2.

[32] H.H. Chan and M. Razavy, Can. J. Phys. 42, 1017 (1964).

[33] See for instance, E. Hewitt and K. Stromberg, *Real and Abstract Analysis*, (Springer-Verlag, New York, 1965).

[34] C.H. Pugh, *Real Mathematical Analysis*, (Springer-Verlag, 2003) p. 163.

[35] M. Razavy and R.J.W. Hodgson, Nucl. Phys. A 149, 65 (1970).

Chapter 14

The Inverse Problem of Torsional Vibration

In Chapter 1 we studied few inverse problems of classical dynamics and in particular the inverse problem of vibrating systems composed of masses connected to each other by springs. Among the vibrating systems studied extensively and with considerable interest in engineering and applied physics are the torsional motion of rigid bodies [1]–[4]. The torsional system similar to the mass-spring one is the system where a number of massive wheels with moments of inertia I_j, are mounted on a rigid rod with a fixed distance from each other, the bar is fixed at one end and at the other end it can be free or attached to a fixed wall and an initial small rotational displacement a small torque applied to it initially, and execute small oscillations Fig. 14.1. The inverse problem is similar to the problem of masses attached to springs [3]. Another related problem is that of forced torsional vibration of an elastic slab of finite thickness. The direct problem here is to determine the contact impedance of such a system. A simple version of this problem is as follows: Consider a slab of thickness h with its face at $z = d$ fixed to a rigid foundation and is excited by means of a rigid circular disc of radius $r = a$ attached to its upper surface at $z = 0$ [4]. The continued fraction expansion method which we discussed in the previous chapter can be applied to a number of other problems of classical physics including the construction of shear modulus and density profiles from torsional vibration data.

    Let us consider a torsional vibrator of the form of a circular plate of radius b which is mounted on the free flat surface of a layered Earth at $z = 0$, and let us assume that this vibrator generates a torsional displacement $u(r, z, \omega)e^{i\omega t}$ at the depth z. The displacement caused by the motion of the vibrator can be described by a second order partial differential equation, which in cylindrical coordinates can

285

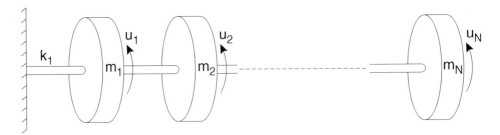

Figure 14.1: A system composed of wheels connected with rods of negligible weights is an analogue of the mass-spring system shown Fig. 1.3. The inverse problem for this system can be formulated and solved as that of mass-spring system which we discussed in Sec. 1.6. Details of the solution can be found in Gladwell's book [4].

be written as [5], [6]

$$\left\{\frac{\partial^2}{\partial r^2} + \frac{1}{r}\frac{\partial}{\partial r} - \frac{1}{r^2} + \frac{\partial^2}{\partial z^2} + L(z)\frac{\partial}{\partial z} + \frac{\omega^2}{c^2(z)}\right\} u(r, z, \omega) = 0, \qquad (14.1)$$

where

$$c(z) = \sqrt{\frac{\mu(z)}{\rho(z)}} \qquad (14.2)$$

is the shear velocity, $\rho(z)$ is the density, $\mu(z)$ is the shear density and $L(z)$ is the logarithmic derivative of the shear modulus

$$L(z) = \frac{d}{dz}\ln\mu(z). \qquad (14.3)$$

In this problem we can express the torsional stress due to the vibrator as

$$T(r, \omega) = \begin{cases} \mu(z)\frac{\partial}{\partial z}u(r, z, \omega)|_{z=0} & \text{if } 0 \le r \le b \\ 0 & \text{if } r > b \end{cases}. \qquad (14.4)$$

Let us suppose that we have information about $T(r, \omega)$ and $u(r, z, \omega)$ on the surface, i.e. at $z = 0$. From this information we want to determine the two profiles of $c(z)$ and $\rho(z)$. By taking the Hankel transform of (14.1) we obtain

$$\left\{\frac{d^2}{dz^2} - q^2 + \frac{\omega^2}{c^2(z)} + L(z)\frac{d}{dz}\right\}\tilde{u}(z, q, \omega) = 0, \qquad (14.5)$$

where

$$\tilde{u}(z, q, \omega) = \int_0^\infty u(r, z, \omega)[rJ_1(qr)]dr. \qquad (14.6)$$

As in the case of inverse problem of acoustic wave propagation we assume that we know the logarithmic derivative of $\tilde{u}(z, q, \omega)$ on the surface, but to simplify the problem further we assume that $\tilde{u}$ does not have any reflected wave. Thus altogether we want to make the following assumptions:

(1) That after a depth d, both the shear modulus and the density become constant, i.e. for $z \geq d$, $L(z) = 0$, and $c(z) = c(d)$.

(2) That the data is available as a function of the wave number $q \geq \frac{\omega}{c(d)}$.

(3) That the differential equation with z as the variable can be approximated by a finite difference equation.

Now we write the ordinary differential equation (14.5) in the following form:

$$\left[\frac{d^2}{dz^2} - \eta^2 + k^2 w(z) + L(z) \frac{d}{dz} \right] \tilde{u} = 0, \tag{14.7}$$

where we have introduced the new symbols η, $V(z)$, and $w(z)$ by

$$\eta^2 = q^2 - \frac{\omega^2}{c^2(d)}, \quad \text{and} \quad V(z) = \frac{c(z)}{c_0}, \tag{14.8}$$

and

$$w(z) = \frac{1}{V^2(z)} - \frac{1}{V^2(d)}, \quad \text{and} \quad k = \frac{\omega}{c_0}. \tag{14.9}$$

Here c_0 is an arbitrary constant introduced in order to make $V(z)$ a dimensionless function. Also from assumption (1) it follows that for $z \geq d$ we have

$$\tilde{u} = T e^{-\eta z}, \tag{14.10}$$

where T is a constant. Let us also define $\psi(z)$ by

$$\psi(z) = e^{\eta z} \frac{\tilde{u}(z)}{T}, \tag{14.11}$$

then by replacing $\tilde{u}$ by $\psi(z)$ we find the differential equation for this new function to be

$$\left\{ \frac{d^2}{dz^2} + [L(z) - 2\eta] \frac{d}{dz} + \left[k^2 w(z) - \eta L(z) \right] \right\} \psi(z) = 0. \tag{14.12}$$

Now with these changes the differential equation for $\psi(z)$ satisfies simple boundary conditions at $z = d$

$$\psi(d) = 1, \quad \text{and} \quad \left(\frac{d\psi(z)}{dz} \right)_{z=d} = 0. \tag{14.13}$$

We can transform (14.12) to a Riccati differential equation by introducing the torsional impedance at the surface $z = 0$ (for the corresponding function in acoustic waves see Eq. (13.177))

$$\Omega(q, \omega) = \frac{d}{dz} \ln \psi(z)|_{z=0} = \eta + \left(\frac{d}{dz} \ln \tilde{u} \right)_{z=0}$$

$$= \eta + \frac{\tilde{T}(q, \omega)}{\mu(0)\tilde{u}(0, q, \omega)}, \tag{14.14}$$

where $\tilde{T}(q,\omega)$ is the Hankel transform of $T(r,\omega)$, Eq. (14.4) [7]

$$\tilde{T}(q,\omega) = \int_0^\infty T(r,\omega) r J_1(qr) dr. \tag{14.15}$$

This last quantity can be found from $\tilde{T}(q,\omega)$, $\tilde{u}(z=0)$ and $c(d)$. Following the same technique that we used to solve the inverse problem of acoustic wave propagation, we introduce $\phi(z)$ by

$$\phi(z) = \frac{1}{\psi(z)} \frac{d\psi(z)}{dz}. \tag{14.16}$$

By substituting $\psi(z)$ from (14.16) in (14.12) we find

$$\frac{d\phi(z)}{dz} + \phi^2(z) + [L(z) - 2\eta]\phi(z) + k^2 w(z) - \eta L(z) = 0. \tag{14.17}$$

For the solution of the direct problem we integrate (14.17) with the boundary condition

$$\phi(d) = 0, \tag{14.18}$$

and obtain the torsional impedance at the point $z = 0$, i.e.

$$\phi(0) = D(q,\omega). \tag{14.19}$$

This is the exact solution of the direct problem.

The Inverse Problem of Torsional Vibration — To solve the inverse problem by the continued fraction method, we replace the differential equation (14.12) by the difference equation

$$(1 - \eta\Delta)\left(1 + \frac{1}{2}L_n\Delta\right)\psi_{n+1} + (1 + \eta\Delta)\left(1 - \frac{1}{2}L_n\Delta\right)\psi_{n-1} - U_n\psi_n = 0, \tag{14.20}$$

where N is the total number of divisions used in this approximation, and we have introduced the following quantities:

$$U_n = 2 - k^2\Delta^2 w_n, \quad \Delta = \frac{d}{N}, \quad \text{and} \quad \psi_n = \psi(z = n\Delta), \tag{14.21}$$

where the boundary conditions (14.13) for the finite difference equation takes the following form:

$$\psi_N = \psi_{N+1} = 1. \tag{14.22}$$

To bring this difference equation to a form which is suitable for continued fraction expansion, we introduce the following changes in the function and the variables:

$$F_n = \frac{a_n\psi_n}{(1 + \eta\Delta)^n}, \quad y = \eta^2\Delta^2 - 1, \tag{14.23}$$

$$a_{n+1} = \frac{1 + \frac{1}{2}L_n\Delta}{1 - \frac{1}{2}L_n\Delta} a_{n-1}, \tag{14.24}$$

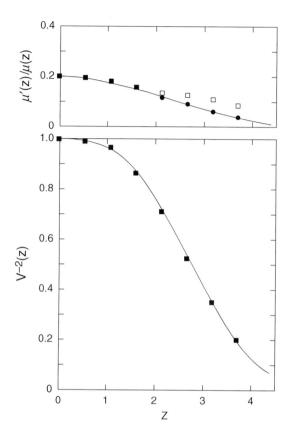

Figure 14.2: In this figure the input data, consisting of the logarithmic derivative of the shear modulus (the upper graph) and $V^{-2}(z)$, i.e. the inverse square of the velocity profile (the lower graph) are shown when both of these profiles are smoothly varying functions of z. The results of inversion when the surface data for two frequencies are used are shown by $\bullet$ and when a single frequency is used by $\square$. When the results overlap the points are shown by $\blacksquare$.

and

$$\Omega_n = \frac{U_n a_{n-1}}{a_n \left(1 - \frac{1}{2} L_n \Delta\right)}.$$ (14.25)

Substituting these in (14.20), we obtain a simple difference equation

$$F_{n-1} = \Omega_n F_n + y F_{n+1}.$$ (14.26)

This equation can be solved for $\frac{F_0}{F_1}$ by successive elimination of $\frac{F_n}{F_{n+1}}$,

$$\frac{F_0}{F_1} = G(y, \omega) = \Omega_1 + \frac{y}{\Omega_2 +} \frac{y}{\Omega_3 +} \cdots \frac{y}{\Omega_{N-2} +} \frac{y}{\Omega_{N-1} +} \frac{y}{\frac{y}{(1+\sqrt{y+1})} \frac{a_{N-1}}{a_N}}.$$ (14.27)

Also from (14.14) we can find $G(y, \omega)$ directly from the surface data using the finite difference analogue of $\Omega(q,\omega)$:

$$G(y,\omega) = \left(\frac{a_0}{a_{-1}}\right) \frac{\left(1 - \frac{1}{2}\eta L_0\Delta^2\right)}{\left(1 + \frac{1}{2}L_0\Delta\right)\left[\Delta D(q,\omega) + \frac{1}{2}\frac{U_0}{(1+\eta\Delta)\left(1-\frac{1}{2}L_0\Delta\right)}\right]}. \tag{14.28}$$

The term $\left(\frac{a_0}{a_{-1}}\right)$ in this equation is the normalization constant which can be set equal to one. Now let us consider the Thiele expansion [3] (see Appendix B)

$$1 + \sqrt{1+y} = 2 + \frac{y}{2+}\frac{y}{2+}\cdots\frac{y}{2+}\frac{y}{1+\sqrt{1+y}}. \tag{14.29}$$

Now the similarity between (14.29) and the equation for $G(y,\omega)$ suggests that Eq. (14.27) must have a Thiele expansion about $y = 0$ which looks like (14.29) [3], i.e.

$$G(y,\omega) = G_0 + \frac{y}{sG_0+}\frac{y}{2ss_1G_0+}\frac{y}{3ss_2G_0+}\cdots, \tag{14.30}$$

where $G_0 = G(0,\omega)$, and s_nG is the nth the reciprocal derivatives of $G(y)$, which is defined by

$$s_0G(y) = G(y), \tag{14.31}$$

$$s_1G(y) = \frac{1}{\frac{dG(y)}{dy}}, \tag{14.32}$$

and (see the Appendix B) [2], [3].

$$s_nG(y) = s_{n-2}G(y) + nss_{n-1}G(y). \tag{14.33}$$

From Thiele's theorem, (Appendix B), and Eq. (14.27) for $\frac{F_0}{F_1}$ we arrive at the interesting result that within our approximate scheme, $c(z)$ and $L(z)$ are related to the reciprocal derivatives of $G(y,\omega)$ at $y = 0$, i.e. to Ω_1 and Ω_{n+1}

$$\Omega_1 = G(0,\omega), \quad \text{and} \quad \Omega_{n+1} = nss_{n-1}G(0,\omega). \tag{14.34}$$

Now we want to find $G(y,\omega)$ and its reciprocal derivatives at $y = 0$, therefore if the data are given for a number of points about q_0, then Δ can be related to q_0 by

$$y = \left(q_0^2 - \frac{\omega^2}{c^2}\right)\Delta^2 - 1 \tag{14.35}$$

a result that follows from Eqs. (14.8) and (14.24). Therefore we choose Δ to be

$$\Delta = \left(q_0^2 - \frac{\omega^2}{c^2(d)}\right)^{-\frac{1}{2}}. \tag{14.36}$$

Returning to Eq. (14.28) for $G(y,\omega)$, we observe that if we know the value of $G(y,\omega)$ and its derivatives at $y = 0$, then from these, the reciprocal derivatives of G can be

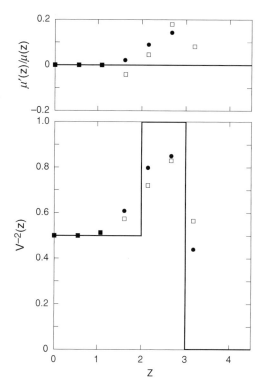

Figure 14.3: Same as the previous figure except here the input $\frac{d}{dz} \ln \mu(z)$ is continuous, whereas $V^{-2}(z)$ is discontinuous.

obtained. These reciprocal derivatives are in turn related to Ω_n s Eq. (14.34), and therefore these relations enable us to determine $c(z)$ and $L(z)$ Eq. (14.25). The most general case for inversion is the one where the data are known at two frequencies and these lead to two sets of Ω_n s. From these Ω_n s both $c(z)$ and $L(z)$ can be found [6].

For the calculation of derivatives, $G(y, \omega)$ was fitted by a polynomial in y and from this polynomial fit the derivatives of $G(y, \omega)$ were approximately calculated. From the definition of the derivatives of $G(y, \omega)$ its reciprocal derivatives were found by the following a method due to Nörlund (Appendix B) [2]:

$$(n + 1)ss_n G = \frac{(-1)^n p_n^2}{p_{n-1}p_{n+1}} \qquad (14.37)$$

with

$$p_0 = 1, \quad p_1 = a_1, \quad p_2 = a_2, \qquad (14.38)$$

$$p_{2n} = p_{2,n}, \quad p_{2n+1} = p_{1,n+1}, \qquad (14.39)$$

where the quantities $p_{r,n+1}$ are defined by the determinant;

$$p_{r,n+1} = \begin{vmatrix} a_r & a_{r+1} & \cdots & a_{r+n} \\ \cdots\cdots\cdots\cdots\cdots\cdots \\ \cdots\cdots\cdots\cdots\cdots\cdots \\ a_{r+n} & a_{r+n+1} & \cdots & a_{r+2n} \end{vmatrix}. \tag{14.40}$$

The elements of this determinant are related to the derivatives of $G(y, \omega)$

$$a_n = \frac{1}{n!}\left(\frac{d^n G(y,\omega)}{dy^n}\right)_{y=0}. \tag{14.41}$$

Numerical Results Obtained by Thiele's Method — To test the reliability and the accuracy of this method we will consider two shear modulus and two density profiles, one where both of these variables are smoothly varying functions of the depth z, and one where $\mu(z)$ is a continuous function of z, whereas $V^{-2}(z)$, defined by Eq. (14.8) changes abruptly Fig. 14.2. In this figure black circles, •, are the points found from the inversion of the surface data using two frequencies, and squares, □, are those obtained for one frequency. When the two overlap, they are shown by black squares ■.

References

[1] A.W. Leissa and M.S. Qatu, *Vibrations of Coninuous Systems*, (McGraw-Hill, New York, 2013) Chapter 3.

[2] G.M.L. Gladwell, J. Sound and Vibration, 211, 309 (1999).

[3] G.M.L. Gladwell, *Inverse Problems in Vibration*, Second Edition, Volume 119 of *Solid Mechanics and Its Applications*, (Dordrecht, Kluwer, 2004).

[4] An early paper giving a detailed account of the forced torsional vibration of an elastic medium is by G.M.L. Gladwell, Intl. J. Engin. Sci. 7, 1011 (1969).

[5] S. Coen, J. Math. Phys. 22, 2338 (1981).

[6] M.A. Hooshyar and M. Razavy in *Conference on Inverse Scattering – Theory and Appliction*,

[7] I.N. Sneddon, *Fourier Transform*, (McGraw-Hill, New York, 1951). Chapter 2 of this book presents a rigorous account of the Hankel transform and its application. Edited by J.B. Bednar, (SIAM, Philadelphia, 1983).

[8] L.M. Milne-Thomson, *The Calculus of Finite Differences*, (McMillan, London, 1965) Chapter V.

[9] N.E. Nörlund, *Differenzenrechnung*, (Springer-Verlag, Berlin, 1924) Chapter 15.

Chapter 15

Local Nucleon-Nucleon Potentials Found from the Inverse Scattering Problem at Fixed Energy

Nucleon-Nucleon Scattering — Consider the scattering of two nucleons, and let us denote the spin states of the incoming and outgoing waves by χ_{m_s} and $\chi_{m'_s}$ respectively. If the matrix $M^{m_s,m'}$ in spin space describes the scattering of an incident particle with spin $\chi_{m'_s}$ into the final spin χ_{m_s}, then we can express the asymptotic form of the wave function as

$$\psi(r) \to e^{i\mathbf{k}\cdot\mathbf{r}}\chi_{m_s} + \frac{1}{r}e^{ikr}\sum_{\chi_{m'_s}} M^{m_s,m'_s}(\theta,\phi)\chi_{m'_s}. \tag{15.1}$$

There are nine complex amplitudes M^{m_s,m'_s} for triplet and one M^{m_s,m'_s} for the singlet state for a given isotopic spin state. Since there are two states of isotopic spin, $T = 0$ and $T = 1$, therefore altogether there are twenty amplitudes. Due to the unitarity of the S matrix and also as a consequence of the symmetries of the interaction, the number of independent quantities required for a complete determination of the scattering amplitude reduces to ten complex or twenty real terms at a given energy and angle. However one needs only five complex coefficients over the angular range zero to $\frac{1}{2}\pi$ at a fixed energy to obtain the scattering amplitude. For energies below the threshold of pion production ≈ 350 MeV, the summation is over real phase shifts and mixing parameter. There are four sets of phase shifts: δ_J for the singlet, and the phases δ_{JJ} and $\delta_{J\pm 1,J}$ for the triplet, plus the mixing parameter ϵ_J for the triplet in each isotopic spin state. These five real functions of

J, in principle, can be generated by five different potential functions. A detailed analysis of the two nucleon interaction shows that in addition to the central and the spin-orbit force, there are other forces including the tensor force and quadratic spin-orbit force [1]–[5]. The most general form of the interaction between two nucleons is of the form

$$V(r) = V_1(r) + V_2(r)\mathbf{L} \cdot \mathbf{S} + (\boldsymbol{\sigma}_1 \cdot \boldsymbol{\sigma}_2)V_3(r) + S_{12}V_4(r)$$
$$+ (\boldsymbol{\sigma}_1 \cdot \mathbf{L})(\boldsymbol{\sigma}_1 \cdot \mathbf{L})V_5(r). \tag{15.2}$$

In this relation $\boldsymbol{\sigma}_1$ and $\boldsymbol{\sigma}_2$ denote the spin of the two nucleons and $\mathbf{L}$ and $\mathbf{S}$ are measured in units of $\hbar$. We can write the two nucleon potential in a slightly different form, viz,

$$V(r) = V_c(r) + S_{12}V_T + V_{so}(r)\mathbf{L} \cdot \mathbf{S} + W_{12}V_W + \mathbf{L}^2 V_{LL}, \tag{15.3}$$

where W_{12} is related to spin-orbit coupling and is given by

$$W_{12} = \frac{1}{2}\left[(\boldsymbol{\sigma}_1 \cdot \mathbf{L})(\boldsymbol{\sigma}_2 \cdot \mathbf{L}) + (\boldsymbol{\sigma}_2 \cdot \mathbf{L})(\boldsymbol{\sigma}_1 \cdot \mathbf{L}) - \frac{1}{3}\boldsymbol{\sigma}_1 \cdot \boldsymbol{\sigma}_2\mathbf{L}^2\right]. \tag{15.4}$$

In addition to the terms shown in (15.2), in some phenomenological models (e.g. Bonn potential model [6] or Paris potential model [7]) there is an additional potential which is a quadratic function of the relative momentum p of the two nucleons

$$V_1\left(r, p^2\right) = \mathbf{p}^2 V_p(r) + V_p \mathbf{p}^2 \tag{15.5}$$

or

$$V_2\left(r, p^2\right) = \mathbf{p} \cdot V_p(r)\mathbf{p}. \tag{15.6}$$

This potential was initially introduced to replace the hard core in the nucleon-nucleon potentials, and in this way the application of perturbation theory to nuclear structure calculation became more acceptable [8].

15.1   Constructing the S Matrix from Empirical Data

Having discussed some of the important analytic properties of the S matrix in Chapter 4, we want to outline a method for constructing the S matrix from empirical data in such a way that these analytic properties are preserved and the formulation is stable for rational fraction representation [9]. We start with the assumption that we have a knowledge of the bound states energies, phase shifts, resonances and their normalization constants. From these we want to find a rational form for the S matrix or for the Jost function. Let us also note that these quantities are not completely independent. For example Chadan's sum rule indicates the existence of a relation between the bound state energy, the phase shift and the range of the force [10]–[12].

Even when all these information are available, one cannot construct the S matrix uniquely. But as Kranopolsky and Kukulin argue, if the basic analytic properties are correctly taken into account, the problem of construction of the scattering matrix has a unique solution and is stable [9].

For simplicity we consider a single channel scattering where for the ℓth partial wave we write $S_\ell(k)$ as a rational function of k

$$S_\ell(k) = e^{2i\delta_\ell(k)} = \frac{g_\ell(k) + ik^{2\ell+1}}{g_\ell(k) - ik^{2\ell+1}}. \tag{15.7}$$

In this equation $\delta_\ell(k)$ is the phase shift and $k^2 = \frac{2\mu E}{\hbar^2}$, where E is the energy in the center-of-mass, and μ is the reduced mass. For short range potentials, $g_\ell(k^2)$ which is an analytic function of k^2 can be expanded for small k (effective-range expansion)

$$g_\ell(k^2) = -\frac{1}{a_\ell} + \frac{1}{2}r_\ell k^2 - \frac{1}{4}P_\ell k^4 + \cdots, \tag{15.8}$$

where a_ℓ and r_ℓ are the scattering length and the effective range respectively.

The power expansion of $g_\ell(k^2)$ cannot be used for all energies, since if for the energy E_0, the phase shift passes through zero, $g_\ell(E_0)$ becomes infinite and the effective range expansion diverge. Such a situation happens, for instance, for S wave $n - p$ scattering, since in this scattering in the energy range 300–350 MeV, the phase shift $\delta_0(E)$ passes through zero. Thus the expansion (15.8) is valid as long as $E < E_0$.

A possible way of avoiding the power expansion of $g_\ell(k^2)$ is to consider Padé approximation which approximates a function by a ratio of two polynomials of order N and M respectively i.e. [13];

$$g_\ell(E) \approx g_\ell^{N,M}(E) = \frac{P_N(E)}{Q_M(E)} = \frac{a_0 + a_1 E + \cdots + a_N E^N}{1 + b_1 E + \cdots + b_M E^M}. \tag{15.9}$$

In the whole region where $g_\ell(E)$ is analytic, this expansion converges with increasing N and M. Even in the vicinity of cuts which are simulated by the zeros and poles of $g_\ell^{N,M}(E)$ this expansion is convergent. By substituting (15.9) in (15.7) we find a simple analytic form for $S_\ell(k)$

$$S_\ell(k) = \frac{P_N(E) + ik^{2\ell+1}Q_M(E)}{P_N(E) - ik^{2\ell+1}Q_M(E)}. \tag{15.10}$$

Now an analytic continuation of the S matrix allows one to calculate the poles of the S matrix by setting the denominator of $S_\ell(E)$ equal to zero

$$P_N(E) - ik^{2\ell+1}Q_M(E) = 0. \tag{15.11}$$

The root $k = i\gamma$, $\gamma > 0$ of this equation corresponds to a bound state with an energy $E = -\frac{\hbar^2\gamma^2}{2\mu}$, while the root $k = \alpha + i\beta$ with $\alpha > 0$, $\beta > 0$ corresponds to a resonant energy $E = E_r + \frac{i}{2}\Gamma$, where $E_r = \alpha^2 - \beta^2$ and $\Gamma = 4\alpha\beta$. Once we have found a zero of (15.11), we can easily find the residue at the position of this

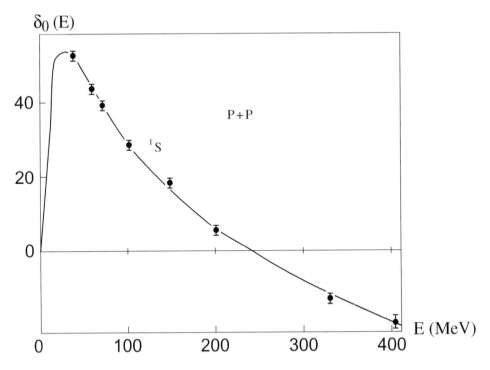

Figure 15.1: This figure shows how accurate the Padé approximation with $[M = 1, N = 1]$ is for the $p - p$ scattering phase shift over a wide range of energies. The phase shift is given in degrees and the energy in the laboratory system is in MeV [9].

pole. Thus if we obtain the coefficients a_j and b_j in (15.11) we can complete the construction of $S_\ell(k)$.

Construction of the Padé Approximants — Let us assume that the phase shifts are measured at discrete energies $\{E_j\}$ with the corresponding errors $\{\varepsilon_j\}$, $j = 1, 2, \cdots K$. Using these quantities we calculate the set $\{g_j\}$, i.e. the values of $g_j(E)$ and the errors $\{\varepsilon_j\}$. Thus the problem becomes that of constructing a the rational approximation of $g_j(E)$ from the given set $\{E_j, g_j, \varepsilon_j\}$. There are several ways of carrying out this programme, but here we describe one particular method where one minimizes χ^2 defined by [9]

$$\chi^2 = \sum_{l=1}^{K} \frac{1}{\varepsilon_j^2} \left| \left(\frac{P_n(E_j)}{Q_M(E_j)} - g_j \right) \right|^2 . \tag{15.12}$$

In this method the resulting equations are nonlinear and complicated. However one can linearize these nonlinear set of equations. By solving the linear equations we will not find the absolute minimum of χ^2 (15.12), but from its initial value we can find improved values systematically.

Let us start with the inequality

$$\chi^2 \leq \frac{1}{|Q_{min}|^2} \sum_{j=1}^{K} \left| \frac{1}{\varepsilon_j^2} \left(P_N(E_j) - g_j Q_M(E_j) \right) \right|^2, \tag{15.13}$$

where Q_{min} is the minimum of all $Q_M(E_j)$ s, assuming that this minimum is not zero. Thus instead of minimizing (15.12) we try to minimize the functional

$$\chi_0^2 = \sum_{j=1}^{K} \left| \frac{1}{\varepsilon_j^2} \left(P_N(E_j) - g_j Q_M(E_j) \right) \right|^2. \tag{15.14}$$

By minimizing this functional, χ_0^2, we get a set of linear equations, but with this functional χ_0^2 will not be the absolute minimum of χ^2. Thus by minimization of χ_0^2 from (15.14) we find $P_N^0(E_j)$ and $Q_M^0(E_j)$. But now we have a new set of errors given by

$$\varepsilon_j^{(1)} = \varepsilon_j Q_M^{(0)}(E_j). \tag{15.15}$$

By substituting these in (15.14) we find χ_1^2

$$(\chi_1)^2 = \sum_{j=1}^{K} \left| \frac{1}{(\varepsilon_j^{(1)})^2} \left(P_N(E_j) - g_j Q_M(E_j) \right) \right|^2. \tag{15.16}$$

From this expression we find $P_N^{(1)}(E_j)$ and $Q_M^{(1)}(E_j)$, and from these we can compute χ_1^2 etc. This iteration scheme converges rapidly and can be used for nucleon-nucleon or nucleon-nucleus scattering. As an example consider the $n - \alpha$ scattering in two different states. The results are shown in Table 15.1 where for the $n - \alpha$ scattering in two different states, $P_{3/2}$ and $P_{1/2}$ the Padé approximants for the S matrix are given for increasing orders $[N, N]$ of this approximation.

We can also include the information about the bound state energy in the Padé approximation. To this end let

$$f_\ell(k) = \frac{P_N(E) - ik^{2\ell+1} Q_M(E)}{Q_M(E)}, \tag{15.17}$$

be the Jost function, then the bound state with the energy E_0 corresponds to a zero of the function $f_\ell(k)$ at $k_0 = i\gamma$ where γ is related to E_0 by $E_0 = -\frac{\hbar^2 \gamma^2}{2\mu}$. If we know k_0, then we can include it into the functional to be minimized

$$\chi_k^2 = \chi^2 + \lambda |f_\ell(k_0)|^2, \tag{15.18}$$

where χ^2 is given by (15.12) and λ is a positive constant. For this case we start with the linearized form of the functional χ^2:

$$\chi_{PL}^2 = \sum_{j=1}^{K} \left| \frac{1}{\varepsilon_j^2} \left(P_N(E_j) - g_j Q_M(E_j) \right) \right|^2$$
$$+ \lambda \left| P_N(E_0) - ik_0^{2\ell+1} Q_M(E_0) \right|^2. \tag{15.19}$$

Table 15.1: Results showing the rapid convergence for $P_{3/2}$ and $P_{1/2}$ in $n - \alpha$ resonance energy with increasing order of the Padé approximants [9].

	$P_{3/2}$		$P_{1/2}$	
$P[N, N]$	Re E_0(MeV)	-Im E_0 (MeV)	Re E_0(MeV)	-Im E_0(MeV)
$P[1, 1]$	0.77255	0.32227	1.8310	3.1357
$P[2, 2]$	0.77259	0.32246	1.9874	2.5988
$P[3, 3]$	0.77273	0.32233	1.9882	2.6057
$P[4, 4]$	0.77271	0.32253	1.9710	2.6103
$P[5 - 5]$	0.77273	0.32247	1.9757	2.6102

Even when we take M and N in the Padé approximant to be small, e.g. $N = 2$, and $M = 1$ still we can get a very good fit for the phase shifts for all energies below 500 MeV, which in this case the fit contains only four parameters [9]. Now if we change the order to $M = 2$ and $N = 1$, then the Padé approximant not only describes the phase shift for a wide range of energies, but also gives an accurate analytic continuation into the pole region, i.e. the poles corresponding to the deuteron bound ^3S state and virtual ^1S state. For instance, the effective range parameters (see Eq. (4.137)) for the ^1S and ^3S states of $n - p$ scattering we find $a_0 = 5.43$ fm, $r_0 = 1.68$ fm, $P_0 = -0.806$ and $E_0 = -2.14$ MeV. These should be compared with the empirical values of $a_0 = 5.399 \pm 0.011$ fm, $r_0 = 1.732 \pm 0.012$ fm, and $E_0 = -2.22$ MeV [14]. For the ^1S scattering from Padé approximant one finds $a_0 = -23$ fm $r_0 = 2.4$ fm, $P = -1.2$ and $E_0 = -0.068$ MeV [9]. From the analysis of the experimental results Arndt *et al.* give us $a_0 = -23 \pm 0.028$ fm $r_0 = 2.4 \pm 0.12$ fm and $E_0 = -0.066$ MeV. These numbers again show the excellent agreement between the empirical values and the simplest forms of the Padé approximants.

15.2   A Method for the Numerical Calculation of the Local Potential Using the Gel'fand–Levitan Formulation

In Chapter 4 we have discussed the Gel'fand–Levitan method of inversion in detail. Here we just outline one way of finding the nucleon-nucleon potential for a given partial wave from the information that we have about the dependence of the phase shift or Jost function on momentum (or k). But first let us consider the k or ℓ dependence of the phase shift $\delta_\ell(k)$ and decide on whether it is better to do the calculation for fixed energy (or k) or for the fixed partial wave ℓ. The nonlocality in the angular momentum dependence of nuclear forces dictate the use of fixed angular momentum inversion [15]–[19]. Thus the potentials that we get are local, energy independent, i.e. no explicit momentum dependence, however it can be state dependent. In fact some potential models, such as Reid's, independent (different) potentials are assumed for each state of distinct isotopic spin [20]. Following the work of Kirst et $al.$ we study the inversion of the empirical phase shifts with fixed angular momentum to obtain the potential for each state. We start with the Gel'fand–Levitan equation for the ℓth partial wave and we write it as (see Sec. 4.4);

$$K_\ell(r,r') = g_\ell(r,r') + \int_0^r K_\ell(r,r'') g_\ell(r'',r') \, dr'', \qquad (15.20)$$

where $g_\ell(r,r')$ which is the input kernel is related to the Jost function by

$$g_\ell(r,r') = -\frac{2}{\pi} \int_0^\infty j_\ell(kr) \left[\frac{1}{|f_\ell(k)|^2} - 1 \right] j_\ell(kr') \, dk, \qquad (15.21)$$

and the Jost function which is related to the phase shift by

$$f_\ell(k) = e^{-i\delta_\ell(k)}. \qquad (15.22)$$

Thus for finding $g_\ell(r,r')$ we need the phase shift at all energies plus the bound state data. The result of the phase shift analysis is available up to an energy of about 1.4 GeV [21], however for energies greater than 300 MeV particle production becomes important and the concept of a static potential becomes invalid.

For the purpose of inversion we choose a rational function of the phase shift $\delta_\ell(k)$ subject to the symmetry condition $\delta_\ell(k) = -\delta_\ell(-k)$, with the asymptotic form $\delta_\ell(k) \to k^{-1}$ for large k (see Eq. (4.44)). The information about the phase shift has to be supplemented by the energies and the number of bound states in any given partial wave. To this end we try to to find either the Padé approximation to the phase shift

$$\delta_\ell(k) = \frac{\sum_{n=1}^{\frac{N}{2}} a_{2n-1} k^{2n-1}}{1 + \sum_{n=1}^{\frac{N}{2}} b_{2n} k^{2n}}, \qquad (15.23)$$

or to the Jost function. This expression for $\delta_\ell(k)$ guarantees that $\lim \delta_\ell(k) \to k^{-1}$ as $k \to \infty$ and that $\delta_\ell(k) = -\delta_\ell(-k)$. From this expression we can find $S(k)$ or the

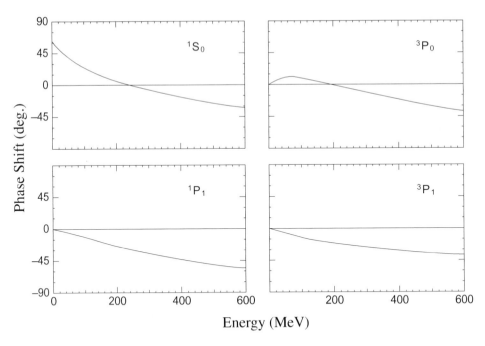

Figure 15.2: Nucleon-nucleon phase shifts for S and P waves calculated for the Reid's potential are shown by solid lines. The rational approximation for the same phase shifts used for inversion are indistinguishable from the calculated phase shifts [16].

Jost function, or we can also write the latter as a Padé approximant

$$f_\ell(k) = \prod_{j=1}^{\frac{J}{2}} \frac{k - \sigma_j^\downarrow}{k - \sigma_j^\uparrow}, \tag{15.24}$$

where $\sigma_j^\downarrow$ and $\sigma_j^\uparrow$ are defined by

$$\sigma_j = \begin{cases} \sigma_j^\uparrow & \text{if Im } \sigma_j > 0 \\[2mm] \sigma_j^\downarrow & \text{if Im } \sigma_j < 0 \end{cases}, \tag{15.25}$$

where arrows indicate the poles in the lower and upper half of k-plane respectively.

    As we have seen before a rational representation of $f_\ell(k)$ enables one to express $g_\ell(r, r')$ and consequently $K_\ell(r, r')$ as solutions of a linear set of equations. Let us write the expanded form of $K_\ell(r, r')$ as

$$K_\ell(r, r') = \sum_{i=1}^{J/2} \Phi_\ell(i, r) j_\ell\left(\sigma_i^\uparrow, r'\right). \tag{15.26}$$

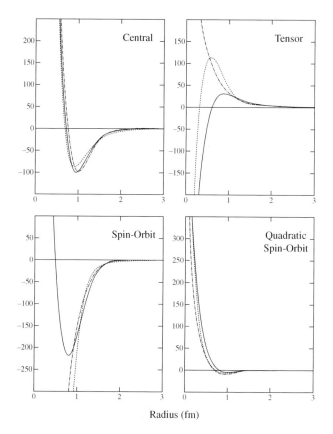

Figure 15.3: Results of construction of the potential from the phase shifts. From Arndt's phase shifts the resulting potentials are shown by solid lines, Paris's potential by dashed curves and Bonn's by dotted curves [16].

Then the Gel'fand–Levitan equation reduces to a set of linear equations

$$h_\ell^- \left(\sigma_j^\downarrow, r \right) + \sum_{n=1}^{N} w_{jn} \Phi_\ell(n, r) = 0, \tag{15.27}$$

where the matrix elements, w_{jn} are given by

$$w_{j,n}(r) = \frac{1}{\sigma_n^{\uparrow 2} - \sigma_j^{\downarrow 2}}$$

$$\times \left[\sigma_j^\downarrow \, j_\ell \left(\sigma_n^\uparrow, r \right) h_{\ell-1}^- \left(\sigma_j^\downarrow, r \right) - \sigma_j^\uparrow \, j_{\ell-1} \left(\sigma_n^\uparrow, r \right) h_{\ell-1}^- \left(\sigma_j^\downarrow, r \right) \right]. \tag{15.28}$$

In Eqs. (15.26)–(15.28), $j_\ell(\sigma_j r)$ and $h_\ell^-(\sigma_j r)$ are spherical Bessel and Hankel functions respectively [22].
We first solve (15.27) for the unknown quantities $\Phi_\ell(n, r)$ and then substitute these in (15.26) to obtain the Gel'fand–Levitan kernel $K(r, r')$. Once this kernel is found,

then we calculate the potential from

$$v_\ell(r) = -2\frac{d}{dr}K_\ell(r,r).$$ (15.29)

Inversion by Marchenko's Method — In the last subsection we outlined how one can construct local energy independent potentials from the Gel'fand–Levitan formulation. We could have used Marchenko's method where instead of fitting the Jost function by a rational function of momentum k we would have used a rational function fit of the $S(k)$ matrix of the form

$$S(k) = \prod_{j=1}^{J}\frac{k+\sigma_j}{k-\sigma_j} = \prod_{j=\frac{J}{2}+1}^{J}\frac{k+\sigma_j^\uparrow}{k-\sigma_j^\uparrow}\prod_{j=1}^{J/2}\frac{k+\sigma_j^\downarrow}{k-\sigma_j^\downarrow},$$ (15.30)

and then by substituting this $S(k)$ matrix in Marchenko's equation obtained the potential. One of the advantages of using either the Gel'fand–Levitan or Marchenko s formulations is that the input information, i.e. a rational function fit to the phase shift data can be determined to any desired accuracy [16]. In the case of Reid's soft core potentials, the phase shifts were found by the numerical integration of the Schrödinger equation. Then the corresponding Jost function were fitted with a rational function of k like the one given in (15.30). The potential found from inversion agreed well with the original one for most of the states of the nucleon-nucleon system.

Given the fact that both the Paris and the Bonn potentials were constructed to produce the same set of empirical phase shifts very accurately, it is surprising to observe differences between various potentials found from inversion and particularly in distances of the order of one fermi [15]–[17].

15.3   Direct and Inverse Problems for Nucleon-Nucleon Scattering Using Continued Fraction Formulation

Let us start with a discrete version of the wave equation and solve the direct and the inverse problems for the two nucleons scattering [23], [24].

We denote the reduced mass of the system by μ and introduce $v_i(r)$ by $v_i(r) = \frac{2\mu}{\hbar^2}V_i(r)$, then we have the following Schrödinger equation for the uncoupled states:
(a) For the singlet state we have

$$\psi_J''(r) + \left[k^2 - \frac{J(J+1)}{r^2} - v_c - J(J+1)v_{LL}\right]\psi_J(r) = 0.$$ (15.31)

(b) For triplet state with $L = J$ the wave equation becomes

$$\psi_J''(r) + \left[k^2 - \frac{J(J+1)}{r^2} - (v_c + 2v_T - v_{so} + v_w) \right.$$

$$\left. - J(J+1) \left(v_{LL} - \frac{4}{3}v_w \right) \right] \psi_J(r) = 0. \tag{15.32}$$

The asymptotic solutions of these equations yield the phase shifts δ_J and δ_{JJ} respectively.

For the coupled states the wave equation can be written as

$$\psi_J''(r) + \left[k^2 - \frac{L(L+1)}{r^2} - \tilde{v}_J(r) \right] \psi_J(r) = 0, \tag{15.33}$$

where $\psi_J(r)$ has two components

$$\psi_J(r) = \begin{bmatrix} u_J(r) \\ w_J(r) \end{bmatrix} \tag{15.34}$$

and $\tilde{v}_J$ and $L(J)$ are defined by the following matrices:

$$\tilde{v}_J = \begin{bmatrix} \tilde{v}_{11} & \tilde{v}_{12} \\ \tilde{v}_{21} & \tilde{v}_{22} \end{bmatrix} = \begin{bmatrix} v_c + v_1 - \frac{2(J-1)}{2J+1}v_T & \frac{6}{2J+1}[J(J+1)]^{\frac{1}{2}}v_T \\ \frac{6}{2J+1}[J(J+1)]^{\frac{1}{2}}v_T & v_c + v_2 - \frac{2(J+2)}{2J+1}v_T \end{bmatrix}, \tag{15.35}$$

and

$$L(J) = \begin{bmatrix} J-1 & 0 \\ 0 & J+1 \end{bmatrix}. \tag{15.36}$$

In Eq. (15.35), $v_c(r)$ and $v_T(r)$ are the central and the tensor parts of the potential and $v_1(r, T)$ and $v_2(r, T)$ are the other components of the potential arising from $v_{LL}(r)$ and are given by (see Eq. (15.3))

$$v_1(r, J) = (J-1)v_{LS} + \frac{1}{3}(J-1)(2J-3)v_W + J(J-1)v_{LL}, \tag{15.37}$$

and

$$v_2(r, J) = -(J+2)v_{LS} + \frac{1}{3}(J+2)(2J+5)v_W$$

$$+ (J+1)(J+2)v_{LL}. \tag{15.38}$$

Matrix Riccati Equation for the Coupled States — Since ψ_J has two components, Eq. (15.34), therefore for a given J there corresponds two eigenstates of ψ_J, say $\psi_{J\alpha}$ and $\psi_{J\beta}$, and we have four wave functions $u_{J\alpha}$, $w_{J\alpha}$, $u_{J\alpha}$ and $w_{J\beta}$. The boundary conditions for these wave functions at the origin are complicated and depend on the form of the potential and will be studied later [24], [25] On the other hand for large r, the asymptotic behaviour of the wave functions are simple and are expressible in terms of the phase shifts and the mixing parameter;

$$\lim u_{J,\alpha} \rightarrow -2iA_{J\alpha}e^{i\delta_{J\alpha}}\sin\left[kr - \frac{\pi}{2}(J-1) + \delta_{J\alpha}\right] \quad \text{as} \quad r \rightarrow \infty, \quad (15.39)$$

$$\lim w_{J,\alpha} \rightarrow -2iA_{J\alpha}\tan\epsilon_J e^{i\delta_{J\alpha}}\sin\left[kr - \frac{\pi}{2}(J+1) + \delta_{J\alpha}\right] \quad \text{as} \quad r \rightarrow \infty, \quad (15.40)$$

$$\lim u_{J,\gamma} \rightarrow +2iA_{J\gamma}\tan\epsilon_J e^{i\delta_{J\gamma}}\sin\left[kr - \frac{\pi}{2}(J-1) + \delta_{J\gamma}\right] \quad \text{as} \quad r \rightarrow \infty, \quad (15.41)$$

and

$$\lim w_{J,\gamma} \rightarrow -2iA_{J\gamma}e^{i\delta_{J\gamma}}\sin\left[kr - \frac{\pi}{2}(J+1) + \delta_{J\gamma}\right] \quad \text{as} \quad r \rightarrow \infty. \quad (15.42)$$

In these relations $A_{J\alpha}$ and $A_{J\gamma}$ are the amplitudes, δ_J s are the eigenphases and ϵ_J is the mixing parameter.

As in the case of uncoupled Schrödinger equation we want to define the variable $\mathcal{R}_J$ matrix, and for this we write ψ_J as a 2×2 matrix

$$\psi_J = \begin{bmatrix} u_{J\alpha} & u_{J\gamma} \\ w_{J\alpha} & w_{J\gamma} \end{bmatrix}, \quad (15.43)$$

and then define the $\mathcal{R}_J$ matrix by

$$\mathcal{R}_J = r\psi'_J(\psi_J)^{-1}, \quad (15.44)$$

where prime denotes differentiation with respect to r. From Eqs. (15.39)–(15.42) it follows that the matrix elements of $\mathcal{R}_J$ matrix are dependent on δ_J s and ϵ_J. There is a problem about the boundary conditions at the origin, since it depends on the form of the tensor potential as $r \rightarrow 0$. To simplify the problem we assume that near the origin the tensor force tends to zero. This assumption guarantees that u_J and w_J tend to zero as r^J and r^{J+2} respectively. Therefore as $r \rightarrow 0$, $\psi_J(r)$ tends to zero as

$$\psi_J(r) \rightarrow T_J(r)c_J, \quad \text{as} \quad r \rightarrow 0, \quad (15.45)$$

where

$$T_J(r) = \begin{bmatrix} r^J & 0 \\ 0 & r^{2J+2} \end{bmatrix}, \quad (15.46)$$

and c_J is a constant 2×2 matrix. This c_J matrix is the limit of a matrix $\phi_J(r)$ as $r \rightarrow 0$ where

$$\psi_J(r) = T_J(r)\phi_J(r). \quad (15.47)$$

To find the matrix equation satisfied by $\phi_J(r)$, we substitute (15.47) in (15.33) and simplify the result using (15.46) and thus we have

$$\phi''_J(r) + \frac{2(L+1)}{r}\phi'_J(r) + \left(k^2 - U_J(r)\right)\phi_J(r) = 0, \quad (15.48)$$

where $U_J(r)$ is defined by

$$U_j(r) = T_J^{-1}(r)\tilde{v}_J(r)T_J(r) = \begin{bmatrix} \tilde{v}_{11} & r^2\tilde{v}_{12} \\ \frac{1}{r^2}\tilde{v}_{21} & \tilde{v}_{22} \end{bmatrix}. \tag{15.49}$$

Solution of the Direct Problem with Matrix Riccati Equation — To obtain the phases and the mixing parameter we introduce the dimensionless matrix $Z_J(r)$ by

$$Z_J(r) = r\phi'_J(r)\phi_J^{-1}(r). \tag{15.50}$$

This matrix satisfies the matrix Riccati differential equation

$$Z'_J(r) + \frac{2L+1}{r}Z_J(r) + \frac{1}{r}Z_J^2(r) + r\left(k^2 - U_J(r)\right) = 0, \tag{15.51}$$

where $Z_J(r)$ is subject to the boundary condition

$$Z_J(0) = 0. \tag{15.52}$$

From the asymptotic forms of the wave functions (15.39)–(15.42) and the definitions of $\mathcal{R}_J$ and Z_J matrices we obtain the following relations:

$$\mathcal{R}_{11} = Z_{11} + J = \frac{r}{D}\left(u'_{J\alpha}w_{J\gamma} - u'_{J\gamma}w_{J\alpha}\right), \tag{15.53}$$

$$\mathcal{R}_{12} = \frac{1}{r^2}Z_{12} = -\frac{r}{D}\left(u'_{J\alpha}u_{J\gamma} - u'_{J\gamma}u_{J\alpha}\right), \tag{15.54}$$

$$\mathcal{R}_{21} = r^2 Z_{21} = -\frac{r}{D}\left(w'_{J\gamma}w_{J\alpha} - w'_{J\alpha}w_{J\gamma}\right), \tag{15.55}$$

and

$$\mathcal{R}_{22} = Z_{22} + (J+2) = \frac{r}{D}\left(u_{J\alpha}w'_{J\gamma} - u_{J\gamma}w'_{J\alpha}\right), \tag{15.56}$$

where in these relations D is the determinant of the 2×2 matrix wave function ψ_J (see Eq. (15.43));

$$D = u_{J\alpha}w_{J\gamma} - u_{J\gamma}w_{J\alpha}. \tag{15.57}$$

Equations (15.39)–(15.42) and (15.53)–(15.56) show that for $r \geq R$, the three quantities $\delta_{J\alpha}$, $\delta_{J\gamma}$ and ϵ_J completely specify the asymptotic forms of $\mathcal{R}_J$ and Z_J matrices.

15.4 Inverse Problem of Scattering in the Presence of the Tensor Force

The method of continued fraction expansion that we studied for the problems of acoustic and the torsional waves can be applied to the matrix $\phi_J(r)$, Eq. (15.48). In writing the discrete form of the matrix equation for $\phi_J(r)$, we suppress the index J in (15.48) and divide the range of integration $0 \le r \le R$ into N equal parts, each of length Δ, i.e. we take $r = n\Delta$ and $\Delta = \frac{R}{N}$. The simplest difference equation which approximates (15.48) to the order Δ^4 is given by

$$\frac{1}{\Delta^2}(\phi_{n+1} + \phi_{n-1} - 2\phi_n) + \frac{1}{n\Delta^2}\begin{bmatrix} J & 0 \\ 0 & J+2 \end{bmatrix}(\phi_{n+1} - \phi_{n-1})$$

$$+ \left[k^2 - U_n\right]\phi_n = 0. \tag{15.58}$$

By rearranging the terms in (15.58) we can write this equation as a 2×2 matrix equation for ϕ_{n+1};

$$\phi_{n+1} = B_n A_n \phi_n + C_n \phi_{n-1}. \tag{15.59}$$

In this relation A_n, B_n and C_n are defined by

$$A_n = 1 - \Delta^2 \left[k^2 - U_n\right], \tag{15.60}$$

$$B_n = \begin{bmatrix} \frac{n}{n+J} & 0 \\ 0 & \frac{n}{n+J+2} \end{bmatrix}, \tag{15.61}$$

and

$$C_n = \begin{bmatrix} \frac{J-n}{J+n} & 0 \\ 0 & \frac{J+2-n}{n+J+2} \end{bmatrix}. \tag{15.62}$$

From the matrix difference equation for ϕ_n we can determine $\mathcal{R}_J$ or Z_J matrix at $R = N\Delta$. Thus according to (15.50) we have

$$Z_J(R) = R\phi'_J(R)\phi_J^{-1}(R) = N\Delta\phi'_J(R)\phi_J^{-1}(R). \tag{15.63}$$

Next we replace $\phi'_J(R)$ by

$$\phi'(R) \approx \frac{1}{2\Delta}(\phi_{N+1} - \phi_{N-1}), \tag{15.64}$$

in (15.63) to get

$$Z_J(R) = \frac{N}{2}\left[B_N A_N + (C_N - 1)\Omega_N^{-1}\right], \tag{15.65}$$

where

$$\Omega_n \approx \phi_n \phi_{n-1}^{-1}. \tag{15.66}$$

Now Eq. (15.65) can be inverted and Ω_N can be expressed in terms of $Z_J(R)$ and A_N, both of these are known quantities. We observe that

$$\Omega_N(J) = \frac{C_N - 1}{\frac{2}{N}Z_J(R) - B_N A_N}. \tag{15.67}$$

Thus from (15.59) and (15.66) it follows that

$$\Omega_n(J) = B_{n-1}A_{n-1} + C_{n-1}\Omega_{n-1}^{-1}(J), \tag{15.68}$$

and consequently

$$\Omega_N(J) = B_{N-1}A_{N-1} + C_{N-1}\Omega_{N-1}^{-1}(J). \tag{15.69}$$

The last two equations give a matrix generalization of Thiele's continued fraction representation of Ω_N. We can use these relations recursively to solve the forward problem, viz, to find $Z_J(R)$ if the potentials are known. The same set of relations can be used to solve the inverse problem and determine the interaction potential if $Z_J(R)$ is known for values of $J = 1, 2, \cdots N - 3$. We note that the information which is gained from data for $J = 0$, (in this case $w_{J\gamma}$ is of physical importance) amounts to a constraint on the potential near the origin, but it does not determine the shape of the potential about $r = 0$. Now, because of the recursive nature of constructing the potential, we assume that $\Omega_n(J)$ is derivable from $Z_J(R)$ for values of $J = 1, 2, \cdots n - 3$ for some arbitrary n, $n \leq N$, and we note that

$$C_{n-1}(J = n - 3) = \begin{bmatrix} \frac{-1}{n-2} & 0 \\ 0 & 0 \end{bmatrix}, \tag{15.70}$$

therefore according to (15.68), the second row of Ω_n is directly related to the central and tensor potentials, i.e.

$$V_c(r') = \frac{\hbar^2}{\mu\Delta^2} \left[\Omega_n^{(2,2)}(J = n - 3) - 1 + \frac{1}{2}\Delta^2 k^2 \right] \tag{15.71}$$

$$- V_2(r', n - 3) + \frac{2(n-1)}{2n-5}V_T(r'), \tag{15.72}$$

and

$$V_T(r') = \frac{\hbar^2}{6\mu} \frac{(2n-5)(n-1)^2}{[(n-2)(n-3)]^{\frac{1}{2}}}\Omega_n^{(2,1)}(J = n - 3), \tag{15.73}$$

where $r' = (n-1)\Delta$, and $\Omega_n^{(2,1)}$ denotes the ijth matrix element of Ω_n. Therefore if we know $\Omega_n(J = n - 3)$ and $V_2(r', J)$, then we can find V_c and V_T at $r' = (n-1)\Delta$. Once we have found these, A_{n-1} for different values of J can be calculated. From the calculated values we can determine $\Omega_{n-1}(J)$ for values of $J = 1, 2 \cdots n - 4$, with the help of the relation

$$\Omega_{n-1}(J) = [-B_{n-1}A_{n-1} + \Omega_n(J)]^{-1}C_{n-1}. \tag{15.74}$$

We repeat this process, and thus we can determine the form of the potential up to a point $n - 3$.

An interesting aspect of this method is that although the information about $\Omega_n^{(1,1)}$ and $\Omega_n^{(1,2)}$ do play an important role in the calculation of Ω_{n-1}, as can be seen in Eq. (15.74), but unlike the matrix elements $\Omega_n^{(2,1)}$ and $\Omega_n^{(2,2)}$ they may not directly be helpful in the determination of the potentials, as can be seen from Eqs. (15.72) and (15.73). The fact is that this method does not use all the information contained

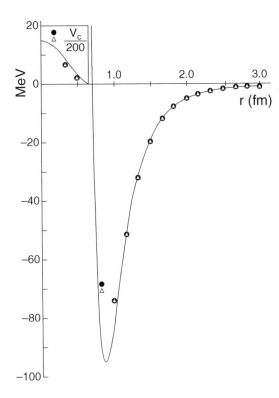

Figure 15.4: The solid line is a plot of OPEG potential for the singlet state. The results of inversion for two different laboratory energies are shown by △ for $E_L = 10$ MeV and ● for $E_L = 100$ MeV. The OPEG nuclear potential is described in the text.

in the matrix elements $\Omega_n^{(1,1)}$ and $\Omega_n^{(1,2)}$. However it is possible to generalize this method so that four sets of quantities rather than two sets discussed here can be used to find the shape of the local potentials. But because of the symmetric nature of the potential matrix $\tilde{V}_J = \frac{\hbar^2}{2\mu}\tilde{v}_J$, this set of four quantities cannot be independent and there must be a relation between them. We must note that the eigenphases $\delta_{J\pm1,J}$ and the mixing parameter ϵ_J all given at a fixed laboratory energy, E_L, yield three independent set of relations for the potentials. Clearly these relations are not sufficient for determining all the terms in $\tilde{V}_J$, but we can supplement them with the force laws obtained from the scattering phase shifts for the uncoupled states, i.e. the singlet state given by (15.31) and the triplet given by (15.32). We also observe that Eqs. (15.31) and (15.32) are identical in form as far as dependence on J is concerned. That is, the uncoupled triplet state depends on two potentials; one which looks like a central potential and the other which acts like an $\mathbf{L}^2$-dependent potential. We can relate these to the five different terms in the potential operator (15.3) by the combinations $V_c + 2V_T - V_{LS} + V_W$ and $V_{LL} - \frac{4}{3}V_W$. These effective potentials can be obtained if the scattering data are known at two different energies [24].

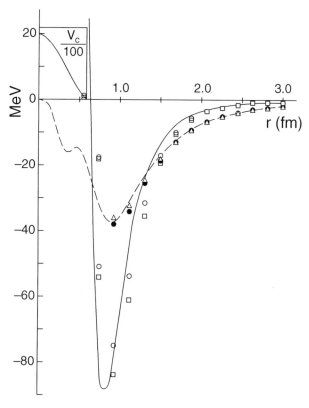

Figure 15.5: Solid and dashed lines in this figure represent the central and the tensor input potentials (OPEG) [27]. For this inversion we have assumed that the spin-orbit potential is known. The results of inversion for the central potential for two different laboratory energies are shown by ○ for $E_L = 10$ MeV and by □ for $E_L = 100$ MeV. For the tensor potential and $E_L = 10$ MeV, the points found from inversion are shown by △ for $E_L = 10$ MeV and by ● for $E_L = 100$ MeV.

15.5   Potential Model for Generating the Input Data for Testing the Inversion Method

Rather than using the phase shifts and the mixing parameter obtained from the empirical results, just as in our earlier discussion in this chapter, we will use a local potential model which produces these phase shifts and mixing parameter. In this way we will not be concerned with the questions of (a) the compatibility of the scattering data with the general form of the local potential (15.2) or (15.3) and (b) the nonzero value of the tensor force at the origin. Other advantages of starting with a local potential model are discussed in Ref. [26].

In the following discussion we will use one of the potential models proposed by Tamagaki [27]. The model that we will use has the following interesting features:
(1) For large separation between the two nucleons the tail of the potential is of the one-pion-exchange form.
(2) In the intermediate range the potential is chosen by utilizing the result found from one-boson-exchange model.
(3) For small distances the potential is assumed to be a superposition of three Gaussian terms with different ranges [27]. In Figs. 15.4–15.6 the input data are the Tamagaki potential (in his paper these are denoted by OPEG, $^1E - 2$) shown by solid, dashed and dashed dotted curves. These potentials do not have a simple exchange character, therefore here we assume the (non-exchange) type for nuclear forces, i.e. Wigner's (see Ref. [28]).

For the singlet state, the solution of this problem is similar to the inverse scattering problem of nuclear-nucleon which will be discussed in the Chapter 16. Now let us consider the general case when the tensor force is present. Since the main purpose of our study is to test the inversion for different potentials, we consider the simplest case, i.e. when $\mathcal{R}$ matrices for both even (odd) values of J with the even (odd) states of the potential model are given [27]. In the potential model that we are using, V_{LL} and V_W are either zero or are very small compared to the other terms, and we ignore them completely.

In Fig. 15.4 the result of inversion is shown in the case of scattering in singlet state. For the triplet state we observe that there are three unknown potentials V_c, V_T and V_{LS}. Now we can assume that one of these functions is known and determine the other two. In Fig. 15.5 we show the results when V_{LS} is known and the object is to find V_c and V_T by inversion. In Fig. 15.6 we assume V_c is known and we calculate V_{LS} and V_T.

Results of Inversion Obtained from the Thiele expansion of the $\mathcal{R}_J$ Matrix — The input data is found by the integration of the Riccati matrix equation (15.51) from the origin to the point $r = R$. From the scattering data the same $\mathcal{R}_J$ can be found provided that the potential for $r > R$ is known, and then Eq. (15.51) can be integrated, from some large value of r where $U(r)$ is negligible, to the point $r = R$. If we assume that the central potential is known, then we can obtain the result of inversion for the tensor and the spin-orbit potentials. Figure 15.6 is a plot of the input potential and the points found by solving the inverse problem. By comparing the input potential and the potential recovered by inversion we reach the following conclusions:
(a) For separation greater than one fermi between the two nucleons, the results are generally good and reliable, whether we consider a singlet state or a triplet state, and they are not very sensitive to the scattering energy. In the case of the triplet state, no matter which of the the two, V_c or V_{LS} is assumed to be known (see Figs. 15.5 and 15.6) the results are acceptable. Finally, we can solve the inverse problem for the triplet channel $L = J$ to find the sum $V_c + 2V_T - V_{LS}$, and combine this result with the two relations obtained for the potentials calculated in the coupled channels and in this way determine V_c, V_{LS} and V_T. In Fig. 15.7 shows the triplet even potentials of Eikenmeir and Hackenbriech [29]. These potentials have

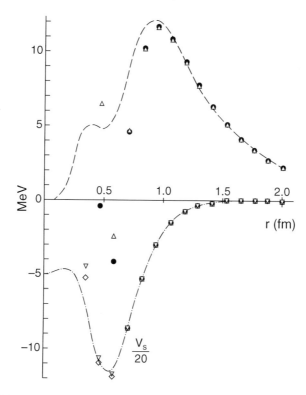

Figure 15.6: If the central potential is known, then the results of the inversion yields the tensor and the spin-orbit force. In this figure the input tensor force is shown by dashed line and the spin-orbit potential by dashed-dotted curve. The input force is assumed to be given by OPEG ($^3O-2$). For the laboratory energies of $E_L = 10$ MeV and $E_L = 100$ MeV, the calculated results for spin-orbit potential are shown by ∇ and by $\Diamond$ respectively ($V_{LS} = V_s$). The same symbols as those of Fig. 15.5 are used here.

soft core and are superpositions of Gaussian potentials. They are shown by dashed line (tensor), dashed-dotted line (spin-orbit) and solid line (central) in this figure. The results of inversion are also shown, and are in good agreement with the starting potentials for larger separation. For smaller separation $r < 1$ fm the potentials are not accurately reproduced, but they show some of the features of force laws, e.g. where they are attractive and where they are repulsive.

(b) The main cause of discrepancy at close distances is the accumulation of the round off errors in the continued fraction expansion and the limitation placed on the size of Δ just as in the other problems that we have studied with this method of inversion.

(c) Since we cannot increase N much larger than 20, a way of improving the result is to make Δ small by choosing smaller values of R. For this we assume that the tail of the potential is known exactly, and then by integrating Eq. (15.51) inwards we find the $\mathcal{R}_J$ matrix for smaller values of r. As we can see in Fig. 15.6, by changing R from 3 to 2 fm, the potential can be reproduced rather accurately to a distance

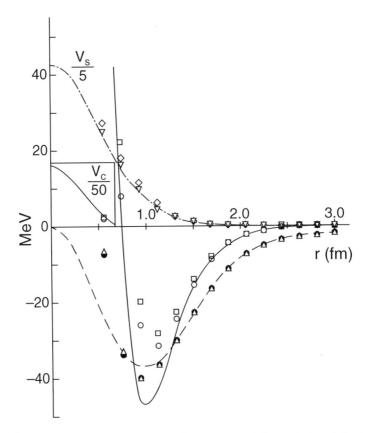

Figure 15.7: Solid and dashed lines in this figure represent the central and the tensor input potentials (Eikenmeir and Hackenbriech) [29] triplet even potential. Symbols are the same as those used in Fig. 15.5.

of about 0.5 fm. For other examples of this type of inversion the reader is referred to the original paper Ref. [24].

15.6 Inverse Method of Nucleon-Nucleon Phase Shift and the Calculation of Nuclear Structure

In nuclear structure calculation we usually work with a complete set of independent particle harmonic oscillator wave function, i.e. we use the matrix elements of nucleon-nucleon of the relative motion to calculate the desired nuclear properties [30], [31].

If we denote these matrix elements for the relative motion by

$$\langle n'\ell's|V(r)|n\,\ell,\,s\rangle,$$

then apart from n and n', the labelling of the matrix elements are the same as that of the phase shifts in the two-body problem. Abbreviating $\ell\ell'$s by the symbol α, denoting the potential for the set $(\ell,\,s=\alpha)$ by $V_\alpha(r)$, and taking the matrix elements of $V_\alpha(r)$, we find that we need the information about the weighted integrals of the potential, viz,

$$I(\alpha,\ell,p,b) = \int_0^\infty V_\alpha(r)\exp\left(-\frac{r^2}{2b^2}\right)r^{2\ell+p+2}dr, \tag{15.75}$$

where $p \geq 0$ and is an even integer. The constant b here is the oscillator strength parameter.

The phase shifts for $\ell > 0$ do not exceed a magnitude of 0.5 radians and are generally smaller than this value. We know that for small $\delta_\ell(k)$ the Born approximation is valid, i.e. the phase shift for the ℓth partial wave is given by

$$\tan\delta_\ell(k) = -\frac{mk}{\hbar^2}\int_0^\infty V_\alpha(r)j_\ell^2(kr)r^2dr, \tag{15.76}$$

where $E_{\text{lab}} = \frac{2\hbar^2 k^2}{m}$, and m is the mass of a nucleon. From Eq. (15.76) we can find $V_\alpha(r)$ if $\tan\delta_\ell(k)$ is known for all values of k. Let us denote the inverse function of $j_\ell^2(kr)$ by $g_\ell(kr)$, i.e.

$$\int_0^\infty j_\ell^2(kr)g_\ell(kr')\,dk = \delta(r-r'). \tag{15.77}$$

The special case of this inversion for $\ell = 0$ was considered in Eq. (13.259). Here we want to find $g_\ell(kr')$ which is the inverse of $j_\ell^2(kr)$ as defined by Eq. (15.77). By multiplying (15.76) by $\frac{1}{k}g_\ell(k')$ and integrating over k, using Eq. (15.77) we find the potential $V(r)$ to be

$$V_\ell(r) = -\frac{\hbar^2}{2mr^2}\int_0^\infty \frac{1}{k}\tan\delta_\ell(k)g_\ell(kr)dk. \tag{15.78}$$

Thus we have reduced the problem to that of finding $g_\ell(kr)$. We can find this inverse with the help of Mellin's transform, and the result is [30]–[33]

$$g_\ell(kr) = -\left[\frac{4(kr)^2}{\pi\left(\ell+\frac{1}{2}\right)}\right]{}_1F_2\left(\frac{3}{2};\frac{1}{2}-\ell;\ell+\frac{3}{2};-kr^2\right)$$

$$= \frac{8(kr)^2}{\pi}\frac{d}{dr}\left[kr^2 n_\ell(kr)j_\ell(kr)\right], \tag{15.79}$$

where ${}_1F_2(\alpha;\beta;\gamma;-kr^2)$ is the hypergeometric function [15]. In fact it is possible to eliminate $V(r)$ between (15.75) and (15.76) by substituting $V_\ell(r)$ found from

(15.78) in (15.75). This can be achieved by noting that [31]

$$r^{2\ell} \exp\left(-\frac{r^2}{2b^2}\right) = -\frac{2}{\pi} \left(2b^2\right)^{\ell+\frac{3}{2}} 2\Gamma\left(\ell + \frac{1}{2}\right) \int_0^\infty \left\{j_\ell^2(kr)\exp\left(-2k^2b^2\right)\right.$$

$$\times \left. {}_1F_1\left(-\ell - 1; \frac{1}{2} - \ell; 2k^2b^2\right)k^2\right\} dk, \tag{15.80}$$

and formulae similar to this. For instance if $n' = n = 0$ and $\ell' = \ell = 1$, we find

$$\langle 0, s\ell j|0, s\ell j\rangle = \frac{2}{3}\hbar\omega \int_0^\infty e^{-E}\left(1 + 4E - 4E^2\right)\tan\delta_\alpha(E)dE, \tag{15.81}$$

where

$$E = 2\hbar^2 b^2 = \frac{1}{\hbar\omega}E_{\text{lab}} \tag{15.82}$$

is the laboratory scattering energy in units of the oscillator strength $\hbar\omega$. While we observe that the integral in (15.81) is over all energies, and the phase shifts for energies greater than 350 MeV are not well-defined, but the exponential term in the integral of (15.81) cuts off the integral well below $E_{max} = 350$ MeV. Elliott and collaborators, using this method, have shown that the spin-orbit splittings in several nuclei obtained in this way agree well with the experimentally observed results [31].

References

[1] A detailed discussion of the role of the tensor force in nucleon-nucleon interaction is given in the classic work of J.M. Blatt and V.F. Weisskopf *Theoretical Nuclear Physics*, (Springer, New York, 1979) Chapter III.

[2] S. Okubo and R.E. Marshak, Ann. Phys. (NY), 4, 166 (1958).

[3] A complete review of the two nucleon interaction with an extensive list of references up to 1972 can be found in M.J. Moravcsik's report in Rep. Prog. Phys. 35, 587 (1972).

[4] G.E. Brown and A.D. Jackson, *The nucleon-nucleon interaction*, (North-Holland, 1976).

[5] For a recent survey of nucleon-nucleon interaction see the review article by M. Naghdi, Phys. Par. Nucl. 45, 924 (2014).

[6] R. Machleidt, K. Holinde and Ch. Elster, Phys. Rep. 149, 1 (1987).

[7] M. Lacombe, B. Loiseau, J.M. Richard, R. Vinh Mau, J. Côté, P. Pirès and R. de Tourreil, Phys. Rev. C 21, 861 (1980).

[8] M. Razavy, G. Field and J.S. Levinger, Phys. Rev. 125, 269 (1962).

[9] V.M. Kranopolsky, V.I. Kukulin, and J. Horáček, Czech. J. Phys. B 35, 805 (1985).

[10] K. Chadan, Nuovo Cimento, 40, 1195 (1965).

[11] K. Chadan and A. Montes Lozano, Phys. Rev. 164, 1762 (1967).

[12] M. Razavy and E.S. Krebes, Nucl. Phys. A 184, 533 (1972).

[13] G.A. Baker, Jr., *Essentials of Padé Approximants*, (Academic Press, New York, 1975).

[14] R.A. Arndt, L.D. Roper and R.I. Shotwell, Phys. Rev. C 5, 2100 (1971).

[15] Th. Kist, H. Kohlhoff and V. Geramb in *Inverse Methods in Action*, P.C. Sabatier (Ed.) (Springer-Verlag, New York, 1990).

[16] Th. Kirst, K. Amos, L. Berge, M. Coz and H.V. von Geramb, Phys. Rev. C 40. 912 (1989).

[17] L. Jäde, M. Sander and H.V. von Geramb, in *Inverse and Algebraic Quantum Scattering Theory*, Edited by B. Apagyi *et al.*, (Springer-Verlag, Berlin 1997) p. 124.

[18] H.V. Geramb and H. Kohlhoff in *Quantum Inversion Theory and Applications*, Edited by H.V. Geramb, (Springer-Verlag, Berlin, 1994) p. 285.

[19] H. Kohlhoff, M. Küker, H. Freitag and H.V. von Geramb, Phys. Scripta, 48, 238 (1993).

[20] R.V. Reid, Ann. Phys. 50, 411 (1968).

[21] R.A. Arndt, J.S. Hyslop III and L.D. Roper, Phys. Rev. D 35, 128 (1986).

[22] P.M. Morse and H. Feshbach, Methods of Theoretical Physics, Part 1 (McGraw-Hill, New York, 1953) p. 622.

[23] M.A. Hooshyar and M. Razavy, Can. J. Phys. 59, 1627 (1981).

[24] M.A. Hooshyar and M. Razavy, Phys. Rev. C 20, 29 (1984).

[25] The difficulty of defining the regular solution by the boundary condition at $r = 0$ when the tensor force is present, is discussed in a number of books. For instance see R.G. Newton's *Scattering Theory of Waves and Particles*, Second Edition (Springer, New York, 1982) p. 462.

[26] M.A. Hooshyar, J. Math. Phys. 21, 1695 (1980).

[27] R. Tamagaki, Prog. Theor. Phys. 39, 91 (1968).

[28] J.M. Blatt and V.F. Weisskopf, *Theoretical Nuclear Physics*, (Springer, New York, 1979) p. 137.

[29] H. Eikenmeir and H.H. Hakenbroich, Nucl. Phys. A169, 407 (1971).

[30] H.A. Mavromatis and K. Schilcher, J. Math. Phys. 9, 1627 (1968).

[31] J.P. Elliott, H.A. Mavromatis and E.A. Sanderson, Phys. Lett. 42B, 358 (1967).

[32] W. Sollfrey, J. Math. Phys. 10, 1429 (1969).

[33] E.C. Titchmarch, *Introduction to the Theory of Fourier Integrals*, (Oxford University Press, London, 1948) p. 212.

[34] I.S. Gradshteyn and I.M. Ryzhik, *Tables of Integrals, Series, and the Products*, (Academic Press, New York, 1965) p. 1039.

Chapter 16

The Inverse Problem of Nucleon-Nucleus Scattering

Starting with a many channel Schrödinger equation, Feshbach showed that the low energy nucleon-nucleus collision can be reduced to that of a nucleon interacting with an optical potential [1], [2]. This potential is, in general, nonlocal and has negative-definite imaginary part. Now a detailed study of the solution of the Schrödinger equation with nonlocal potentials has shown that in many problems one can replace the nonlocal interaction by an equivalent, but energy dependent potential [3]. Using this equivalence, one can apply the methods of inverse scattering theory and construct local optical potentials at fixed energy.

It is instructive first to consider the WKB inversion technique. This method has been successfully applied by Kujawski to construct the optical potential for $\alpha -^{12} C$ and $\alpha -^{90} Zr$ scatterings [4]. We know that the WKB approximation works well at higher energies, therefore the inversion is also more accurate at higher energies. A formal but exact procedure has been advanced by Newton and by Sabatier, but in practice this method works well also at high energies [5]. Coudray who has used this technique to find the optical potential for neutron-α scattering has found acceptable results for incident energies of about 300 MeV [6]. An alternative method advanced by Lipperheide and Fiedeldey, which we discussed in Sec. 6.5 can also be applied to study the inverse problem for heavy-ion collision [7], [8]. In addition to these, important contributions to the application of the inverse method for construction of the optical potentials have been made by Mackintosh and collaborators [9]–[12].

The discrete method based on continued fraction expansion that we will be discussing can easily be extended to proton-nucleus scattering as well as to the low energy heavy ion scattering. Here we assume not only a central force but also we include the spin-orbit potential. In the present formulation the $\mathcal{R}$ matrix turns out

317

---

to be a complex rational function of the angular momentum ℓ. In solving the direct problem, we integrate the radial Schrödinger equation for the ℓ-th partial wave from the origin to the point $r = R$. At this point we calculate $\mathcal{R}_\ell = \left|\frac{d}{dr}\ln\psi_\ell(r)\right|_{r=R}$. Here R denotes the range beyond which both V_c and V_{so} become negligible. We can also use the discrete form of the Schrödinger equation that we will be using for the inverse problem to determine the $\mathcal{R}_\ell$ matrix. Once $\mathcal{R}_\ell$ is given at the exterior of the potential, then we can find the potential at N points, where N is the number of partial waves.

Solution of the Inverse Problem when the $\mathcal{R}$ Matrix Is Known — Let us start with the wave equation for the scattering of a neutron of mass m from a spin zero target of mass M, i.e.

$$\nabla^2\psi(r) + \left[k^2 - v_c(r) - \frac{4}{\hbar^2}\mathbf{L}\cdot\mathbf{S}\,v_{so}(r)\right]\psi(r) = 0, \tag{16.1}$$

where

$$k^2 = \frac{2mM^2E_L}{\hbar^2(m+M)^2}, \tag{16.2}$$

$$v_c(r) = \frac{2mM}{\hbar^2(m+M)}V_c(r), \tag{16.3}$$

and

$$v_{so}(r) = \frac{2mM}{\hbar^2(m+M)}V_{so}(r). \tag{16.4}$$

Here E_L is the energy of the neutron in the laboratory frame, and $V_c(r)$ and $V_{so}(r)$ are the central and the spin orbit potentials respectively. The differential cross section for scattering depends on two amplitudes $f(\theta)$ and $g(\theta)$ defined in terms of partial wave amplitudes by [3]

$$f(\theta) = \sum_{\ell=0}^{\infty}\left[(\ell+1)f_\ell^+ + \ell f_\ell^-\right]P_\ell(\cos\theta), \tag{16.5}$$

and

$$g(\theta) = \sum_{\ell=1}^{\infty}\left[f_\ell^+ - f_\ell^-\right]P_\ell(\cos\theta). \tag{16.6}$$

Therefore when $f(\theta)$ and $g(\theta)$ are given, then the two amplitudes f_ℓ^+ and f_ℓ^- can be determined. Let us note that $\psi(r)$ in (16.1) is a two component column vector which we write as $\begin{bmatrix}\psi^+(r)\\\psi^-(r)\end{bmatrix}$ and this vector can also be written in terms of the partial waves $\psi_\ell^\pm(r)$. For our description of the inverse problem it is convenient to change $\psi_\ell^\pm(r)$ to $\phi_\ell^\pm(r)$ where $\phi_\ell^\pm(r)$ is defined by

$$\phi_\ell^\pm(r) = (kr)^{-\ell-1}\psi_\ell^\pm(r), \tag{16.7}$$

and is a solution of the differential equation

$$\frac{d^2}{dr^2}\phi_\ell^\pm(r) + \frac{2(\ell+1)}{r}\frac{d}{dr}\phi_\ell^\pm(r) + \left[k^2 - v(r) \mp (2\ell+1)Q(r)\right]\phi_\ell^\pm(r) = 0. \quad (16.8)$$

Here $v(r)$ and $Q(r)$ are related to $v_c(r)$ and $v_{so}(r)$ by [5]

$$v(r) = v_c(r) - v_{so}(r), \quad (16.9)$$

and

$$Q(r) = v_{sc}(r). \quad (16.10)$$

Let us assume that the range beyond which both $v_c(r)$ and $v_{so}(r)$ are negligible is R, then the $\mathcal{R}_\ell$ matrix is given by

$$\mathcal{R}_\ell^\pm = \left[\frac{\psi_\ell^\pm(r)}{\psi_\ell'^\pm(r)}\right]_{r=R} = R\left[\ell+1+R\left(\frac{\phi_\ell'^\pm(r)}{\phi_\ell^\pm(r)}\right)_{r=R}\right]^{-1}, \quad (16.11)$$

where prime denotes derivative with respect to r. We can also write $\mathcal{R}_\ell^\pm$ using the asymptotic form of $\psi_\ell^\pm(r)$ and thus relate $\mathcal{R}_\ell^\pm$ to $f_\ell^\pm$:

$$\mathcal{R}_\ell^\pm = R\left\{\frac{j_\ell(kr) + ikf_\ell^\pm h_\ell^{(1)}(kR)}{[rj_\ell(kr)]_R' + ikf_\ell^\pm\left[rh_\ell^{(1)}(kr)\right]_R'}\right\}. \quad (16.12)$$

The last two equations show that if we know $f_\ell^\pm$, then we can find $\left(\frac{\phi_\ell'^\pm(kR)}{\phi_\ell^\pm(kr)}\right)_{r=R}$. Then this logarithmic derivative of $\phi_\ell^\pm(kr)$ at $r = R$ will be used as the input in our approach.

Discrete Form of the Wave Equation — We want to use a discrete form of the Schrödinger equation, therefore we divide R into N equal parts, $R = N\Delta$, so that r is equal to $n\Delta$, $n = 0,\ 1\ 2\cdots N$. Now we replace Eq. (16.8) by the following difference equation [13], [14]

$$\phi_{n+1}^\pm = A_n^\pm(\ell)B_n(\ell)\phi_n^\pm + C_n(\ell)\phi_{n-1}^\pm, \quad (16.13)$$

where

$$B_n(\ell) = \frac{n}{\ell+1+n}, \quad (16.14)$$

$$C_n(\ell) = \frac{\ell+1-n}{\ell+1+n}, \quad (16.15)$$

and

$$A_n^\pm(\ell) = 2 - \Delta^2 k^2 + \Delta^2 v_n \pm (2\ell+1)\Delta^2 Q_n. \quad (16.16)$$

In Eq. (16.13) we have suppressed the dependence of ϕ_n s on ℓ. Now let us examine the boundary condition for this difference equation. Assuming that the potential is less singular than r^{-2} at the origin, then the asymptotic form of $\psi_\ell^\pm(r)$ has the form

$$\psi_\ell^\pm(r) \to (kr)^{\ell+1}, \quad \text{as} \quad r \to 0, \tag{16.17}$$

and thus the initial condition for the difference equation becomes

$$\phi_0 = 1. \tag{16.18}$$

For the solution of direct problem we iteration procedure, and then for $n = N$ we find

$$\frac{\phi_N^\pm}{\phi_{N-1}^\pm} = A_{N-1}^\pm(\ell)B_{N-1}(\ell) + \cfrac{C_{N-1}}{A_{N-2}^\pm(\ell)B_{N-2}(\ell)+} \cdots$$

$$\frac{C_3(\ell)}{A_2^\pm(\ell)B_2(\ell)+} \frac{C_2(\ell)}{A_1^\pm(\ell)B_1(\ell)\left[\frac{C_\ell(\ell)}{\phi_1^\pm}\right]}, \tag{16.19}$$

and a similar relation for $\left(\frac{\phi_{N+1}^\pm}{\phi_N^\pm}\right)$.

From Eq. (16.19) it may appear that $\left(\frac{\phi_{N+1}^\pm}{\phi_N^\pm}\right)$ is dependent on $\phi_\ell^\pm$, but as (16.15) shows this is not the case for the physical values of the angular momentum. Once $\left(\frac{\phi_{N+1}^\pm}{\phi_N^\pm}\right)$ and $\left(\frac{\phi_{N-1}^\pm}{\phi_N^\pm}\right)$ are found then we can obtain $\mathcal{R}_\ell^\pm$ from

$$\frac{\Delta}{\mathcal{R}_\ell^\pm} = \frac{\ell+1}{N} + \frac{1}{2}\left(\frac{\phi_{N+1}^\pm - \phi_{N-1}^\pm}{\phi_N^\pm}\right). \tag{16.20}$$

By substituting this relation in (16.12) we find $f_\ell^\pm$ and hence $f(\theta)$ and $g(\theta)$. It is important to note that both $\mathcal{R}_\ell^\pm$ matrix and $\left(\frac{\phi_{N+1}^\pm}{\phi_N^\pm}\right)$ are given as ratios of polynomials.

16.1 Solving the Inverse Nucleon-Nucleus Problem

We assume that from the collision cross section for nucleon-nucleus we can deduce the two amplitudes $f(\theta)$ and $g(\theta)$ and in addition we know the range R of the forces between the two systems. Our object is to construct the two potentials v_c and v_{so}. This can be done in three steps:
(1) First we decompose $f(\theta)$ and $g(\theta)$ according to Eqs. (16.5) and (16.6). Then by substituting these results in (16.12) we find $\mathcal{R}_\ell^\pm$ which then yields $\left(\frac{\phi_{N+1}^\pm - \phi_{N-1}^\pm}{\phi_N^\pm}\right)$ as a function of ℓ.

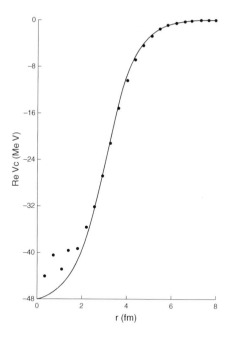

Figure 16.1: The real part of the central potential for $n - O$ scattering at $E_L = 14$ MeV. In this plot the input potential is shown by a solid line and the circles, • s, indicate the output for $N = 22$ partial waves.

(2) We also assume that only a finite number of partial waves, N, contribute to $f(\theta)$ and $g(\theta)$, and the rest are negligible. In principle we can choose N arbitrarily large, however, as we have seen before there is an optimal value say N_0, for which the method works best.

(3) We also need to know the potential at the boundary, i.e. $v(R) = v_N$ and $Q(R) = Q_N$, (either of these may or may not be zero). In addition we need to find $Z_N^{\pm}(\ell_n)$ at $N - 1$ points where

$$Z_N^{\pm}(\ell) = N \left(\frac{\phi_{N+1}^{\pm} - \phi_{N-1}^{\pm}}{2\phi_N^{\pm}} \right), \tag{16.21}$$

and

$$\ell_n = N - 1 - n, \qquad n = 1, \, 2 \cdots N - 1. \tag{16.22}$$

Since v_n and Q_n are assumed to be known, therefore we can find $A_N^{\pm}(\ell)$ and consequently $A_N^{\pm}(\ell_N)$ from (16.16). Also from (16.13) and (16.21) we have

$$\frac{\phi_N^{\pm}(\ell)}{\phi_{N-1}^{\pm}(\ell)} = \frac{C_N(\ell) - 1}{\frac{2}{N} Z_N^{\pm}(\ell) - A_N^{\pm}(\ell) B_N(\ell)}. \tag{16.23}$$

This ratio is known for $\ell = \ell_n$. But according to (16.15), $C_n(\ell_{N-n}) = 0$, therefore

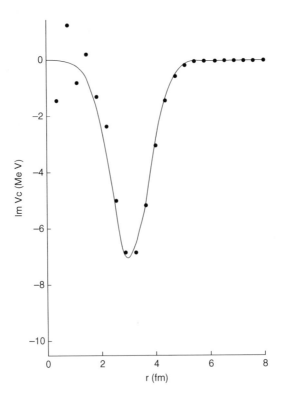

Figure 16.2: The imaginary part of the central potential for $n - O$ scattering at $E_L = 14$ MeV. In this plot the input potential is shown by a solid line and the circles • s indicate the output for $N = 22$ partial waves.

from (16.13) it follows that

$$A_{N-1}^{\pm}(\ell_1) = \frac{\phi_N^{\pm}(\ell_1)}{\phi_{N-1}^{\pm}(\ell_1)B_{N-1}(\ell_1)}, \tag{16.24}$$

and

$$A_{N-j}^{\pm}(\ell_j) = \frac{1}{B_{N-j}(\ell_j)}\left[\frac{C_{N+1-j}(\ell_j)}{-A_{N+1-j}^{\pm}(\ell_j)B_{N+1-j}(\ell_j)+} \cdots \right.$$

$$\left. \frac{C_{N-2}(\ell_j)}{-A_{N-2}^{\pm}(\ell_j)B_{N-2}(\ell_j)+} \frac{C_{N-1}(\ell_j)}{-A_{N-1}^{\pm}(\ell_j)B_{N-1}(\ell_j) + \frac{\phi_N^{\pm}(\ell_j)}{\phi_{N-1}^{\pm}(\ell_j)}}\right],$$

$$j = 2,\, 3,\, \cdots N - 1. \tag{16.25}$$

To calculate $A_{N-j}^{\pm}(\ell)$ from the previous terms we solve Eq. (16.16) for v_{N-j} and Q_{N-j} with $\ell = \ell_j$ and $n = N - j$;

$$v_{N-j} = \frac{1}{2\Delta^2}\left[A_{N-j}^{+}(\ell_j) + A_{N-j}^{-}(\ell_j) - 4 + 2\Delta^2 k^2\right], \tag{16.26}$$

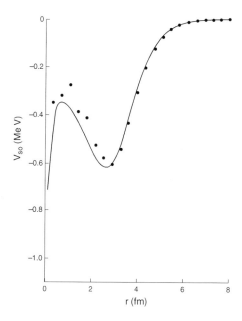

Figure 16.3: The spin-orbit potential shown by solid line for $n - O$ scattering at $E_L = 14$ MeV with $N = 22$, and the results of inversion are shown by circles.

and

$$Q_{N-j} = \frac{\left[A^{+}_{N-j}(\ell_j) - A^{-}_{N-j}(\ell_j)\right]}{2(2\ell_j + 1)\Delta^2}. \tag{16.27}$$

Once v_n s and Q_n s are determined, we can find the central and the spin-orbit potentials from Eqs. (16.9) and (16.10). In this way we get these potentials at the discrete points $(r = n\Delta)$.

$$\text{Re } v_c(r = n\Delta) = \text{Re } v_n + Q_n, \tag{16.28}$$

$$\text{Im } v_c(r = n\Delta) = \text{Im } v_n, \tag{16.29}$$

and

$$v_{so}(r = n\Delta) = Q_n. \tag{16.30}$$

Construction of the Optical Potential — In principle, the test of the inverse nuclear scattering theory should be the use of the experimental results for the cross section and polarization as the input, and the determination of the possible potential models as the outcome of the calculation. But in practice there are problems which prevents us from achieving this goal. First we have the uncertainty in the experimental data, and then the uncertainty in the partial wave analysis. Added to this is the numerical errors arising in the determination of the coefficients of the continued fraction which introduces large accumulative errors when N is large. For this reason it is preferable to start with potential models which produce experimental results accurately, and use these to test the validity of solving the inverse problem. Now let us consider the specific example of the neutron-oxygen collision at 14 MeV. The empirical result can be best fitted with an optical potential of the form [15]–[18]

$$V(r) = -V_0 f(r) - iW_0 g(r) - V_{so} h(r) \mathbf{L} \cdot \mathbf{S}, \qquad (16.31)$$

where $f(r)$ has the Wood–Saxon form

$$f(r) = \frac{1}{1 + \exp\left(\frac{r-R}{a}\right)}, \qquad (16.32)$$

and $g(r)$ is the surface-centered imaginary potential

$$g(r) = \exp\left[-\left(\frac{r-R}{a}\right)^2\right]. \qquad (16.33)$$

The spin-orbit term $h(r)$ is given by

$$h(r) = -\left(\frac{\hbar}{m_\pi c}\right)^2 \frac{1}{r} \frac{df(r)}{dr}, \qquad (16.34)$$

where, in these equations, R denotes the nuclear radius, $R = r_0 A^{\frac{1}{3}}$, A being the mass number of the target nucleus.

We have used these potentials, shown by solid lines in Figs. 16.1–16.3 and calculated $(R)_\ell^\pm$ by integrating the Schrödinger equation from the origin to the point $r = R$. Then using $\mathcal{R}_\ell^\pm$ as the input we have solved the inverse problem as discussed earlier. In these figures the results of inversion are shown by solid circles. The optimum number of divisions N for this inversion is about 22. Again we notice that for large N, the round-off errors become large. In all three figures we notice that the results of inversion is very good for large separation between the target and projectile, i.e. for $r > 1.5$ fm, but for smaller separation they do not follow the details of the potentials. By applying this method to other systems, e.g. the case of neutron-carbon scattering similar results have been obtained with more or less the same degree of accuracy [14].

16.2   Inverse Scattering Theory Incorporating Both Coulomb and Nuclear Forces

Alam and Malik have extended the inversion method that we described in this chapter to find the local potential for $\alpha - {}^{12}C$ when the Coulomb force is present [19]. In this generalization, the $\mathcal{R}(\ell)$ matrix has to be modified by the presence of the long range Coulomb potential. Thus we replace the $\mathcal{R}(\ell)$, Eq. (16.12) by

$$\mathcal{R}(\ell) = \left[\frac{\chi_\ell(r)}{\chi_\ell'(r)}\right]_{r=R}, \qquad (16.35)$$

where
$$\chi_\ell(r) = r\left\{F_\ell(\eta, kr) + T(\ell)[G_\ell(\eta, kr) + iF_\ell(\eta, kr)]\right\}. \tag{16.36}$$
In this relation η is the Sommerfeld parameter;
$$\eta = \frac{Z_1 Z_2 e^2}{4\pi\epsilon_0 \hbar v} = \alpha Z_1 Z_2 \sqrt{\frac{\mu c^2}{2E}}, \tag{16.37}$$
e is the elementary charge, Z_1 and Z_2 are the atomic numbers of the two nuclei, v is the magnitude of their relative velocity in the center-of-mass frame, and α is the fine structure constant. The two wave unctions $F_\ell(\eta, kr)$ and $G(\eta, kr)$ are defined by the following relations:

For $r > R$, we use the asymptotic form of a linear combination of the regular and irregular Coulomb wave functions [20], [21]. Thus if we define $Y_\ell(r)$ by
$$Y_\ell(r) = i\exp\left\{-i\left[kr + \eta\ln(2kr) - \frac{\pi\ell}{2} + \sigma_\ell\right]\right\}, \tag{16.38}$$
then the regular solution is
$$F_\ell(\eta, kr) = \sqrt{\frac{1}{2\pi}}\,[Y_\ell(r) + Y_\ell^*(r)]$$
$$\rightarrow \sqrt{\frac{2}{\pi}}\sin\left[kr - \frac{\pi\ell}{2} - \eta\ln(2kr) + \sigma_\ell\right], \quad \text{as } r \to \infty, \tag{16.39}$$
and the irregular solution is
$$G_\ell(\eta, kr) = \sqrt{\frac{1}{2\pi}}\,[Y_\ell(r) - Y_\ell^*(r)]$$
$$\rightarrow -\sqrt{\frac{2}{\pi}}\cos\left[kr - \frac{\pi\ell}{2} - \eta\ln(2kr) + \sigma_\ell\right], \quad \text{as } r \to \infty, \tag{16.40}$$
where
$$e^{2i\sigma_\ell} = \frac{\Gamma(\ell + 1 + i\eta)}{\Gamma(\ell + 1 - i\eta)}. \tag{16.41}$$
In Eq. (16.36) $T(\ell)$ matrix is related to the ℓth partial phase shift $\delta(\ell)$ by
$$T(\ell) = \frac{1}{2i}\left(e^{2i\delta(\ell)} - 1\right) = \frac{S(\ell) - 1}{2i}. \tag{16.42}$$
Thus knowing the phase shifts $\delta(\ell)$ and R enable us to determine $\mathcal{R}(\ell)$.

For the scattering of two nuclei, the potential is, in general complex, therefore $\delta(\ell)$ is also complex. To solve the direct problem we can use the finite difference formulation, but we must replace $A_n(\ell)$ of Eq. (16.16) by
$$A_n(\ell) = 2 - k^2\Delta^2 + \Delta^2(v_n + v_n^c). \tag{16.43}$$
Here v_n and v_n^c are the nuclear and Coulomb potentials respectively.
$$v_n = \frac{2\mu}{\hbar^2}V_n, \quad v_c = \frac{2\mu}{\hbar^2}V_n^c, \tag{16.44}$$
where μ is the reduced mass $\mu = \frac{mM}{m+M}$.

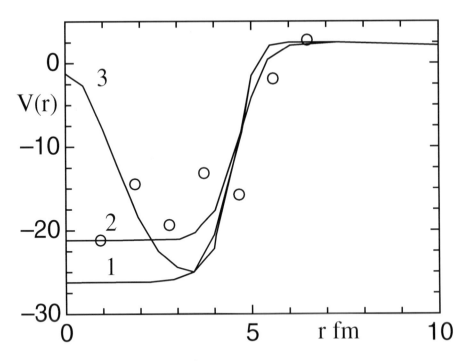

Figure 16.4: By solving the Schrödinger equation we can find a set of potentials 1, 2, and 3 each generating a good fit for the empirical phase shifts obtained by Plaga *et al* for $\alpha - {}^{12}C$ scattering [24]. On the other hand the circles in this graph show the result of inversion using the phase shifts directly as the input [19].

Rather than using a potential model to test the reliability of the inversion as we did earlier, Alam and Malik use the data and the phase shift analysis of Plaga *et al.* [24] and try to determine the unknown potential (or potentials) which will produce the empirical data. Now in their analysis Plaga *et al.* assumed that the imaginary part of the phase shift is negligible. This is not an unreasonable assumption since most of the incident energies (which is 6.458 MeV) in the center-of-mass system is below the first excited state of ^{12}C [19].

Based on the knowledge of a finite number of partial waves, the inverse method will not yield a unique nuclear potential v_n or $v(r)$. But by assuming that the potential is energy independent, only one of these non-unique set of potentials is acceptable and is expected to reproduce the approximate phase shifts at adjacent energies. At these nearby energies the angular distributions are calculated with the chosen potential and if the angular distributions are not well reproduced, then the potential set is discarded and another set that fits the data at the selected energies is chosen. This process is continued until an acceptable potential which reproduces the overall features of the cross section over a wide range of energies is found.

In the problem of $\alpha - {}^{12}C$ scattering, we choose the data for scattering at the fixed energy of 6.458 MeV in laboratory frame since it provides information about the maximum number of partial wave [24]. For this incident wave the total number of phase shifts for a given k is seven, $\ell = 0,\ 1\cdots 6$. From these partial waves we find $T(\ell)$, Eq. (16.42) and then from Eqs. (16.36) and (16.35) we calculate $\mathcal{R}(\ell)$. A continued fraction expansion of $\mathcal{R}(\ell)$ gives us v_n, since v_n^c is known, Eq. (16.44). The points that has been found for v_n are shown by circles in Fig. 16.4. By passing a curve through these points we get an idea about the shape of the potential $v(r)$, and as we see in this figure these points do not show a smooth potential. An indication of the nonuniqueness of the result of inversion can be provided by the construction of potentials which exactly gives these phase shifts once used in the Schrödinger equation. Three such potentials with the analytic form

$$U(r) = V_0 \frac{1}{1 + \exp\left(\frac{R-a}{a}\right)} + V_1 \exp\left[-\left(\frac{r}{a_1}\right)^2\right], \qquad (16.45)$$

with V_0, V_1, a, a_1 and R being parameters, have been obtained by Alam and Malik [19].

16.3   Inverse Scattering Method for Two Identical Nuclei at Fixed Energy

As we had seen earlier for the continued fraction method of constructing the potential, or at least finding a number of points of a local potential function, we required an input consisting of phase shifts for all partial waves at a given energy (or wave number k). Therefore our method has to be changed slightly when we want to determine the potential for scattering of two identical particles such as $p - p$, $\alpha - \alpha$, $^{12}C - ^{12}C$ and $^{16}O - ^{16}O$ scatterings, where only even states contribute to the cross section. Alam and Malik have modified the continued fraction method of inversion so that only the knowledge of even ℓ partial waves is needed for the construction of the potential. Following their formulation, we start with the Schrödinger equation but we change $\psi_\ell(r)$ to $\phi_\ell(r)$ not as in (16.7) but as

$$\phi_\ell(r) = (kr)^{-\left(\frac{\ell}{2}+1\right)} \psi_\ell(r), \qquad (16.46)$$

noting that $\chi_\ell(r)$ has the asymptotic form (16.36). Thus the resulting wave equation for $\phi_\ell(r)$ becomes

$$\phi_\ell''(r) + \frac{\ell+2}{r}\phi_\ell' + \left[k^2 - U(r) - \frac{\ell(3\ell+2)}{4r^2}\right]\phi_\ell(r) = 0. \qquad (16.47)$$

We can approximate this differential equation with the difference equation for ϕ as in Eq. (16.13), and we will suppress the ℓ dependence of ϕ_ℓ. The difference equation now becomes

$$\phi_{n+1} = A_n(\ell)B_n(\ell)\phi_n + C_n(\ell)\phi_{n-1}. \qquad (16.48)$$

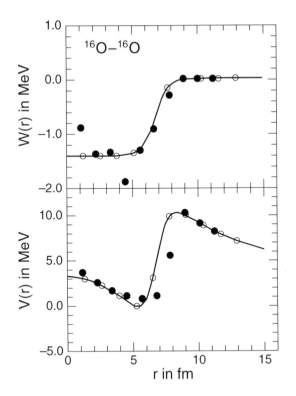

Figure 16.5: The complex potential $V(r) + iW(r)$ which gives a good fit to the $^{16}O\,{-}^{16}O$ scattering data at the energy of 20 MeV (lab) is shown by solid line. Open circles represent the points found by the continued fraction inverse scattering method with $Z_N(\ell)$ as the input. Solid circles are the points found from the knowledge of the phase shifts [22].

But now the coefficients $A_n(\ell)$, $B_n(\ell)$ and $C_n(\ell)$ are different from those given by Eqs. (16.14)–(16.16);

$$B_n(\ell) = \frac{2n}{2n + \ell + 2}, \tag{16.49}$$

$$C_n = \frac{\ell + 2 - 2n}{2n + \ell + 2}, \tag{16.50}$$

and

$$A_n(\ell) = 2 - \Delta^2 k^2 + \Delta^2 v_n + \frac{\ell(3\ell + 2)}{4n^2}. \tag{16.51}$$

Now we expand $\frac{\phi_{N+1}}{\phi_N}$ as a continued fraction just as we did in Eq. (16.19). For the physical values of ℓ, (i.e. for any even ℓ) the actual values of ϕ_0 and ϕ_1 do not enter in the calculation of $\frac{\phi_{N+1}}{\phi_N}$, since $C_n(\ell) = 0$, for $n = \frac{1}{2}(\ell + 2)$ (all ℓ s are even). Defining ℓ_n by

$$\ell_n = 2(N - n - 1), \quad \text{for} \quad n = 1, 2, \cdots N - 1, \tag{16.52}$$

we calculate

$$Z_N(\ell) = \frac{N}{2}\left(\frac{\phi_{N+1} - \phi_{N-1}}{\phi_N}\right). \tag{16.53}$$

By combining (16.48) and (16.53) we find

$$\frac{\phi_N(\ell)}{\phi_{N-1}(\ell)} = \frac{C_N(\ell) - 1}{\frac{2}{N}Z_N(\ell) - A_N(\ell)B_N(\ell)}. \tag{16.54}$$

Since $C_n(\ell)$ is zero for $2n = (\ell+2)$ and since n has to be an integer, ℓ must be even. Starting from (16.54) we can iterate inwards, noting that according to (16.50) and (16.52), $C_n(\ell_{N-n}) = 0$. Then from the continued fraction of $\frac{\phi_{N+1}}{\phi_N}$, i.e.

$$\frac{\phi_{N+1}}{\phi_N} = A_N(\ell)B_N(\ell) + \frac{C_N(\ell)}{A_{N-1}(\ell)B_{N-1}(\ell)+} \frac{C_{N-1}(\ell)}{A_{N-2}(\ell)B_{N-2}(\ell)+}$$

$$\cdots \frac{C_2(\ell)}{A_1(\ell)B_1(\ell) + C_1(\ell)/(\phi_1/\phi_0)}, \tag{16.55}$$

and Eq. (16.52) we find $A_{N-1}(\ell_1)$ to be

$$A_{N-1}(\ell_1) = B_{N-1}(\ell_1)\left(\frac{\phi_N(\ell_1)}{\phi_{N-1}(\ell_1)}\right). \tag{16.56}$$

Other $A_{N-j}(\ell_j)$ s are given by

$$A_{N-j}(\ell_j) = \frac{1}{B_{N-j}}\left[\frac{C_{N+\ell-j}}{-A_{N+1-j}(\ell_j)B_{N+1-j}+}\right.$$

$$\left.\cdots \frac{C_{N-1}(\ell_j)}{-A_{N-1}(\ell_j)B_{N-1}(\ell_j) + \left(\frac{\phi_N(\ell_j)}{\phi_{N-1}(\ell_j)}\right)}\right], \quad j = 2, 3, \cdots N - 1. \tag{16.57}$$

Once all $A_{N-j}(\ell_j)$ s are determined, we can find v_{N-j} and hence V_{N-j} from v_{N-j}:

$$v_{N-j} = \frac{1}{\Delta^2}\left[A_{N-j}(\ell_j) - 2 + \Delta^2 k^2 - \frac{\ell_j(3\ell_j + 2)}{4(N - j)\Delta}\right], \tag{16.58}$$

and

$$V_{N-j} = \frac{\hbar^2}{2\mu}v_{N-j}. \tag{16.59}$$

The results of inversion for $^{16}O - ^{16}O$ are shown in Fig. 16.5. Other results for $\alpha - \alpha$ and $^{12}Ca - ^{12}Ca$ scattering are given in Ref. [22], and the results are good.

As we have mentioned before, in this as well as other problems of fixed energy type using continued fraction expansion, the success of the method depends on how many partial waves are used for the input data. For large step sizes, replacing the differential equation by difference equation is not a good approximation. At the same time a large number of steps enhances the propagation of error. For optimal energy range, we have to choose energies where one has information about 8 to 20 partial waves are best suited for this type of expansion.

The inverse scattering theory at fixed energy that we have discussed in this chapter has been extended to the scattering of spinless particles at relativistic energies governed by the Klein-Gordon equation. As an application of this extension, Shehadeh *et al.* have determined the complex potential in the pion-^{40}Ca scattering and found acceptable results [23].

References

[1] H. Feshbach, Ann. Phys. (NY), 5, 357 (1958).

[2] L.L. Foldy and J.D. Walecka, Ann. Phys. (NY), 54, 447 (1969).

[3] P.E. Hodgson, *The Optical Model of Elastic Scattering*, (Oxford University Press, London, 1963), Chapter 2.

[4] E. Kujawski, Phys. Rev. C8, 100 (1973).

[5] K. Chadan and P.C. Sabatier, *Inverse Problems in Quantum Scattering Theory*, (Springer, New York, 177), Chapter 15.

[6] C. Coudray, Lett. Nuovo Cimento, 19, 319 (1977).

[7] Lipperheide and Fiedeldey, Z. Phys. A286, 45 (1978).

[8] Lipperheide and Fiedeldey, Z. Phys. A301, 81 (1981).

[9] S.C. Cooper and B.S. Mackintosh, Inverse Problems, 5, 707 (1989).

[10] S.C. Cooper and B.S. Mackintosh, Phys. Rev. C43, 1001 (1991).

[11] V.I. Kukulin and R.S. Mackintosh, J. Phys. G 30, R1 (2004).

[12] V.M. Kranopolsky, V.I. Kukulin, and J. Horáček, Czech. J. Phys. B 35, 805 (1985).

[13] M.A. Hooshyar and M. Razavy, Can. J. Phys. 59, 1627 (1981).

[14] M.A. Hooshyar, R. Nadeau, and M. Razavy, Phys. Rev C25, 1187 (1982).

[15] H.F. Lutz, J.B. Mason, and M.D. Karvelis, Nucl. Phys. 47, 521 (1963).

[16] A.J. Frasca, R.W. Finlay, R.D. Koshel and R.L. Cassola, Phys. Rev. 144, 854 (1966).

[17] R.D. Wood and D.S. Saxon, Phys. Rev. 95, 577 (1954).

[18] C.M. Perey and F.G. Perey, At. Data Nucl. Data Tables 17, 1 (1976).

[19] M.M. Alam and F.B. Malik, Phys. Lett. B 237, 14 (1990).

[20] M.L. Goldberger and K.M. Watson, *Collision Theory*, (John Wiley & Sons, New York, 1964) p. 261.

[21] L.D. Landau and E.M. Lifshitz, *Quantum Mechanics*, (Addison-Wesley, Reading, 1958) p. 126.

[22] M.M. Alam and F.B. Malik, Nucl. Phys. A524, 88 (1991).

[23] Z.F. Shehadeh, M.M. Alam and F.B. Malik, Phys. Rev. C59, 826 (1999).

[24] R. Plaga, H.W. Becker, A. Redder, C. Rolfs and H.P. Trautvetter, Nucl. Phys. A465, 291 (1987).

[25] H.V. von Geramb and H. Kohlhoff, in *Quantum Inversion Theory and Applications*, (Springer-Verlag, 1994) p. 266.

Chapter 17

Two Inverse Problems of Electrical Conductivity in Geophysics

17.1 Inverse Problem of Electrical Conductivity in One-Dimension

Among a number of different approaches to the inverse problem of the Earth's conductivity, the mathematical method developed by Gel'fand and Levitan and we discussed it in Chapter 4 seems to have received special attention [1]–[6]. This is partly due to the great success of this method in other fields, notably in quantum theory of scattering, and partly due to the adaptability to different sets of available information input. However in applying this method, one encounters certain difficulties and limitations, among them the most serious is the instability of the resulting profile against the variation of the initial data which is common to nearly all of the inverse problems. As an alternative, we can use Goupillaud's method, discussed in Sec. 13.4, and consider layered medium with layers of constant conductivity to formulate and solve the inverse conductivity problem.

As we will see in this and the next sections we can use similar formulation for the determination of conductivity in one spatial dimension or in a three-dimensional problem provided that in the latter case, we assume, that the conductivity is dependent only on the radial coordinate.

We will first consider the inverse problem which is of special interest to geophysicists namely the determination of the electrical conductivity of the Earth from the geomagnetic induction data when the total response (or impedance) of the medium at the surface is known as a function of frequency of the electromagnetic field [1]–[7]. The simplest case is a one-dimensional model where the electromagnetic wave is normally incident on a flat Earth [8]. For solving this inverse problem we make the following assumptions:

(1) That the electrical conductivity varies only with the depth z (positive z direction is taken to be downwards).

(2) That the permeability of the medium is the same as the permeability of the vacuum.

(3) That the displacement current can be ignored.

Now let us consider a plane electromagnetic wave normally incident on the flat earth with the time dependence of $e^{i\omega t}$. Then the horizontal electric and magnetic fields E and H (in the directions of y and x respectively) in the medium will satisfy the following equations:

$$\frac{d}{dz}H(z,\omega) = \sigma(z)E(z,\omega),\tag{17.1}$$

$$\frac{d}{dz}E(z,\omega) = i\omega\mu_0 H(z,\omega),\tag{17.2}$$

where ω is the angular frequency, μ_0 is the vacuum permeability and $\sigma(z)$ is the electric conductivity. By eliminating $H(z,\omega)$ between (17.1) and (17.2) we find a second order differential equation for $E(z,\omega)$ where $\sigma(z)$ appears as a coefficient of $E(z,\omega)$ similar to the potential function in the Schrödinger equation. We want to restrict the class of conductivities to those satisfying the integrability condition

$$\alpha = \int_0^{\bar{z}} \sqrt{\sigma(z)}dz < \infty,\tag{17.3}$$

where $\bar{z}$ is an arbitrary but finite depth z where at this depth we assume that $E(z,\omega)$ vanishes.

Next we define the impedance $Z(z,\omega)$ by

$$Z(z,\omega) = \frac{E(z,\omega)}{H(z,\omega)} = i\omega\mu_0\left(\frac{E(z,\omega)}{E'(z,\omega)}\right),\tag{17.4}$$

where prime denotes the derivative with respect to z. From Eqs. (17.1), (17.2) and (17.4) it follows that the impedance satisfies the Riccati equation [9]

$$\frac{dZ(z,\omega)}{dz} + \frac{1}{\rho(z)}Z^2(z,\omega) = i\omega\mu_0,\tag{17.5}$$

where

$$\rho(z) = \frac{1}{\sigma(z)},\tag{17.6}$$

is the resistivity of the medium. Also from (17.4) we find the boundary condition for $Z(z,\omega)$ to be

$$Z(\bar{z},\omega) = 0.\tag{17.7}$$

To solve the direct problem, knowing the conductivity $\sigma(z)$, we integrate (17.5) from $z = \bar{z}$ to $z = 0$ with the boundary condition (17.7) and we find $Z(0,\omega)$ at the surface of the Earth as a function of ω. For the inverse problem we want to find the conductivity $\sigma(z)$ in the range $0 \leq z \leq \bar{z}$ from the information on $Z(0, \omega)$ for different values of ω. Again we will use the method of continued fraction expansion to find $\sigma(z)$ at N points. For this purpose we divide the range $0 \leq z \leq \bar{z}$ into N finite layers of constant conductivity with equal wave attenuation in each layer [8]. Now within each layer, from Eqs. (17.1) and (17.2) for constant σ, we find the following differential equation

$$\frac{d^2 E_n(z)}{dz^n} = K_n^2 E_n(z), \quad z_n \leq z \leq z_{n+1}, \quad n = 1, \, 2 \cdots N, \tag{17.8}$$

where

$$K_n = \gamma \sqrt{\sigma_n}, \quad \gamma = (1 + i)\sqrt{\frac{\mu_0 \omega}{2}}, \quad z_1 = 0,$$

$$z_{n+1} = z_n + \frac{\lambda}{\sqrt{\sigma_n}}, \quad z_{N+1} = \bar{z}, \quad \sigma_n = \sigma(z_n), \tag{17.9}$$

and $\lambda = \frac{\alpha}{N}$ which is the constant attenuation and is the same for all layers. We also have the condition

$$E_N(\bar{z}) = 0. \tag{17.10}$$

The solution of the differential equation (17.8) for $z_n \leq z \leq z_{n+1}$ is given by

$$E_n(z) = C_n \exp[K_n(z - z_{n+1})] + D_n \exp[-K_n(z - z_{n+1})], \tag{17.11}$$

and this solution is subject to the boundary condition

$$C_N + D_N = 0. \tag{17.12}$$

Here the quantity of interest is Ω_n which is the ratio $\frac{C_n}{D_n}$ and is a function of z and ω:

$$\Omega_n = \frac{D_n}{C_n} = \frac{\sqrt{\sigma_n} Z_n(z, \omega) - \gamma}{\sqrt{\sigma_n} Z_n(z, \omega) + \gamma} \exp[2K_n(z - z_{n+1})], \tag{17.13}$$

where $Z_n(z, \omega)$ is the impedance at the point z which is located in the nth layer. Since $E_n(z, \omega)$ and its derivative are continuous functions of z we have

$$Z_n(z_{n+1}, \omega) = Z_{n+1}(z_{n+1}, \omega). \tag{17.14}$$

Equations (17.13) and (17.14) together give us the following recurrence relation from which Ω_n can be obtained

$$\Omega_n = \frac{1}{p_{n+1}} + \frac{p_{n+1} - p_{n+1}^{-1}}{1 + y p_{n+1} \Omega_{n+1}}, \tag{17.15}$$

where

$$p_n = \frac{\sqrt{\sigma_{n-1}} - \sqrt{\sigma_n}}{\sqrt{\sigma_{n-1}} + \sqrt{\sigma_n}}, \tag{17.16}$$

and

$$y = e^{2\gamma\lambda} = \exp[2K_n(z_{n+1} - z_n)]. \tag{17.17}$$

Since the layers are chosen in such a way that y is constant for all the layers, therefore y is independent of n, and is only a function of ω. Now let us define a quantity $G(\omega)$ by the following relation

$$G(\omega) = \frac{\sqrt{\sigma(0)}Z(0,\omega) - \gamma}{\sqrt{\sigma(0)}Z(0,\omega) + \gamma}. \tag{17.18}$$

We note that by definition $G(\omega)$ is a measurable quantity since it only depends on the values of the conductivity and the impedance both on the surface of the Earth. We also note that for large N, $y\Omega$ will tend to $G(\omega)$. Therefore using (17.15) we can write $G(\omega)$ as a continued fraction

$$G(\omega) = y\Omega_1 = yc_1 + \frac{yd_1}{1 + yc_2 +}\frac{yd_2}{1 + yc_3 +}\cdots\frac{yd_{N=2}}{1 + yc_{N-1} +}\frac{yd_{N-1}}{1 + yp_N\Omega_N}, \tag{17.19}$$

where

$$c_n = \frac{p_n}{p_{n+1}}, \tag{17.20}$$

$$d_n = p_n p_{n+1} - c_n, \tag{17.21}$$

$$p_1 = 1, \quad \text{and} \quad \Omega_N = -1. \tag{17.22}$$

From Eq. (17.19) it follows that we can write $G(\omega)$ as the ratio of two polynomials;

$$G(\omega) = \frac{P_N(y)}{Q_{N-1}(y)} = \frac{-y^N + A_{N-1}y^{N-1} + \cdots A_1 y}{B_{N-1}y^{N-1} + \cdots B_1 y + 1}. \tag{17.23}$$

We find a remarkable relation between A_n and B_n by considering the relation between the inverses of the quantities that we have introduced so far. Let us define $\bar{G} = G^{-1}$, $\bar{y} = y^{-1}$, and $\bar{\Omega}_n = \Omega_n^{-1}$. From (17.15) we conclude that

$$\bar{\Omega}_n = p_{n+1}^{-1} + \frac{p_{n+1} - p_{n+1}^{-1}}{1 + \bar{y}p_{n+1}\bar{\Omega}_{n+1}}. \tag{17.24}$$

Furthermore

$$\bar{G} = \bar{y}\bar{\Omega}_1 = \bar{y}c_1 + \frac{\bar{y}d_1}{1 + \bar{y}c_2 +}\frac{\bar{y}d_2}{1 + \bar{y}c_3 +}\cdots\frac{\bar{y}d_{N-1}}{1 + \bar{y}p_n\bar{\Omega}_N}. \tag{17.25}$$

By comparing (17.25) with (17.19) and noting that $\bar{\Omega}_N = \Omega_N = -1$, we conclude that $\bar{G}(\omega)$ can be represented as the ratio of two polynomials in $\bar{y}$:

$$\bar{G}(\omega) = \frac{P_N(\bar{y})}{Q_{N-1}(\bar{y})} = \frac{-\bar{y}^{(N)} + A_{N-1}\bar{y}^{(N-1)} + \cdots A_1\bar{y}}{B_{N-1}\bar{y}^{(N-1)} + \cdots B_1\bar{y} + 1}. \tag{17.26}$$

If we compare (17.26) and (17.23), and note that $G = G^{-1}$, we find that

$$A_j = -B_{N-j}, \quad j = 1, 2, \cdots N - 1. \tag{17.27}$$

From this relation we can obtain $G(\omega)$, Eq. (17.23), if the values of G for $N-1$ values of ω are given. In that case (17.17) is valid for different values of y_n;

$$y_n = \exp(2\gamma_n\lambda),\qquad(17.28)$$

where

$$\gamma_n = (1+i)\sqrt{\frac{\mu_0\omega_0}{2}}.\qquad(17.29)$$

It then follows from (17.26) that

$$\sum_{m=1}^{N-1} \bar{K}_{n,m}B_m = \bar{F}_n,\qquad(17.30)$$

where

$$\bar{F}_n = -y_N^{N-1} - \frac{1}{y_n}G(\omega_n),\qquad(17.31)$$

and

$$\bar{K}_{n,m} = y_n^{N-1-m} + G(\omega_n)y^{m-1}.\qquad(17.32)$$

Noting that all B_n s are real we can simplify the numerical calculation by only working with the real (or imaginary) part of (17.30), i.e.

$$K\mathbf{B} = \mathbf{F},\qquad(17.33)$$

where the elements of the matrix K and the vector $\mathbf{F}$ are given by

$$K_{nm} = \mathrm{Re}\,(\bar{K}_{n,m}),\qquad(17.34)$$

and

$$F_n = \mathrm{Re}\,(\bar{F}_n).\qquad(17.35)$$

By solving Eq. (17.33) for the polynomial constants B_n we find

$$\mathbf{B} = \bar{K}^{-1}\mathbf{F}.\qquad(17.36)$$

Once B_n s are known we can calculate A_n from (17.27) and determine the polynomials $P_n(y)$ and $Q_n(y)$ and their ratio $G(\omega)$ from (17.23). From the rational fraction representation of G in the form of continued fraction (17.19) can be found (see Appendix A) [10]. From this continued fraction representation the coefficients c_n, p_n and d_n can be successively obtained and therefore the conductivity of each layer can be specified. Thus we have a complete technique for the inversion of electrical conductivity of the one-dimensional problem [8].

Let us summarize the steps needed for this method of inversion. The input consists of the information on the impedance for $N-1$ different values of angular frequency (or the corresponding periods T_n) and the conductivity at the surface of the Earth plus the total attenuation factor defined in (17.3). From these information, the quantities $G(\omega_n)$ and y_n, Eqs. (17.18) and (17.28) are found and used

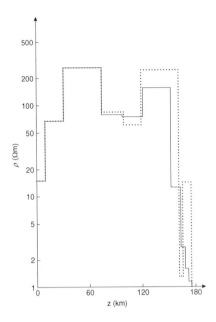

Figure 17.1: A resistivity profile for stratified medium of constant attenuation shown by solid line. This resistivity profile is used as a model for computing the input data (the response function). The dotted line indicates the result obtained from inversion. In this calculation the following parameters have been used. Number of layers $N = 10$, lowest period $T_0 = 10^3$ s and the maximum depth $\bar{z} = 174.4$ km.

in defining the matrix K and the vector $\mathbf{F}$, Eqs. (17.34) and (17.35). Next the matrix K is inverted and the coefficients of $P_N(y)$ and $Q_{N-1}(y)$ are determined. From these, a continued fraction representation of $G(\omega)$ is obtained. Knowing $G(\omega)$ enables us to find the resistivity of the medium.

Figures 17.1–17.3 show the input and the output for three different profiles of resistivity. The solid lines in all three cases show the input, with the analytic form of the resistivity given, and the dotted lines are the numerical results obtained from inversion. In all these calculations the periods used are determined from the relation

$$T_n = T_0 \times 10^{\frac{2(n-1)}{N-1}}, \quad n = 1, \, 2, \cdots N, \tag{17.37}$$

where N is the number of layers, and T_0 is some starting value for the period. Now for a medium of arbitrary resistivity this method enables us to find an approximate profile. As we expect for larger number of layers, N, our approximation improves up to a point. We again note that we can use the same method to solve the direct problem, i.e. if we are given $\sigma(z)$ or $\rho(z)$, we can get the solution of (17.7) through the continued fraction representation (17.19). Thus we find that for a wide range of frequencies and profiles the agreement up to four significant figures could easily be achieved, if we choose the number of layers to be about 500. By calculating the impedance on the surface for different conductivity profiles and using these as input data, shown in the figures by solid line, and subsequent inversion we obtain

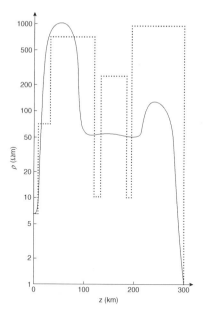

Figure 17.2: In this figure a different profile for resistivity is shown by solid line. This is a plot of $\rho(z) = 6 - 0.02z + 50\exp\left[-\left(\frac{z-150}{100}\right)^4\right] + 100\exp\left[-\left(\frac{z-250}{30}\right)^4\right] + 1000\exp\left[-\left(\frac{z-60}{30}\right)^4\right]$ which is shown with the predicted profile obtained from continued fraction expansion (dotted line) ($N = 13$, $T_0 = 10^3$ s, $\bar{z} = 300$ km).

the output shown by dotted lines. As we can see in each case the agreement between the input data and the final output is better for the layers closer to the surface, and as we look at the deeper depths the results become less reliable. However we can see certain qualitative features such as the increase or decrease in resistivity of the profiles in these examples. Obviously if we use more accurate method of numerical computation and reducing the round off errors we can get better results.

17.2   The Inverse Problem of Geomagnetic Induction at a Fixed Frequency

A second inverse problem for the determination of the electrical conductivity of the Earth which is of interest in geophysics is the one where the components of the response function are given on the basis of spherical harmonics at fixed frequency. Thus this problem is analogous to the quantum mechanical problem of finding the potential if all the phase shifts for a given energy are known (Chapter 6). For solving the geomagnetic induction problem one can, in principle, use the Newton–Sabatier method discussed in that chapter, and as we observed there this method works better at higher energies (or frequencies). For lower frequencies, we can

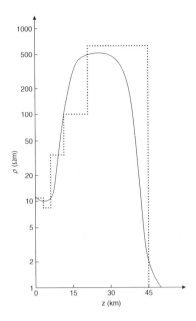

Figure 17.3: A model profile with high resistivity in the intermediate region, $\rho(z) = 11 - 0.2z + 500 \exp\left[-\left(\frac{z-25}{12}\right)^4\right]$, is compared with the predicted layers of constant attenuation shown by dotted line ($N = 11$, $T_0 = 100$ s and $\bar{z} = 50$ km).

follow closely the continued fraction technique which was successfully employed for the inverse problem of quantum scattering theory.

Let us consider a spherical Earth and let a and σ denote the radius and the conductivity of the Earth respectively. In discussing this problem we assume that σ is finite everywhere and is a function of the radial distance r only. The lateral variations of σ which persists to a depth of about 400 km can be included in the calculation, but for the sake of simplicity here we ignore these variations [12]. Now following the work of Lahiri and Price, [13] for the formulation of this problem we choose spherical polar coordinates (r, θ, ϕ) with its origin at the center of the Earth, and we assume that (a) the source field varies as $e^{i\omega t}$ and (b) that the displacement current is negligible. With these assumptions we can express the magnetic field outside the Earth in terms of a scalar potential Ω:

$$\mathbf{B} = -\nabla\Omega, \quad r \geq a, \tag{17.38}$$

where

$$\Omega(r, \theta, \phi) = \sum_{\ell=1}^{\infty} \sum_{m=-\ell}^{\ell} \left[\mathcal{E}_{\ell m} r^\ell + \frac{\mathcal{J}_{\ell m}}{r^{\ell+1}}\right] Y_\ell^m(\theta, \phi), \tag{17.39}$$

and $Y_\ell^m(\theta, \phi)$ are the spherical harmonics. The electric and the magnetic fields inside the Earth satisfy the following equations

$$\nabla \times \nabla \times \mathbf{E} - i\omega\mu_0\sigma(r)\mathbf{E} = 0, \quad r < a, \tag{17.40}$$

$$\mathbf{B} = \frac{1}{i\omega} \nabla \times \mathbf{E}. \tag{17.41}$$

We can write the electric field as

$$\mathbf{E} = \nabla \times (\mathbf{r}\psi(\mathbf{r})), \tag{17.42}$$

where

$$\psi(\mathbf{r}) = \sum_{\ell=1}^{\infty} \sum_{m=-\ell}^{\ell} R_{\ell m}(r) Y_\ell^m(\theta, \phi). \tag{17.43}$$

In this relation $R_{\ell m}$ is the regular solution of the differential equation

$$\frac{d}{dr}\left(r^2 \frac{d}{dr} R_{\ell m}(r)\right) + \left[i\omega\mu_0\sigma(r)r^2 - \ell(\ell+1)\right] R_{\ell m}(r) = 0 \tag{17.44}$$

with the boundary condition

$$\left(\frac{R_{\ell m}}{r^\ell}\right) \to c_{\ell m}, \quad \text{as} \quad r \to 0. \tag{17.45}$$

From the continuity of the fields at $r = 0$, it follows that $c_{\ell m}$ are dependent on the coefficients $\mathcal{E}_{\ell m}$ and $\mathcal{J}_{\ell m}$ in (17.39). Now the solution of (17.44) except for a multiplying constant does not depend on m. Thus by matching the magnetic fields inside and outside of the surface of Earth we find that

$$\left(\frac{\mathcal{E}_{\ell m}}{\mathcal{J}_{\ell m}}\right)^{-1} = \frac{\ell\left(aR'_{\ell m} - \ell R_{\ell m}\right)}{(\ell+1)\left[(aR'_{\ell m} + (\ell+1)R_{\ell m})\right]} = \left(\frac{\mathcal{E}_\ell}{\mathcal{J}_\ell}\right)^{-1}, \tag{17.46}$$

is independent of m. In (17.46) prime denotes derivatives with respect to r. To simplify (17.44) further, we change $R_{\ell m}(r)$ to $\phi_\ell(r)$

$$\phi_\ell(r) = \frac{R_{\ell m}(r)}{U_{\ell m}(r)}, \quad U_{\ell m}(r) = c_{\ell m} r^\ell, \tag{17.47}$$

where $U_{\ell m}(r)$ is the regular solution of (17.44) with $\sigma(r) = 0$. This new function, $\phi_\ell(r)$, is independent of m and satisfies the differential equation

$$\phi''_\ell(r) + \frac{2(\ell+1)}{r}\phi'_\ell(r) + i\omega\mu_0\sigma(r)\phi_\ell(r) = 0, \tag{17.48}$$

with the boundary condition

$$\phi_\ell(0) = 1. \tag{17.49}$$

As a number of other problems that we have studied, we transform (17.48) to a Riccati equation by replacing $\phi_\ell(r)$ by $Z_\ell(r)$:

$$Z_\ell(r) = r\left(\frac{\phi'_\ell(r)}{\phi_\ell(r)}\right), \tag{17.50}$$

where the function $Z_\ell(r)$ is a solution of

$$Z_\ell'(r) = -\frac{1}{r}Z_\ell(r)(Z_\ell(r) + 2\ell + 1) - i\omega\mu_0 r\sigma(r), \tag{17.51}$$

with

$$Z_\ell(0) = 0. \tag{17.52}$$

At the surface of the Earth, $r = a$, we have

$$Z_\ell(a) = (\ell + 1)\left(\frac{\ell}{W_\ell} - 1\right), \quad \ell = 1, 2\cdots, \tag{17.53}$$

where W_ℓ is the response function [14]

$$W_\ell = \frac{\ell\mathcal{E}_{\ell m} - (\ell + 1)\mathcal{J}_{\ell m}}{\mathcal{E}_{\ell m} + \mathcal{J}_{\ell,m}}. \tag{17.54}$$

Now we replace the differential equation (17.48) by the difference equation

$$\phi_{n+1} = A_n B_n(\ell)\phi_n + C_n(\ell)\phi_{n-1}, \quad n = 1, 2, \cdots N, \tag{17.55}$$

i.e. we have divided the radius r into N equal parts each of length $\Delta = \frac{a}{N}$, as we have done before. In writing Eq. (17.55) we have introduced the following symbols:

$$\phi_n = \phi_\ell(n\Delta), \tag{17.56}$$

$$A_n = 2 - i\mu_0\omega\Delta^2\sigma_n, \tag{17.57}$$

$$B_n = \frac{n}{\ell + 1 + n}, \tag{17.58}$$

$$C_n = \frac{\ell + 1 - n}{\ell + 1 + n}, \tag{17.59}$$

and

$$\sigma_n = \sigma(n\Delta). \tag{17.60}$$

Equation (17.55) is a difference equation similar to (16.13). Therefore we follow the steps given after (16.13), but now B_n and A_n in Eqs. (16.15) and (16.16) must be replaced by (17.57) and (17.58). Note that in Eq. (17.55) we have suppressed the dependence of ϕ_n on ℓ. To find the discrete form of $Z_\ell(r)$, using (17.50), we have

$$Z_{\ell,n} = \frac{n(\phi_{n+1} - \phi_{n-1})}{2\phi_n}. \tag{17.61}$$

Let us again emphasize that the replacement of (17.48) by (17.55) is not unique and there are difference equation with faster convergence rate than (17.55). However the difference equation (17.55) lends itself in a natural way to the solution of the inverse problem.

For the solution of the direct problem, iteration can be used to obtain the response function. Thus from (17.55) and (17.61) we find $Z_{\ell,N}$ to be

$$Z_{\ell,N} = \frac{1}{2}N\left[A_N B_N(\ell) + \frac{C_N - 1}{\left(\frac{\phi_N}{\phi_{N-1}}\right)}\right],\tag{17.62}$$

where $\left(\frac{\phi_N}{\phi_{N-1}}\right)$ is given by the continued fraction expansion of (17.55):

$$\frac{\phi_N}{\phi_{N-1}} = A_{N-1}B_{N-1}(\ell) + \frac{C_N - 1}{A_{N-2}B_{N-2}(\ell)+} \cdots$$
$$\cdots \frac{C_3(\ell)}{A_2 B_2(\ell)+} \frac{C_2(\ell)}{A_1 B_1(\ell)\left[\frac{C_1(\ell)}{\phi_1}\right]}.\tag{17.63}$$

Once σ_n s are given, then all of the other terms in the right-hand side of (17.63) are known and therefore $Z_{\ell,N}$ can be calculated. Let us note that in this formulation $Z_{\ell,N}$ is independent of the value of ϕ_1 for the physical values of ℓ. This follows from the fact that $C_m(\ell) = 0$, for $m = \ell + 1$.

If we test this approximation for large values of N, say $N = 500$, we find that $Z_{\ell,N}$ is in agreement with the exact $Z_\ell(a)$ found by integrating (17.51) to at least three significant figures.

Solution of the Inverse Problem — The great advantage of working with Eq. (17.55) is that the corresponding response function has a Thiele type expansion (see Appendix B) and we can find the coefficients of the expansion provided that $Z_{\ell,N}$ s are known for $N - 1$ consecutive values of ℓ [15]. Since W_n s at fixed frequency are found from measurements, the function $Z_\ell(a) = Z_{\ell,N}$ can be obtained from (17.53). Therefore for the input we assume that $Z_{\ell_n,N}$ s are known, and in addition the conductivity at $r = a$ is given. Then by inverting (17.62) we find

$$\frac{\phi_N(\ell)}{\phi_{N-1}(\ell)} = \frac{C_N(\ell) - 1}{\frac{2}{N}Z_{\ell,N} - A_N B_N(\ell)},\tag{17.64}$$

where we have written the dependence of the left-hand side on ℓ explicitly. We also note that $C_n(\ell_{N-n}) = 0$ and $\left(\frac{\phi_N \ell_n}{\phi_{N-1}(\ell_n)}\right)$ are known quantities, therefore from (17.63) it follows that

$$A_{N-1} = \frac{\phi_N(\ell_1)}{\phi_{N-1}(\ell_1)B_{N-1}(\ell_1)},\tag{17.65}$$

and

$$A_{N-j} = \frac{1}{B_{N-j}(\ell_j)}\left[\frac{C_{N+1-j}(\ell_j)}{-A_{N+1-j}(\ell_j)B_{N+1-j}(\ell_j)+}\cdots\right.$$
$$\left.\cdots\frac{C_{N-2}(\ell_j)}{-A_{N-2}(\ell_j)B_{N-2}(\ell_j)+} \frac{C_{N-1}(\ell_j)}{-A_{N-1}(\ell_j)B_{N-1}(\ell_j) + \frac{\phi_N(\ell_j)}{\phi_{N-1}(\ell_j)}}\right]$$
$$j = 2, 3, \cdots N - 1.\tag{17.66}$$

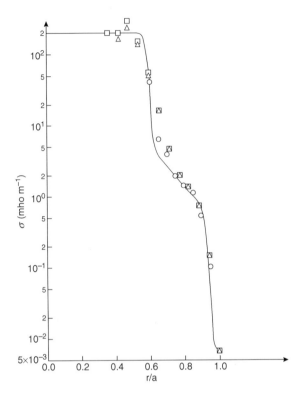

Figure 17.4: The Banks and Parker model for Earth's conductivity with the core conductivity of 200 mho m^{-1} which is displayed by solid line. This model is used for the input. The results of inversion for several periods are shown in this graph for ($\square$ = 30 yr, N = 17), ($\triangle$ = 11.2 yr, N = 17) and ($\circ$ = 3 yr, N = 20).

Thus from the set of ratios $\frac{\phi_N(\ell_n)}{\phi_{N-1}(\ell_n)}$ for $n = 1,\ 2 \cdots N-1$ we can find the set of A_n s by iteration. From these A_n s and (17.57) we obtain σ_n:

$$\sigma_n = \frac{1}{\omega\mu_0\Delta^2}\text{Im}\,(2 - A_n).\tag{17.67}$$

As in other direct problems involving continued fraction expansion, as $N \to \infty$, σ_n converges to the actual conductivity.

Accuracy of the Results of Inversion — To test the accuracy of this method we consider the analytical model of the Earth conductivity suggested by Banks and by Parker [4], [14]

$$\sigma_{BP}(\rho) = 193\exp\left\{-\left[\frac{1}{0.04}(\rho - \rho_c)\right]^2\right\},$$

$$+ \exp\left\{5.39 - 6.3\rho - 4.09\exp\left[-\frac{1}{0.06}(1-\rho)\right]^3\right\},\quad \rho \geq \rho_c\tag{17.68}$$

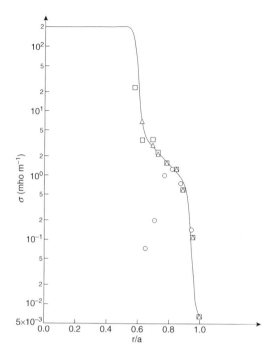

Figure 17.5: The conductivity model here is also the Banks and Parker model, (shown by solid line) which is used for the input. But now the calculation is carried out for shorter periods ($\square = 1$ yr, $N = 19$), ($\triangle = 0.5$ yr, $N = 19$) and ($\circ = 27$ days, $N = 17$).

where

$$\rho_c = 0.546, \quad \text{and} \quad \rho = \frac{r}{a}. \tag{17.69}$$

For the complete set of input data, $\sigma(\rho)$ has to be defined for all values of ρ, but due to the unavailability of information about the conductivity of the core, we choose this conductivity to be a constant of about 10^5 mho m^{-1}.

Now let us briefly review the result of inversion. For an analysis of the noisy data and how the method predicts the actual conductivity, the reader is referred to Ref. [11]. Here we want to test the accuracy of this inversion technique for the case of error-free data. Therefore we use $N \leq 20$ and assume the conductivity profile of Banks and Parker $\sigma(r) = \sigma_{BP}(r)$ which is consistent with the existing partial information about Earth's conductivity. Having $\sigma(r)$ we calculate $Z_\ell(r)$, Eq. (17.51), for different frequencies in the range where the measurement of Earth's response function may be experimentally feasible. Then we use these calculated response functions as the input data for inversion. The results of these calculations are shown in Figs. 17.4–17.6, and from these we can see that except for the external source of 11.2 and 30 yr periods the results of inversion are comparable when different values of the core conductivity are used. On the average the core has reduced the probing ability of the inversion by only 100 km. Let us also note that for high enough periods of the order of 100 yr, the inversion method is capable of probing the conductivity associated with Fig. 17.4 all the way to the center. To test the present method with another conductivity profile see Ref. [11]. Among different profiles that has been tested we find that the ones with smaller overall conductivity are better reproduced in inversion [11]. The outcome of the calculation depends not only on the choice of N, but on the fixed frequency for which the response is given. For certain values of N and ω, on account of the accumulation of errors, there is the possibility of getting negative or even complex values for the conductivity at certain points, these unphysical cases are discarded. If we choose N to be between 16 and 21, we find that the resulting conductivity profiles are close to each other, but for clarity of displaying the results in the figures, different N values for different frequencies have been used in the calculation of the conductivity [11]. In general, the result of the inversion of data for higher periods agree better with the input for the upper mantle as can be seen in Figs. 17.4–17.5.

Let us again emphasize that the numerical computation presented here are based on using the exact data, and there has been no attempt to predict the conductivity of the Earth from actual, observed noisy data. The main reason is that in this approach one requires the knowledge of the response function at a fixed frequency for several harmonics which is not available, and this is the major drawback of the present method.

References

[1] For a review of this inverse problem see the text: *Inversion of Magnetotelluric Data for a One-Dimensional Conductivity* by K.P. Whittal and D.W. Oldenburg (Society of Exploration Geophysicists, 1992).

[2] R.C. Bailey, Proc. R. Soc. A, 315, 185 (1970).

[3] I.M. Johnson and D.E. Smylie, Geophys. J.R. astr. Soc. 22, 41 (1970).

[4] R.L. Parker, Geophys. J.R. astr. Soc. 22, 121 (1970).

[5] R.L. Parker and K.A. Whaler, J. Geophs. Res. 22, 9574 (1981).

[6] P. Weidelt, Z. Geophys. 38, 257 (1972).

[7] T.T. Wu, Geophys. 33. 972 (1868).

[8] M.A. Hooshyar and M. Razavy, Geophys. J.R. astr. Soc. 71, 127 (1982).

[9] D.H. Eckardt, J. Geophys. Res. 73, 5317 (1968).

[10] H.S. Wall, *Analytic Theory of Continued Fraction*, (van Nostrand, New York, 1948).

[11] M.A. Hooshyar and M. Razavy, Geophys. J.R. astr. Soc. 71, 139 (1982).

[12] T. Rikkitake, *Electromagnetism and the Earth's Interior*, (Elsevier, Amsterdam, 1966).

[13] B.N. Lahiri and A.T. Price, Phil. Trans. R. Soc. A, 237, 509 (1939).

[14] R.J. Banks, Geophys. J. R. astro. Soc. 17, 457 (1969).

[15] L.M. Milne-Thomson, *The Calculus of Finite Differences*, Macmillan, New York (1951).

Chapter 18

Determination of the Mass Density Distribution Inside or on the Surface of a Body from the Measurement of the External Potential

Among the classical problems of geophysics is the problem of determining the density distribution within a closed surface (e.g. Earth) which is responsible for the observed gravitational potential outside this surface. One of the most influential papers on the subject of the geophysics of the Earth is the work authored by Bakus and Gilbert [1] where a detailed exposition of the direct as well as the inverse problem of the gravitational potential is discussed. This detailed study is outside the scope of this book. Since we are just interested in the inverse problem of gravitational potential we will mention some other works on the inverse problem.

A number of mathematically solvable versions of this inverse problem have been examined and solved by Tarakanov and collaborators, but these, while mathematically interesting, are not applicable to the gravitational problem of the Earth [2], [3]. On the question of the equivalence of the gravitational potentials produced by masses of varying density the work of Balk [4] is a useful work, and so is the work of Meshcheryakov [5] on the shape of body density and its relation to the external potential. On the conditions for determining the density distribution of a vibrating body the contribution of Gottlieb [6] is important. However, in this book, we will confine our attention to the question of finding the density distribution $\rho(r, \theta, \phi)$ within a spherical body of radius R from the measurement of the gravitational potential $V(r, \theta, \phi)$, $r \geq R$.

The law of universal gravitation states that the external potential field produced by a distribution of mass with the density $\rho(\mathbf{r})$ within the Earth (radius R)
is given by

$$V(\mathbf{r}) = G \int_{\mathcal{V}} \frac{\rho(\mathbf{r}_0)}{|\mathbf{r} - \mathbf{r}_0|} d^3 r_0. \tag{18.1}$$

In this relation $\mathcal{V}$ is Earth's volume, G is the gravitational constant and $d^3 r_0 = r_0^2 dr_0 \sin\theta_0 d\theta_0 d\phi_0$ is the element of the volume in spherical polar coordinates. Let us assume that we can find $V(\mathbf{r})$ empirically, the question is that whether it is possible to calculate $\rho(\mathbf{r}_0)$, i.e. the mass density distribution within the Earth? Since $r > r_0$, it is natural to expand $|r - r_0|^{-1}$ as a power series in $\left(\frac{r_0}{r}\right)^l$, i.e.

$$\frac{1}{|\mathbf{r} - \mathbf{r}_0|} = \sum_{l=0}^{\infty} \sum_{m=-l}^{l} \frac{1}{(2l+1)} \left[\int \int \int \left(\frac{r_0^l}{r^{l+1}}\right) \rho(\mathbf{r}_0) Y_{lm}(\Omega_0) d^3 r \right] Y_{lm}^*(\Omega), \tag{18.2}$$

where Y_{lm} is the spherical harmonics of degree l and order m i.e. $Y_{lm}^e = \cos(m\phi) P_l^m(\cos\theta)$ and $Y_{lm}^o = \sin(m\phi) P_l^m(\cos\theta)$. By substituting (18.1) in (18.2) we find a multipole expansion of $V(\mathbf{r})$ [7]

$$V(\mathbf{r}) = G \sum_{l=0}^{\infty} \sum_{m=-l}^{l} \frac{1}{(2l+1) r^{l+1}} \left[\int \int \int \rho(\mathbf{r}_0) r_0^l Y_{lm}(\Omega_0) d^3 r_0 \right] Y_{lm}^*(\Omega). \tag{18.3}$$

We can write the potential (18.3) also in the following form [7]:

$$V(r, \theta, \phi) = \sum_{l=0}^{\infty} \sum_{m=0}^{l} \frac{1}{r^{l+1}} \left[A_{lm} Y_{lm}^e(\theta, \phi) + B_{lm} Y_{lm}^o(\theta, \phi) \right], \tag{18.4}$$

where

$$A_{lm} = \epsilon_m \frac{(l-m)!}{(l+m)!} \int_0^{2\pi} \cos(m\phi_0) d\phi_0 \int_0^{\pi} P_l^m(\cos\theta_0) \sin\theta_0 d\theta_0$$

$$\times \int_0^R \rho(r_0, \theta_0, \phi_0) r_0^{l+2} dr_0. \tag{18.5}$$

The coefficients B_{lm} are similar volume integrals as A_{lm} with $\sin(m\phi_0)$ being replaced by $\cos(m\phi_0)$, and with $B_{l0} = 0$. In (18.5) ϵ_m is the Neumann factor $\epsilon_0 = 1$,
and $\epsilon_m = 2$ for $m = 1, 2, \cdots$ [7].

Since $V(\mathbf{r})$ satisfies the Laplace equation exterior to $\mathcal{V}$, we can also expand $V(\mathbf{r})$ in terms of $\left(\frac{R}{r}\right)^{l+1}$, i.e.

$$V(\mathbf{r}) = \frac{GM}{R} \sum_{l=0}^{\infty} \sum_{m=-l}^{l} \left(\frac{R}{r}\right)^{l+1} P_l^m(\cos\theta) \left[C_{lm}\cos(m\phi) + S_{lm}\sin(m\phi)\right], \quad (18.6)$$

where M is the total mass of the Earth and P_l^m is the associated Legendre polynomial, $Y_{lm}(\Omega) = P_l^m \cos(\theta)e^{im\phi}$, and the two dimensionless coefficients C_{lm} and S_{lm} are the Stokes coefficients of degree n and order m, $l = 0, 1\ 2\cdots\infty$ and $m = 0, 1, 2\cdots, l$. From Eqs. (18.3) and (18.6) it follow that

$$C_{lm} + iS_{lm} = \frac{1}{(2l+1)MR^l} \int\int\int_{\mathcal{V}} \rho(\mathbf{r}_0) r_0^l Y_{lm}(\Omega_0) d^3 r_0, \quad (18.7)$$

i.e. the Stokes coefficients are the normalized multipoles of the density distribution of the body. Equation (18.7) shows that if the infinite set of Stokes coefficiens are known, in principle one should be able to invert this equation and find $\rho(\mathbf{r}_0)$. That this inverse problem does not yield a unique solution can be seen from the fact that the set $r^l Y_{lm}(\Omega)$ does not form a complete set, and any function orthogonal to it can be added to the solution without affecting the Stokes coefficients [8].

To demonstrate this point, let us consider the expansion of $|\mathbf{r} - \mathbf{r}_0|^{-1}$ in rectangular coordinates:

$$\frac{1}{|\mathbf{r} - \mathbf{r}_0|} = \sum_{n=0}^{\infty} \left[\frac{(-1)^n}{n!}\right] \left[x_0 \frac{\partial}{\partial x} + y_0 \frac{\partial}{\partial y} + z_0 \frac{\partial}{\partial z}\right]^n \left(\frac{1}{|\mathbf{r}|}\right), \quad (18.8)$$

where $r^2 = x^2 + y^2 + z^2$. This is an expansion of $\frac{1}{|\mathbf{r}-\mathbf{r}_0|}$ into a multipole series each of which is the product of $(x_0^\lambda y_0^\mu z_0^\nu)$ and

$$\frac{\partial^{\lambda+\mu+\nu}}{\partial x^\lambda \partial y^\mu \partial z^\nu} \left(\frac{1}{r}\right),$$

and a numerical factor. The first of these factors multiplied by $\rho(x_0 y_0 z_0)$ is integrated over the interior of a sphere of volume $\mathcal{V}$ to give the final numerical constant in front of the function of x, y and z.

For an explicit determination of the form of the resulting integrals we use the multinomial theorem, viz,

$$(\alpha + \beta + \gamma)^n = \sum_{\lambda,\mu,\nu} \frac{n!}{\lambda!\mu!\nu!} \alpha^\lambda \beta^\mu \gamma^\nu. \quad (18.9)$$

In this relation the sum is over all different values of λ, μ, ν for which $\lambda+\mu+\nu = n$. From this multinomial expansion we find

$$V(x,y,z) = \sum_{n=0}^{\infty} \sum_{l=0}^{\infty} \sum_{k=0}^{\infty} \frac{(-1)^n}{l!k!(n-l-k)!}$$

$$\times \left[\int\int\int x_0^l y_0^k z_0^{n-l-k} \rho(x_0 y_0 z_0) dx_0 \, dy_0 \, dz_0 \right]$$

$$\times \frac{\partial^n}{\partial x^l \partial y^k \partial z^{n-k-l}} \left(\frac{1}{r} \right), \quad r > R, \tag{18.10}$$

where $\rho(\mathbf{r}_0)$ and all its derivatives are zero on and outside of radius R.

An important consequence of the expansion of $V(x,y,z)$ is that we can get some idea about the mass distribution $\rho(x_0 y_0 z_0)$ inside the sphere of radius R, by measurements of the potential outside the sphere, provided that we know about the magnitudes of the following components [7]:

$$V_{l,k,n-k-l}(x,y,z) = \frac{(-1)^n}{l!k!(n-l-k)!}$$

$$\times \int\int\int x_0^l y_0^k z_0^{n-l-k} \left[\rho(x_0 y_0 z_0) \right] dx_0 dy_0 dz_0, \tag{18.11}$$

or by the other form:

$$V_{l,k,n-k-l}(x,y,z) = \int\int\int \frac{\partial^n}{\partial x_0^l \partial y_0^k \partial z_0^{n-l-k}} \left[\rho(x_0 y_0 z_0) \right] dx_0 dy_0 dz_0. \tag{18.12}$$

We note that we can distribute $\rho(x_0 y_0 z_0)$ inside the sphere in any way that we want as long as the values of the integrals (18.11) or (18.12) remain unchanged. But as (18.11) shows, if we require that $\rho(\mathbf{r}_0)$ be expressible as a convergent power series in x_0, y_0, z_0 within the sphere, then no redistribution is possible at all, because the coefficients of power series are fixed when we fix the values of the integrals.

Let us compare the two expressions, Eqs. (18.6) and (18.10). We observe that in the Taylor series expansion since the form $x_0^l y_0^k z_0^{n-l-k}$ appears there are $1+2+3\cdots(n+1) = \frac{1}{2}(n+1)(n+2)$ terms of the nth order. On the other hand the spherical harmonic series has $(2n+1)$ terms of the nth order. Thus in both cases one has one zeroth order and three first order terms. For the second order terms, Taylor's series has six and the spherical harmonic has five. We can say that either Taylor's series has a redundant term, or we can say that the spherical harmonics series neglects to account for one degree of freedom of the mass distribution. Therefore we conclude that there is a degree of freedom in Taylor's series which does not contribute to any potential outside the sphere, and this is omitted from the spherical harmonic expansion.

What Can We Find about Mass Density Distribution from Inversion?
— From the knowledge of the gravitational potential outside a solid sphere (or the Earth) we cannot determine the mass distribution density $\rho(\mathbf{r}_0)$ uniquely. We

can show this by the following example. The coefficient of the lth order spherical harmonic in Eq. (18.3) are volume integrals over the sphere of radius R. Any change of $\rho(\mathbf{r}_0)$ which will not modify these coefficients will result in the same $V(\mathbf{r})$ for $r > R$. In particular we can add or subtract from $\rho(\mathbf{r}_0)$ any amount of mass distribution of the form $\chi_l(r_0)Y_{lm}(\theta_0, \phi_0)$ if

$$\int_0^R \chi_l(r_0)r_0^{l+2}\, dr_0 = 0. \tag{18.13}$$

Since $\rho(r_0)$ cannot be negative, whereas $\chi(r_0)$ has to change sign, we have to impose the condition $\rho(r_0) + \chi(r_0) \geq 0$ for $0 < r_0 \leq R$.

Now let us expand $\rho(x_0 y_0 z_0)$ in terms of spherical harmonics

$$\rho(x_0 y_0 z_0) = \sum_{l,m} [\rho_{lm}^e(r_0)Y_{lm}^e(\theta_0, \phi_0) + \rho_{lm}^o(r_0)Y_{lm}^o(\theta_0, \phi_0)], \tag{18.14}$$

where

$$\rho_{lm}^e(r_0) = \frac{\epsilon_m(2l+1)}{4\pi}\frac{(l-m)!}{(l+m)!}\int_0^{2\pi} d\phi_0 \int_0^\pi \sin\theta_0 d\theta_0 \rho(r_0, \theta_0, \phi_0)Y_{lm}^e(\theta_0, \phi_0), \tag{18.15}$$

with a similar expression for ρ_{lm}^o.

Since Y_{lm} s form a complete set, the series (18.14), (18.15) corresponds to ρ itself. Now ρ_{00} represents the total mass inside $r = R$ and that all other ρ_{lm} s contribute zero mass. Having found ρ_{lm}, we can change ρ without changing the potential $V(\mathbf{r})$. Thus as long as by changing ρ_{mn}, the integral of $\rho_{lm}r^{l+2}$ remains unchanged the potential for $r > R$ stays the same. In particular consider the following scaling of ρ_{lm} (excluding $l = m = 0$),

$$q_{lm}^e(r_0) = \epsilon^{-l-3}\rho_{lm}^e\left(\frac{r_0}{\epsilon}\right), \quad 0 < \epsilon < 1, \tag{18.16}$$

$$q_{lm}^o(r_0) = \epsilon^{-l-3}\rho_{lm}^o\left(\frac{r_0}{\epsilon}\right), \quad 0 < \epsilon < 1. \tag{18.17}$$

Then defining $R\epsilon$ by

$$\int_0^{R\epsilon} q_{lm}^e(r_0)r_0^{l+2}dr_0 = \epsilon^{-l-3}\int_0^{R\epsilon}\rho_{lm}^e\left(\frac{r_0}{\epsilon}\right)r_0^{l+2}dr_0 = \int_0^R \rho_{lm}^e(x)x^{l+2}dx$$

$$= \frac{2l+1}{4\pi}A_{lm}, \tag{18.18}$$

and a similar result for $q_{lm}^o(r_0)$. By this scaling we can compress each partial density ρ_{lm} into smaller and smaller sphere of radius $R\epsilon$, by reducing the magnitude of the mass density by the factor ϵ^{-l-2}, and this will not change $V(\mathbf{r})$, $(r > R)$. For very small ϵ, the equivalent mass distribution reduces to a sequence of multipoles, the multipole of the lth order being

$$\lim_{\epsilon \to 0}\left\{\sum_{m=0}^n [q_{lm}^e Y_{lm}^e + q_{lm}^o Y_{lm}^o]\right\}. \tag{18.19}$$

The components of these multipoles are all that can be determined by measuring $V(\mathbf{r})$ outside $r = R$ [7].

Inverse Problem of Mass Distribution on a Spherical Shell — The two dimensional problem of determining the mass density on a spherical shell found from the measured potential outside the shell can be formulated as follows: Let $\sigma(\Omega)$ be the surface density on a spherical shell of radius R. For this special case Eq. (18.7) reduces to [8]

$$C_{lm} + iS_{lm} = \frac{R^2}{(2l+1)M} \int \int_{S_0} \sigma(\Omega_0) Y_{lm}(\Omega_0) d\Omega_0. \tag{18.20}$$

If these Stokes coefficients C_{lm} and S_{lm} are known for all l and m values, then the surface mass distribution $\sigma(\Omega_0)$ can be found by inverting (18.20). For the inversion we expand $\sigma(\Omega_0)$ in terms of the spherical harmonics

$$\sigma(\Omega_0) = \sum_{lm} \sigma_{lm} Y_{lm}^*(\Omega), \tag{18.21}$$

and then substitute (18.21) in (18.20) to find

$$\sigma_{lm} = \frac{(2l+1)M}{4\pi R^2}(C_{lm} + iS_{lm}). \tag{18.22}$$

Thus each spherical harmonic component of $\sigma(\Omega)$ can be found from the corresponding component of the external potential or Stokes coefficients.

References

[1] G.E. Bakus and J.F. Gilbert, Geophys. Intl, 13, 247 (1967).

[2] Yu. A. Tarakanov, Measurement Techniques, 50, 461 (2007).

[3] Yu. A. Tarakanov, O.V. Karagioz and M.A. Kudriavitsky, Gravitation & Cosmology, 9, 214 (2003).

[4] P.L. Bark, Izv. Akad. Nauk SSSR, Physics of the Solid Earth, 15, 566 (1979).

[5] G.O. Meshcheryakov, Akad. Nauk Ukrains'koi RSR, 6, 492 (1997).

[6] J. Gottlieb, Z. Angew. Math. Mech. 76, 360 (1996).

[7] P.M. Morse and H. Feshbach, *Methods of Theoretical Physics*, Part II (McGraw-Hill, New York, 1953) p. 1276.

[8] B.A. Chao, J. Geodynamics, 39, 223 (2005).

Chapter 19

The Inverse Problem of Reflection from a Moving Object

In this chapter we want to consider the direct problem of reflection of a scalar classical wave such as acoustic wave from a moving reflector with arbitrary motion, provided that the speed of the reflector is less than the wave velocity [1]–[8]. Then we want to discuss the inverse problem, i.e. study a method to determine the motion of reflector from the dependence of the reflection coefficient on the wave number.

To simplify the problem, we only consider one-dimensional motion of a flat surface in the x-direction, and a scalar wave traveling in the domain $x > D(t)$. Here $D(t)$ is the position of the reflector which we assume to be a perfect mirror. Thus the scalar wave equation $\phi(x,t)$ propagating in a homogeneous medium is incident on this surface

$$\frac{\partial^2 \phi(x,t)}{\partial x^2} = \frac{1}{c^2}\frac{\partial^2 \phi(x,t)}{\partial t^2}, \tag{19.1}$$

and is subject to the boundary condition

$$\phi[x = D(t), t] = 0. \tag{19.2}$$

We assume that the time-dependence of $D(t)$ is known, and we also assume that

$$\frac{dD(t)}{dt} = v(t) < c, \quad \text{for all values of } t, \tag{19.3}$$

where c is the wave velocity. We can write the general solution of (19.1) in terms of the Fourier amplitude for the incident and for the reflected wave (d'Lambert solution) [9]

$$\phi(x,t) = \frac{1}{2\pi} \int_{-\infty}^{+\infty} I(p)\exp[ip(x+ct)]dp$$

$$+ \frac{1}{2\pi} \int_{-\infty}^{+\infty} R(k)\exp[ik(x-ct)]dk, \qquad (19.4)$$

where $I(p)$ and $R(q)$ are the incident and reflected wave amplitudes respectively. Now the boundary condition (19.2) implies that $\phi(x = D(t),t) = 0$. We impose this condition on the solution (19.4). Then by multiplying the resulting expression by

$$\exp(-\epsilon|t|\exp[ik(D(t)-ct)]d(D(t)-ct)), \quad \epsilon > 0,$$

and integrating from $-\infty$ to $+\infty$, we find $R(k)$:

$$R(k) = \frac{1}{2\pi} \int_{-\infty}^{+\infty} I(p)dp \int_{-\infty}^{+\infty} d(D(t)-ct)$$

$$\times \exp[i(p+k)D(t)+ct(p-k)]\exp(-\epsilon|t|), \qquad (19.5)$$

where the limit of $\epsilon \to 0$ of the right-hand side of (19.5) is to be taken after integration.

In order to simplify the above expression for $R(k)$ we integrate (19.5) by parts to get

$$R(k) = -\frac{c}{\pi} \int_{-\infty}^{+\infty} \left[\frac{pI(p)}{p+k}\right] dp \int_{-\infty}^{+\infty} \exp[i(p+k)D(t)+ct(p-k)]\exp[-\epsilon|t|]dt. \quad (19.6)$$

Let us assume that the incoming wave is monochromatic with the wave number k_0, then the amplitude $I(p)$ is a delta function

$$I(p) = \delta(p-k_0). \qquad (19.7)$$

In this case the reflection amplitude obtained from (19.6) has the following form;

$$R(k,k_0) = -\frac{ck_0}{\pi(k+k_0)} \int_{-\infty}^{+\infty} \exp\{i[(k+k_0)D(t)++ct(k_0-k)]\}e^{-i|\epsilon|t}dt. \quad (19.8)$$

This last relation shows that, in general, the reflected wave contains all wave numbers. Now let us consider the two interesting and at the same time simple cases:
(1) The simplest possible motion is the one where the reflector is moving with constant velocity v, then $D(t) = vt + D_0$, with D_0 a constant and with $v < c$. Then from (19.8) we find $R(k,k_0)$ to be

$$R(k,k_0) = -\frac{2ck_0}{\pi(k+k_0)}\{\exp[i(k+k_0)D_0]\}\delta[k_0(c+v) - k(c-v)]. \qquad (19.9)$$

Therefore in this case the reflected wave is Doppler shifted,

$$k = k_0 \left(\frac{c+v}{c-v} \right),$$

(19.10)

and is monochromatic.

(2) The reflector moves sinusoidally, and $D(t)$ is given by

$$D(t) = A_0 \sin \omega t + B_0 \cos \omega t, \quad \omega A_0 < c, \quad \omega B_0 < c.$$

(19.11)

For this motion the reflection coefficient $R(k, k_0)$ is given by [10]

$$R(k, k_0) = \frac{2ck_0}{k+k_0} \sum_{n=-\infty}^{+\infty} \delta[c(k_0 - k) - n\omega]$$

$$\times \{ J_n[(k+k_0)A_0] + i^n J_n[(k+k_0)B_0] \}.$$

(19.12)

Both of these examples show that $R(k, k_0)$ can be a singular function of wave number k. These singularities are related to the asymptotic behaviour of $D(t)$ as $t \to \pm\infty$. In order to define a reflection coefficient which remains finite for all k values, we can subtract the delta functions in the following way:

Let us write $R(k, k_0)$ as

$$R(k, k_0) = \rho(k, k_0) + \frac{2ck_0}{k+k_0} \left\{ \int_{-\infty}^{0} \exp\left[i[(k+k_0)D_-(t) + ct(k_0 - k)]\right] \exp(\epsilon t) dt \right.$$

$$\left. + \int_{0}^{+\infty} \exp\left[i[(k+k_0)D_+(t) + ct(k_0 - k)]\right] \exp(-\epsilon t) dt \right\}$$

(19.13)

where $D_-(t)$ and $D_+(t)$ denote the asymptotic forms of $D(t)$ as $t \to -\infty$ and $t \to +\infty$ respectively, and $\rho(k, k_0)$ is the nonsingular part of the reflection coefficient. From Eqs. (19.8) and (19.13), we can show that $\rho(k, k_0)$ is finite for all k s provided that

$$D(t) - D_\pm(t) \to \mathcal{O}\left(|t|^{-1-\gamma}\right) \quad \text{as} \quad t \to \pm\infty, \quad \gamma > 0.$$

(19.14)

The advantage of introducing the nonsingular part of $R(k, k_0)$ is the fact that for certain types of motion $\rho(k, k_0)$ can be calculated approximately, whereas the in $R(k, k_0)$ have to be evaluated exactly.

The Inverse Problem of Reflection from a Moving Reflector — In this inverse problem we are given $R(k, k_0)$ as a function of k and we want to find the motion of the reflector (mirror). As Eq. (19.8) shows $R(k, k_0)$ is not the Fourier transform of $D(t)$, therefore in (19.8) we change k to q, where $q = k_0 - k$ and then write the result as

$$\frac{2\pi}{c} \left(\frac{q}{2k_0} - 1 \right) R(q, k_0) = \int_{-\infty}^{+\infty} \exp[i(2k_0 - q)D(t) + cqt] dt.$$

(19.15)

The inverse Fourier transform of (19.15) gives us the desired expression

$$[\exp(2ikD(t))]_{t-\frac{1}{c}D(t)=t'} = \int_{-\infty}^{+\infty} \left(\frac{q}{2k_0} - 1\right) R(q, k_0) \exp\left(-icqt'\right) dq, \qquad (19.16)$$

provided that t on the left-hand side of (19.16) be replaced by t' where the latter is the solution of

$$t - \frac{1}{c}D(t) = t'. \qquad (19.17)$$

Therefore (19.16) can be regarded as a functional equation which in general can be solved by iteration.

Now let us consider the following example where $R(k, k_0)$ is given by

$$R(q, k_0) = \frac{-2k_0}{(2k_0 - q)} \sum_{n=-\infty}^{+\infty} J_n[(2k_0 - q)B_0] \left\{q + \frac{n\omega}{c} + \frac{v}{c}(2k_0 - q)\right\}. \qquad (19.18)$$

By substituting (19.18) in (19.16) we find that $D(t)$ is a solution of

$$[\exp[2ik_0D(t)]]_{t-\frac{1}{c}D(t)=t'} = \exp\left(\frac{2ik_0vct'}{c-v}\right) \sum_{n=-\infty}^{+\infty} J_n\left[\frac{(2k_0c + n\omega)B_0}{c-v}\right]$$

$$\times \exp\left(\frac{in\omega ct'}{c-v}\right). \qquad (19.19)$$

One way of solving this equation is to expand the right-hand side of (19.19) in powers of $\frac{1}{c}$ and determine the left-hand side as an infinite series. But for this example, we have a simpler way to find $D(t)$. We take the limit of both sides as $c \to \infty$, and we obtain

$$\exp[2ik_0D(t)] = \exp(2ik_0vt) \sum_{n=-\infty}^{+\infty} J_n(2k_0B_0)[\exp(i\omega t)]^n$$

$$= \exp[2ik_0(vt + B_0 \sin \omega t)]. \qquad (19.20)$$

Thus the position of the reflector (mirror) is given by

$$D(t) = vt + B_0 \sin \omega t. \qquad (19.21)$$

Inversion for the Case when the Asymptotic Forms of the Motion Are Known — Let us consider the reflection from a mirror which is moving non-uniformly, but in addition to the nonsingular part of the reflection coefficient $\rho(k, k_0)$, defined in (19.13), we know the asymptotic forms of $D(t)$

$$D(t) \to D_A(t), \quad \text{as} \quad t \to \pm\infty. \qquad (19.22)$$

Then Eq. (19.16) yields the following result

$$[\exp(2ik_0D(t))]_{t-\frac{1}{c}D(t)=t'} = [\exp(2ik_0D_A(t))]_{t-\frac{1}{c}D(t)=t'}$$

$$+ \int_{-\infty}^{+\infty} \left(\frac{q}{2k_0} - 1\right) \exp\left(-iqct'\right) \rho(q, k_0) dq. \qquad (19.23)$$

In this case there are two different retarded times corresponding to the displacements $D(t)$ and $D_A(t)$, respectively.

Reflection from an Absorbing Mirror — We can also solve the inverse problem when the boundary condition (19.2) is replaced by a more general condition, viz,

$$\left[\frac{\partial\phi(x,t)}{\partial x} - i\beta\phi(x,t)\right]_{x=D(t)} = 0. \tag{19.24}$$

Then following the same procedure outlined above we find $R(k,k_0)$ to be

$$R(k,k_0) = \frac{ck_0(k_0 - \beta)}{\pi(k + k_0)(k + \beta)}\int_{-\infty}^{+\infty} \exp[i(k + k_0)D(t) + ct(k_0 - k)]e^{-\epsilon|t|}dt. \tag{19.25}$$

Here we can use the inverse Fourier transform which again yields a functional equation for $D(t)$.

References

[1] A large number of problems and their solutions when the boundary condition(s) are time-dependent have been reviewed in detail by V.V. Dodonov *Modern Nonlinear Optics*, Part 1 in Advances in Chemical Physics, Vol. 119, p. 309, Edited by Myron Evans. See also, *Classical and Quantum Dissipative Systems* by M. Razavy, Second Edition, (World Scientific, Singapore, 2017) Chapter 19.

[2] For a detaled mathematical and theoretical account of scattering of electromagnetic waves in one, two or three dimensions from a moving target the reader is referred to the following two texts: B. Borden's *Radar Imaging of Airborne Targets*, (CRC Press, Boca Raton, 1999), and

[3] M. Cheney and B. Borden, *Fundamentals of Radar Imaging*, (SIAM, Philadelphia, 2009).

[4] For an extensive treatment of the reflection from a moving target, particularly electromagnetic waves see, J. Van Bladel and D. De Zutter, IEEE Trans. Antennas Prop. 29, 629 (1981).

[5] Nonuniform Doppler effect is discussed at length by J.E. Gray and S.R. Addison, IEE Proceedings: Radar, Sonar and Navigation, Vol. 150, 262 (2003).

[6] M. Razavy, Lett. Nuovo Cimento, 37, 449, (1983).

[7] M. Razavy, Hadronic J. 8, 153 (1985).

[8] M. Razavy and D. Salopek, Europhys. Lett. 2, 161 (1986).

[9] P.M. Morse and H. Feshbach, Methods of Theoretical Physics, Part 1 (McGraw-Hill, New York, 1953) p. 844.

[10] See, for instance, D.C. Champeney, *Fourier Transforms and Their Physical Applications*, (Academic Press, New York, 1973).

Appendix A

Expansion Algorithm for Continued J-fractions

In Chapter 1 of this book we considered Jacobi matrices, and their application to solve certain inverse problems. There is a connection between periodic Jacobi matrices and continued fractions. Here we will not discuss this relation and refer our interested readers to the work of Andrea [1]. Whether it is possible to relate these two apparently different methods of inversion is a question that needs to be investigated.

In this Appendix we follow the classic work of Wall and derive some results which we have used to solve a number of inverse problems [2]–[6].
Continued fractions of the type

$$\cfrac{f(a_1)}{b_1 + z - \cfrac{a_2}{b_2 + z - \cfrac{a_3}{b_3 + z - \ddots - \cfrac{a_N}{b_N + z}}}}, \tag{A.1}$$

in which $a_i \neq 0$, $i = 1, 2 \cdots, a_N$ are called J fractions. If $a_i > 0$ and b_i are real for all i s, then (A.1) is called a real J-fraction.

First we want to establish the connection between J fractions and rational polynomials. We also want to consider the nth approximant which is a rational function whose numerator and denominator are of degrees $n - 1$ and n respectively, and whose coefficients are polynomials in r_n and s_n. Consider the following polynomials:

$$P_0(z) = a_{00}z^N + a_{01}z^{N-1} + \cdots a_{0N}, \tag{A.2}$$

$$P_1(z) = a_{11}z^{N-1} + a_{12}z^{N-2} + \cdots a_{1N}, \tag{A.3}$$

where $P_0(z)$ being of degree N and $P_1(z)$ of degree $N - 1$ respectively. We want to find numbers $r_n \neq 0$ and s_n, $n = 1, 2 \cdots N$ such that

$$\frac{P_1(z)}{P_0(z)} = \cfrac{1}{r_1 z + s_1 + \cfrac{1}{r_2 z + s_2 + \cfrac{1}{r_3 z + s_3 + \cdots + \cfrac{1}{r_N z + s_N}}}} \tag{A.4}$$

is equivalent to the problem of determining polynomials $P_n(z)$ of degrees $N - n$, $n = 1, 2 \cdots N - 1$, with the property that they are related to $P_2(z)$ and $P_1(z)$ by the recurrence relation

$$P_{n-1}(z) = (r_n z + s_n) P_n(z) + P_{n+1}(z), \quad n = 1, 2 \cdots N, \tag{A.5}$$

where $P_{N-1} = 0$ and P_N is a constant not equal to zero. Now from Eq. (A.5) we get

$$\frac{P_n(z)}{P_{n-1}(z)} = \frac{1}{r_n z + s_n + \frac{P_{n+1}(z)}{P_n(z)}}, \quad n = 1, 2 \cdots N \tag{A.6}$$

also if (A.4) is true, then a step by step calculation of P_n starting with $n = N$, $P_{N+1} = 0$ and $P_N = c \neq 0$, where c is a constant, determines P_n up to a constant factor.

In solving our inverse problem it is essential to find unique values for r_n and s_n. To show this, let us assume that (A.5) is true with $r_n \neq 0$ and suppose that $P_0 = \mathcal{P}_0$ and $P_1 = \mathcal{P}_2$, then

$$\mathcal{P}_{n-1} = (\bar{r}_n z + \bar{s}_n) \mathcal{P}_n + \mathcal{P}_{n+1}, \quad n = 1, 2 \cdots, \tag{A.7}$$

where $\mathcal{P}_{N-1} = 0$ and $\mathcal{P}_N$ is a nonzero constant. For $n = 1$, we get

$$P_0 - \mathcal{P}_0 = [(r_1 - \bar{r}_1) z + s_1 - \bar{s}_1] P_1 + (P_2 - \mathcal{P}_2) = 0. \tag{A.8}$$

Now $(P_2 - \mathcal{P}_2)$ is at most of the degree $N - 2$, and since P_1 is of degree $N - 1$, we conclude that $r_1 = \bar{r}_1$, $s_1 = \bar{s}_1$ and $P_2 = \mathcal{P}_2$. Next we choose $n = 2$ and use the same argument to show that $r_2 = \bar{r}_2$, $s_2 = \bar{s}_2$ and $P_3 = \mathcal{P}_3$. This process can be repeated with the result that $r_3 = \bar{r}_3, \cdots r_N = \bar{r}_N$, $s_3 = \bar{s}_3, \cdots$ and $s_N = \bar{s}_N$. Thus the two expressions (A.6) and (A.7) are identical and the expansion (A.4) is unique.

Recurrence Formulae for Finding the Coefficients of the *J*-fraction r_n and s_n — In order to find r_1, we start with the rational expression of z of the form $\frac{P_N(z)}{P_{N-1}(z)}$. The division process goes as follows. We write

$$\frac{a_{00} z^N + a_{01} z^{N-1} + \cdots + a_{0N}}{a_{11} z^{N-1} + a_{12} z^{N-2} + \cdots + a_{1N}} = r_1 z + \frac{b_{11} z^{N-1} + b_{12} z^{N-2} + \cdots + b_{1N}}{a_{11} z^{N-1} + a_{12} z^{N-2} + \cdots + a_{1N}}, \tag{A.9}$$

where

$$r_1 = \frac{a_{00}}{a_{11}}, \tag{A.10}$$

and

$$\begin{cases} b_{11} = a_{01} - r_1 a_{12} \\ b_{12} = a_{02} - r_1 a_{13} \\ \dots\dots\dots\dots\dots \\ b_{1N} = a_{0N} \end{cases} \tag{A.11}$$

Next, the second term on the right-hand side of (A.9) can be expanded, and here we find

$$\frac{b_{11}z^{N-1} + a + b_{12}z^{N-2} + \cdots + b_{1N}}{a_{11}z^{N-1} + a_{12}z^{N-2} + \cdots + a_{1N}} = s_1 + \frac{a_{22}z^{N-2} + a_{23}z^{N-3} + \cdots + a_{2N}}{a_{11}z^{N-1} + a_{12}z^{N-2} + \cdots + a_{1N}}, \tag{A.12}$$

where

$$s_1 = \frac{b_{11}}{a_{11}}, \tag{A.13}$$

and

$$\begin{cases} a_{22} = b_{12} - s_1 a_{12} \\ a_{23} = b_{13} - s_1 a_{13} \\ \dots\dots\dots\dots\dots \\ a_{2N} = b_{2N} - s_1 a_{1N} \end{cases} \tag{A.14}$$

Once r_1, s_1 and $P_2(z) = a_{22}z^{N-2} + a_{23}z^{N-3} + \cdots + a_{2N}$ are determined, the equations for the calculation of r_2, s_2 and $P_3(z)$ may be found from the previous ones by increasing all the subscripts by one and decreasing all of the exponents by one. In this way we can find r_3, s_3 and $P_4(z)$, and continue this process until the coefficients $a_{00}, a_{11}, a_{22} \cdots$ are not zero. Wall has suggested the following table for arranging the coefficients of the continued J-fraction (see Table A.1) [6].

Finding a Rational Approximation from the Input Points — Let us assume that the value of a function $f(z)$ at the set of N points $z_1, z_2 \cdots z_N$ are given, then there are three ways of obtaining a rational approximation to $f(z)$ [7], [8]. Using any one of these three provides us with a rational approximation the form

$$f(z) \approx R_{N,M} = \frac{P_N(z)}{Q_M(z)}, \tag{A.15}$$

where $P_N(z)$ and $Q_M(z)$ are the Nth and the Mth order. The simplest one of these three is a method based on continued J-fraction [8].

Consider the continued fraction

$$C_N = \cfrac{f(z_1)}{1 + \cfrac{(z - z_1)a_1}{1 + \cfrac{(z - z_2)a_2}{1 + \cdots + \cfrac{(z - z_N)a_N}{1}}}}. \tag{A.16}$$

From this expression it is clear that $C_N(z_1) = f(z_1)$ and that the coefficients $a_1 \cdots a_N$ can be determined is such a way that $C_N(z)$ is equal to $f(z)$ at all points $z_1 \cdots z_N$. From (A.16) it follows that the coefficients $a_1 \cdots a_N$ are given by [8]

$$a_1 = \frac{1}{z_j - z_{j+1}} \left\{ 1 + \cfrac{(z_{j+1} - z_{j-1})a_{j-1}}{1 + \cfrac{(z_{j+1} - z_{j-1})a_{j-1}}{1 + \cfrac{(z_{j+1} - z_{j-2})a_{j-2}}{1 + \cdots + \cfrac{(z_{j+1} - z_1)a_j}{1 - \frac{f(z_1)}{f(z_{j+1})}}}}} \right\}, \tag{A.17}$$

and

$$a_1 = \frac{f(x_1)/f(x_2) - 1}{x_2 - x_1}. \tag{A.18}$$

This algorithm yields rational polynomials of order $\left(\frac{N-1}{2}, \frac{N+1}{2}\right)$ for N odd, and $\left(\frac{N}{2}, \frac{N}{2}\right)$ for N even.

Table A.1: Wall's arrangement for finding the constant coefficients r_n and s_n in the continued J-fraction expansion [6].

	a_{00}	a_{01}	a_{02}	$\cdots$
	a_{11}	a_{12}	a_{13}	$\cdots$
$r_1 = \frac{a_{00}}{a_{11}}$	$b_{11} = a_{01} - r_1 a_{12}$	$b_{12} = a_{02} - r_1 a_{13}$	$b_{13} = a_{03} - r_1 a_{14}$	$\cdots$
$s_1 = \frac{b_{11}}{a_{11}}$	$a_{22} = b_{12} - s_1 a_{12}$	$a_{23} = b_{13} - s_1 a_{13}$	$a_{24} = b_{14} - s_1 a_{14}$	$\cdots$
$r_2 = \frac{a_{11}}{a_{22}}$	$b_{22} = a_{12} - r_2 a_{23}$	$b_{23} = a_{13} - r_2 a_{24}$	$b_{24} = a_{14} - r_2 a_{25}$	$\cdots$
$s_2 = \frac{b_{22}}{a_{22}}$	$a_{33} = b_{23} - s_2 a_{23}$	$a_{34} = b_{24} - s_2 a_{24}$	$a_{35} = b_{25} - s_2 a_{25}$	$\cdots$

References

[1] This connection has been studied by S.A. Andrea and T.G. Berry, in Linear Algebra and Its Applications, 161, 117 (1992). These authors suggest an algorithm based on continued fraction expansion to construct a periodic Jacobi matrix from spectral data which characterizes a periodic Jacobi matrix.

[2] For a detailed discussion of continued fraction and Its applications see W.E. Jones and W.J. Thron, *Continued Fractions, Analytic Theory and Applications* in *Encyclopedia of Mathematics and its Applications*, Volume 11 (Addison-Wesley Publishing Company, Reading, 1980).

[3] O. Perron, *Die Lehre von der Kottenbrüchen*, Vol. 1 (Teubner 1954).

[4] C.D. Olds *Continued Fractions*, (Random House, New York, 1963).

[5] C.D. Dough, *Continued Fractions*, (World Scientific, Singapore, 2006).

[6] H.S. Wall, *Analytic Theory of Continued Fractions*, (Chelsea Publishing Company, 1967) Chapter IX.

[7] G.A. Baker Jr. and P. Graves-Morris, *Padé Approximants*, (Cambridge University Press, Cambridge, 1996).

[8] *The Padé Approximant in Theoretical Physics*, Edited by G.A. Baker and G.L. Gammel, Vol. 71 in *Mathematics in Science and Engineering*, (Academic Press, New York, 1970).

Appendix B

Reciprocal Differences of a Quotient and Thiele's Theorem

In our numerical treatment of the solution of inverse problem, in many cases, we have found a rational representation of the input, e.g. Padé approximants. Reciprocal differences introduced by Thiele leads to the approximate representation of a function by a rational function, and therefore a more desirable method of interpolation, than e.g. by polynomials [1]. Now we want to present a brief account of Thiele's work, and list some of the essential properties of this interpolation. For details and proofs the reader is referred to few excellent texts on this subject [1]–[3].

Definition of Reciprocal Differences — Consider a function $f(x)$ with the known values at $x = x_0, x_1 \cdots x_n$. The reciprocal difference of $f(x)$ for the two points x_0 and x_1 is defined by

$$\rho(x_0 x_1) = \frac{x_0 - x_1}{f(x_0) - f(x_1)}, \tag{B.1}$$

and for three points x_0, x_1 and x_2 the reciprocal difference is a generalization of (B.1), i.e.

$$\rho_2(x_0 x_1 x_2) = \frac{x_0 - x_1}{\rho(x_0 x_1) - \rho(x_1 x_2)} + f(x_1). \tag{B.2}$$

Similarly for four points we have

$$\rho_3(x_0 x_1 x_2 x_3) = \frac{x_0 - x_3}{\rho_2(x_0 x_1 x_2) - \rho_2(x_1 x_2 x_3)} + \rho(x_1 x_3). \tag{B.3}$$

Thus for $n + 1$ point the reciprocal difference takes the form of

$$\rho_n(x_0 x_1 \cdots x_n) = \frac{x_0 - x_n}{\rho_{n-1}(x_0 \cdots x_{n-1}) - \rho_{n-1}(x_1 x_2 \cdots x_n)} + \rho_{n-2}(x_1 \cdots x_{n-1}).$$

(B.4)

Thiele Interpolation Formula — In in Eqs. (B.1)–(B.4) we replace x_0 by x we have successively the following results:

$$f(x) = f(x_1) + \frac{x - x_1}{\rho(xx_1)},$$

(B.5)

$$\rho(xx_1) = \rho(x_1 x_2) + \frac{x - x_2}{\rho_2(xx_1 x_2) - f(x_1)},$$

(B.6)

$$\rho_2(xx_1 x_2) = \rho_2(x_1 x_2 x_3) + \frac{x - x_3}{\rho_3(xx_1 x_2 x_3) - \rho(x_1 x_2)}.$$

(B.7)

···

We can also write $f(x)$ as a continued fraction;

$$f(x) = f(x_1) + \cfrac{x - x_1}{\rho(x_1 x_2) + \cfrac{x - x_2}{\rho_2(x_1 x_2 x_3) - f(x_1) + \cfrac{x - x_3}{\rho_3(x_1 \cdots x_4) - \rho(x_1 x_2) + \cdots}}}.$$

(B.8)

An important property of the continued fraction is that if a numerator of one of the constituent partial fraction vanishes, this and all the following constituent do not effect the value. Thus in (B.8) if we put in turn $x = x_1, x_2, x_3 \cdots x_n$ we obtain $f(x_1), f(x_2), f(x_3) \cdots f(x_n)$.

Reciprocal Derivatives — So far in the definition of the reciprocal differences we have assumed that the arguments are distinct. Now we want to see what happens if two or more arguments coincide, i.e. to study the so called "confluent reciprocal differences" [3]. The special case where all the arguments have a common value and is called reciprocal derivative are of special interest.

Let us define the operator r_n by

$$r_n f(x) = \lim_{x_1, x_2 \cdots x_{n+1} \to x} \rho_n(x_1, x_2 \cdots x_{n-1}).$$

(B.9)

In particular we have

$$rf(x) = \lim_{x_1, x_2 \to x} \frac{x_1 - x_2}{f(x_1) - f(x_2)} = \frac{1}{f'(x)},$$

(B.10)

where prime denotes derivative with respect to x. Thus the reciprocal derivative of the first order is the reciprocal of ordinary derivative. Now from (B.4) we have

$$\frac{\rho_2(xxx) - \rho_2(xxy)}{x - y} = \frac{1}{\rho_3(xxxy) - \rho(xx)}$$

(B.11)

$$\frac{\rho_2(xxy) - \rho_2(xyy)}{x - y} = \frac{1}{\rho_3(xxyy) - \rho(xy)} \tag{B.12}$$

and

$$\frac{\rho_2(xyy) - \rho_2(yyy)}{x - y} = \frac{1}{\rho_3(xyyy) - \rho(yy)}. \tag{B.13}$$

By adding these three equations we find

$$\frac{\rho_2(xxx) - \rho_2(yyy)}{x - y} = \frac{1}{\rho_3(xxxy) - \rho(xx)} + \frac{1}{\rho_3(xxyy) - \rho(xy)}$$
$$+ \frac{1}{\rho_3(xyyy) - \rho(yy)}. \tag{B.14}$$

Let us consider the limit of this expression as $y \to x$, then we have

$$\frac{1}{rr_2 f(x)} = \frac{3}{r_3 f(x) - r f(x)}, \tag{B.15}$$

and solving for $r_3 f(x)$ we get

$$r_3 f(x) = r f(x) + 3 r r_2 f(x). \tag{B.16}$$

By considering ρ_3, $\rho_4 \cdots$ we obtain the following recurrence relation:

$$r_n f(x) = r_{n-2} f(x) + n r r_{n-1} f(x). \tag{B.17}$$

This general recurrence relation can be used to find $r f(x)$, $r_2 f(x)$ etc. For instance in addition to (B.10) for $r f(x)$ we have

$$r_2 f(x) = f(x) + 2r \left(\frac{1}{f'(x)} \right) = f(x) - 2 \frac{[f'(x)]^2}{f''(x)}. \tag{B.18}$$

Let us consider two simple examples

$$(a) - \qquad f(x) = x^2, \tag{B.19}$$

and

$$(b) - \qquad f(x) = \left(\frac{ax + b}{cx + d} \right). \tag{B.20}$$

For (a) we have

$$(a) \quad r x^2 = \frac{1}{2x}, \quad r_2 x^2 = -3x^2, \quad r_3 x^2 = 0, \tag{B.21}$$

and for (b) the result is

$$(b) \quad r_2 \left(\frac{ax + b}{cx + d} \right) = \frac{a r_2 x + b}{c r_2 x + d} = \frac{a}{c}. \tag{B.22}$$

In getting the last relation we have used $r_2 x = \infty$.

Thiele's Theorem — Thiele's interpolation formula enables us to express a function as a continued fraction in which the reciprocal derivatives at a point are used. In fact, when $x_1 x_2 \cdots x_n \to x$ we have

$$\lim\{\rho_n(x_1 x_2 \cdots x_{n+1}) - \rho_{n-2}(x_1 x_2 \cdots x_{n-1}) r_n f(x) - r_{n-2} f(x)$$
$$= n r r_{n-1} f(x), \tag{B.23}$$

a result which follows from (B.17).

The Thiele interpolation formula (B.8) gives us the Thiele theorem, namely

$$f(x+h) = f(x) + \cfrac{h}{rf(x) + \cfrac{h}{2rrf(x) + \cfrac{h}{3rr_2 f(x) + \ldots}}}. \tag{B.24}$$

We know that Taylor's series terminates when a function is a polynomial, Thiele's continued fraction terminates when the function is rational. The following example illustrates this point

$$\frac{a(x+h) + b}{c(x+h) + d} = \frac{ax+b}{cx+d} + \cfrac{h}{\frac{(cx+d)^2}{ad-bc} + \cfrac{h}{\frac{ad-bc}{c(cx+d)}}}, \tag{B.25}$$

where the first term on the right-hand side of (B.25) can be written as

$$\frac{ax+b}{cx+d} = \frac{b}{d} + \cfrac{x}{\frac{d^2}{ad-bc} + \cfrac{x}{\frac{ad-bc}{cd}}}. \tag{B.26}$$

If $f(x)$ is not a rational function and its continued fraction expansion does not terminate, then by terminating it we obtain an approximate rational fraction representation for $f(x)$.

References

[1] The original work of T.N. Thiele can be found in *Interpolationrechnung*, (Leipzig, 1909).

[2] N.E. Nörlund, *Differenzrechnung*, (Berlin, 1924).

[3] L.M. Milne-Thomson, *The Calculus of Finite Differences*, (Macmillan, London, 1965) Chapter V.

Index

Printed in the United States
By Bookmasters